Industrial and Applied Mathematics

Industrial and Applied Mathematics, a Scopus-indexed book series, publishes high-quality research-level monographs, lecture notes and contributed volumes focusing on areas where mathematics is used in a fundamental way, such as industrial mathematics, bio-mathematics, financial mathematics, applied statistics, operations research and computer science.

Binod Chandra Tripathy · Hemen Dutta ·
Susanta Kumar Paikray · Bidu Bhusan Jena
Editors

Operators, Inequalities and Approximation

Theory and Applications

Springer

Editors
Binod Chandra Tripathy
Department of Mathematics
Tripura University
Agartala, Tripura, India

Hemen Dutta
Department of Mathematics
Gauhati University
Guwahati, Assam, India

Susanta Kumar Paikray
Department of Mathematics
Veer Surendra Sai University of Technology
Burla, Odisha, India

Bidu Bhusan Jena
Faculty of Science (Mathematics)
Sri Sri University
Cuttack, Odisha, India

ISSN 2364-6837 ISSN 2364-6845 (electronic)
Industrial and Applied Mathematics
ISBN 978-981-97-3237-1 ISBN 978-981-97-3238-8 (eBook)
https://doi.org/10.1007/978-981-97-3238-8

Mathematics Subject Classification: 41A36, 41A25, 40C15, 47S40, 42B05, 40G05, 30C45, 54E05, 54H25

This Springer imprint is published by the registered company Springer Nature Singapore Pte Ltd.
The registered company address is: 152 Beach Road, #21-01/04 Gateway East, Singapore 189721, Singapore

Dedication

In recognition of his significant contributions across various scientific domains, this book is dedicated to Prof. Umakanta Misra on the occasion of his 72nd birthday on 20 July 2024. He stands as one of the most distinguished and globally recognized mathematicians. For the past 28 years, he has served as an esteemed faculty member in the P.G. Department of Mathematics at Berhampur University, Odisha, India. Presently, he holds the esteemed position of Professor Emeritus in the Department of Mathematics at the National Institute of Science and Technology, Berhampur, having joined the faculty there in 2012. Throughout his career, he received several honorary memberships and fellowships from numerous scientific academies and societies nationwide. His research interests span a wide array of pure and applied mathematical sciences, encompassing areas such as real and complex analysis, fractional calculus and its applications, integral equations and transforms, graph theory, and inventory optimization. He boasts an impressive publication record, including 5 books, 30 book chapters, 15 papers in international conference proceedings, and over 200 articles in various peer-reviewed international scientific research journals. Furthermore, Prof. Misra has supervised 3 D.Sc. scholars, 27 Ph.D. scholars, 35 M.Phil. scholars and 5 Ph.D. scholars currently working under his guidance. Professor U.K. Misra is not only an exceptional mathematician but also a remarkable individual. His visionary contributions to mathematical analysis have positioned him as a role model and inspiration for mathematicians worldwide, as well as for countless others whose lives he has touched. Professor Misra is known for his kindness, warmth, and willingness to assist others. Professor Misra is esteemed by all who know him. Finally, we extend our heartfelt wishes to Prof. Misra for a healthy and joyous long life with his esteemed wife, Mrs. Arati Misra, children, and grandchildren. May he continue to guide, encourage, and enlighten researchers in mathematical and allied sciences for many more decades to come.

Editors

Preface

The purpose of this book is to cover some important topics in the areas of operator theory, inequalities and approximation theory. The properties of various types of operators, methods for approximating such operators, proximity point problems, applications of approximation methods in other fields, and some analytic inequalities are the main topics covered. It aims to capture both the pure and applied aspects of the concerned topics, and many of the ideas presented in the book are fundamental to several areas of applied science and engineering. There are nine chapters in the book, and a very brief overview of each chapter is presented below.

Chapter "1" investigates a bivariate extension of the linear positive operators based on multivariate q-Lagrange–Hermite polynomials and determine the degree of approximation with the help of the moduli of continuity and the Peetre's K-functional for bi-dimensionally continuous functions. It also investigates some related generalized operators and obtains the rate of convergence.

Chapter "2" deals with the construction of fuzzy Meyer-König and Zeller operators and fuzzy post quantum Meyer-König and Zeller operators, and investigates their fuzzy Korovkin-type approximation results and inequalities estimating their approximations.

Chapter "3" establishes some results on degree of approximation of functions (signals) belonging to generalized Zygmund class using $(E, s)(\overline{N}, q_n)$ mean.

Chapter "4" deals with approximation of the functions belonging to different classes by a product $(C, 1)(N, p_m, q_m)(E, \theta)$ means of conjugate Fourier series.

Chapter "5" investigates some novel Turán type inequalities using the (p, k)-generalization of the Mittag-Leffler function.

Chapter "6" aims to study multiplicative behaviour of Hardy-Rogers-type F-contraction and best proximity results in the framework of complete multiplicative metric spaces.

Chapter "7" establishes certain new classes of proximal contractive mappings in G-metric spaces, and obtains the best proximity point theorems for such kind of mappings. It also deduces certain new best proximity and fixed point results.

Chapter 8" investigates the class $\mathcal{Q}_{\Sigma}(m, n; \Theta)$ of bi-univalent functions by using Lucas polynomials, subordination and Sălăgean differential operator. It also examines $(\mathcal{P}, \mathcal{Q})$-Lucas polynomial coefficient estimates and Fekete–Szegö inequalities for functions belonging to $\mathcal{Q}_{\Sigma}(m, n; \Theta)$.

Chapter "9" presents sufficient conditions for the families of integral operators containing normalized forms of the Rabotnov functions so that they are univalent in the open unit disk. It also obtains the convexity order of the families of the operators

We are thankful to the contributors and reviewers for their efforts and cooperation in developing and completing this book project. We hope that the book will be a useful source of new information and knowledge in the fields of operators, inequalities, and approximations.

Agartala, India — Binod Chandra Tripathy
Guwahati, India — Hemen Dutta
Burla, India — Susanta Kumar Paikray
Cuttack, India — Bidu Bhusan Jena

Contents

Chapter 1
Approximation by a Double Sequence of Operators Involving Multivariable q-Lagrange–Hermite Polynomials

Behar Baxhaku, Purshottam Narain Agrawal, and Kailash Chandra Lachhwani

Abstract In this chapter, we introduce a bivariate extension of the linear positive operators based on multivariate q-Lagrange–Hermite polynomials and determine the degree of approximation with the help of the moduli of continuity and the Peetre's K-functional for bi-dimensionally continuous functions. Further, we define the corresponding Generalized Boolean Sum (GBS) operators and derive the rate of convergence of these operators for functions in a Bögel space. We also generalize the above operators with the aid of Taylor polynomial and examine the rate of approximation of the generalized operators for functions whose sth derivative is continuous and belongs to the Lipschitz class. Finally, we consider a Kantorovich version of the above bivariate operators involving multivariate q-Lagrange- Hermite polynomials and note that the above study could be extended to these operators.

Keywords Multivariate Lagrange–Hermite polynomials · q-Lagrange–Hermite polynomials · Deferred type statistical convergence · P-summability · Modulus of continuity

MSC(2020) 41A25 · 41A36 · 33C45 · 26A15

B. Baxhaku
Department of Mathematics, University of Prishtina, Prishtina, Kosovo
e-mail: behar.baxhaku@uni-pr.edu

P. N. Agrawal (✉)
Department of Mathematics, Indian Institute of Technology Roorkee, Roorkee 247667, India
e-mail: p.agrawal@ma.iitr.ac.in

K. C. Lachhwani
Department of Applied Science, National Institute of Technical Teachers Training and Research, Sector-26, Chandigarh 160 019, India
e-mail: kailashclachhwani@nitttrchd.ac.in

B. C. Tripathy et al. (eds.), *Operators, Inequalities and Approximation*, Industrial and Applied Mathematics, https://doi.org/10.1007/978-981-97-3238-8_1

1.1 Introduction

Special functions, which include the trigonometric functions, have been used for centuries. The core of special functions is the Gaussian hypergeometric function ${}_2F_1$ and the confluent hypergeometric functions ${}_1F_1$ and ψ. The ${}_2F_1$ includes as special cases the Legendre functions, the incomplete beta function, the complete elliptic functions of the first and second kinds and most of the classical orthogonal polynomials. The confluent hypergeometric functions include as special cases Bessel functions, parabolic cylinder functions, Coulomb wave functions, and incomplete gamma functions. In the past three decades, several new special functions and applications have been discovered. For an overview focusing primarily on the hypergeometric functions along with the associated hypergeometric series and recent developments, we refer the interested reader to the book [14]. Researchers have defined and studied the approximation properties of several positive linear operators by involving special functions. Lupas and Müller [56] introduced Gamma operators to study the approximation of continuous functions on $[0, \infty)$. These operators have been extensively studied and generalized by researchers by using the gamma function, see for instance [1, 2, 23, 39, 51, 70, 71, 76, 85] and the references therein. Gupta and Greubel [42] and Dhamija and Deo [30] determined the moments for the Jain–Durrmeyer operators and Jain–Durrmeyer operators associated with the inverse Pólya–Eggenberger distribution, respectively by using the Tricomi's confluent hypergeometric function and Stirling numbers of the first kind. Lupas [55] proposed another sequence of positive linear operators, known as beta operators, by means of beta distribution and studied its convergence properties. Upreti [75] defined another kind of beta operator and studied the uniform convergence of the operator on compact intervals of $[0, \infty)$. The beta approximating operators of the first and the second kinds were defined by Mihesan [58, 59], respectively.

Erkuş et al. [36] constructed positive linear operators by involving multivariable Lagrange polynomials, known as Chan–Chyan–Srivastava polynomials introduced by Chan et al. [27] and studied their approximation properties via statistical convergence. Multivariable Lagrange polynomials based on q-integers were defined by Altin et al. [13]. Erkuş-Duman [33] studied a q-analogue of the operator defined in [36]. Baxhaku et al. [22] considered a bivariate generalization of the q-analogue of the operators proposed in [36] and investigated the approximation degree by means of the modulus of continuity and the Peetre's K-functional. Agrawal et al. [6] discussed a Kantorovich version of the q- analogue of the operators proposed in [36] and studied Korovkin type theorems in the deferred weighted A-statistical approximation and the power series summability method. Erkuş and Srivastava [37] introduced a family of multivariable polynomials. Erkuş-Duman [32] constructed a q-analogue of these polynomials. Agrawal et al. [5] proposed a sequence of linear positive operators based on the polynomials introduced in [32] and studied the Korovkin type theorems via statistical convergence and the power series method. Further, the authors [5] also discussed the k-th order generalization of the operators and derived a convergence theorem for the bivariate case of these operators via A-statistical approximation.

The pioneering work of Jakimovski and Leviatan [46] on the linking of linear positive operators with orthogonal polynomials triggered a virtual chain reaction in approximation theory. The authors [46] introduced a generalization of Szász operators by means of the Appell polynomials. Motivated by this study, Ismail [45] considered the Szász operators by involving Sheffer polynomials. Subsequently, Varma and Tasdelen [79] proposed a sequence of positive linear operators by linking Szász operators with the Charlier polynomials. Kajla and Agrawal [47] presented the Szász-Durrmeyer type operators involving Charlier polynomials. Varma et al. [77] proposed Szász type operators based on Brenke type polynomials. Sucu et al. [72] defined Szász type operators involving Boas–Buck type polynomials which include Brenke type polynomials, Sheffer polynomials and Appell polynomials. For some other studies along similar lines, we refer to [3, 4, 15–17, 28, 38, 40, 41, 44, 48–50, 61, 62, 64, 65, 69, 74, 77, 78, 82, 84] and the references therein. For more details on other important contributions and recent developments in the application of special functions and orthogonal polynomials, the reader may refer to the book [60] dedicated to Prof. H. M. Srivastava.

In the past twenty years, several well-known orthogonal polynomials have been extended to the multivariable case, for instance, see [9, 10, 13, 35]. Chan et al. [27] defined multivariable Lagrange polynomials as follows:

$$g_n^{(\eta_1,\dots\ \eta_r)}(z_1, z_2, \dots\ \ z_r) = \sum_{l_1+l_1+\dots+l_r=n} \left\{ \prod_{i=1}^{r} (\eta_i)_{l_i} \frac{(z_i)^{l_i}}{l_i!} \right\}, \tag{1.1}$$

where $(\zeta)_l = \zeta(\zeta+1)\dots(\zeta+l-1)$ and $(\zeta)_0 = 1$. The generating function for these polynomials is given by

$$\prod_{i=1}^{r}(1-ty_i)^{-\beta^{(i)}} = \sum_{p=0}^{\infty} g_p^{(\beta^{(1)},\beta^{(2)},\dots\ \beta^{(r)})}(y_1, y_2, \dots\ \ y_r)t^p, \tag{1.2}$$

where $|t| < \min\{|y_1|^{-1}, \dots\ , |y_r|^{-1}\}$. For $\varphi \in C(I) := \{\varphi : I \to \mathbb{R} |\ \varphi\ is\ continuous\ on\ I\}$ with $I = [0, 1]$, Erkuş et al. [36] considered the following operators:

$$\mathfrak{F}_n^{\beta^{(1)},\dots,\beta^{(r)}}(\varphi; x) = \left\{ \prod_{k=1}^{r} \left(1 - x\beta_n^{(k)}\right)^n \right\} \sum_{p=0}^{\infty} \left\{ \sum_{l_1+l_2+\dots+l_r=p} \varphi\left(\frac{l_r}{n+l_r-1}\right) \prod_{s=1}^{r} \left(\beta_n^{(s)}\right)^{l_s} \frac{(n)_{l_s}}{l_s!} \right\} x^p, \tag{1.3}$$

where the sequence $\beta^{(j)} = (0 < \beta_n^{(j} < 1)_{n\in\mathbb{N}}$, (for all $j = 1, 2, \dots\ \ r$). The researchers [36] discussed the Korovkin type theorems in the statistical sense for the operator (1.3) and also determined the statistical convergence rate via modulus of continuity. Erkuş-Duman [33] and Mursaleen et al. [63] independently defined a q-operator associated to the one given in [36]. Later, Baxhaku et al. [22] studied the bivariate generalization of these operators and also investigated the associated GBS (Generalized Boolean Sum) operators. In our further study, we shall need the following definitions: For $q \in (0, 1)$, the q-(basic) analogues of a positive integer $n \in \mathbb{N}$, and Pochhammer symbol are given by

$$[n]_q := \frac{1-q^n}{1-q}; \quad \text{and} \quad (\mu; q)_n = \begin{cases} 1, & \text{if } n = 0 \\ (1-\mu)(1-\mu q)...(1-\mu q^{n-1}), & \text{if } n \in \mathbb{N} \end{cases},$$

respectively, where the parameter μ is arbitrary.

Encouraged by the research in [27], Altin et al. [12] considered the multivariate extension of the Lagrange–Hermite polynomials $h_s^{(\beta^{(1)},\beta^{(2)},...,\beta^{(r)})}(z_1, z_2, ...z_n)$ generated by

$$\prod_{j=1}^{r} \left(1 - t^j z_j\right)^{-\beta^{(j)}} = \sum_{s=0}^{\infty} h_s^{(\beta^{(1)},\beta^{(2)},... \beta^{(r)})}(z_1, z_2, \dots z_r) t^s,$$

with $|t| < \min\{|z_1|^{-1}, |z_1|^{-1/2}, \dots \ |z_r|^{-1/r}\}$. Later, Erkuş [34] proposed q-Lagrange–Hermite polynomials generated by

$$\prod_{i=1}^{r} \frac{1}{\left(z_i t^i; q\right)_{\beta^{(i)}}} = \sum_{s=0}^{\infty} h_{s,q}^{(\beta^{(1)},\beta^{(2)},... \beta^{(r)})}(z_1, z_2, \dots z_r) t^s. \tag{1.4}$$

From (1.4), it follows that

$$h_{p,q}^{(\beta^{(1)},\beta^{(2)},... \beta^{(r)})}(z_1, z_2, \dots z_r) = \sum_{l_1+2l_2+...+rl_r=p} \left\{ \prod_{k=1}^{r} \left(q^{\beta^{(k)}}, q\right)_{l_k} \frac{(z_k)^{l_k}}{(q,q)_{l_k}} \right\}.$$

Hence, using (1.4), we get

$$\prod_{i=1}^{r} \left(z_i t^i; q\right)_{\beta^{(i)}} \sum_{p=0}^{\infty} \left(\sum_{l_1+2l_2+...+rl_r=p} \left\{ \prod_{k=1}^{r} \left(q^{\beta^{(k)}}, q\right)_{l_k} \frac{(z_k)^{l_k}}{(q,q)_{l_k}} \right\} \right) t^p = 1. \tag{1.5}$$

Inspired by the above research, for $\varphi \in C(I)$, Agrawal et al. [8] proposed Lagrange–Hermite type operators as

$$\begin{aligned} \mathfrak{R}_{n,q}^{\alpha^{(1)},...,\alpha^{(r)}}(\varphi; x) = & \left\{ \prod_{i=1}^{r} \left(\alpha_i x^i; q\right)_n \right\} \sum_{p=0}^{\infty} \left\{ \sum_{l_1+2l_2+...+rl_r=p} \left\{ \prod_{k=1}^{r} (q^n, q)_{l_k} \frac{(\alpha_n^{(k)})^{l_k}}{(q,q)_{l_k}} \right\} \right. \\ & \left. \varphi\left(\frac{[l_1]_q}{[n+l_1-1]_q} \right) \right\} x^p, \end{aligned} \tag{1.6}$$

where the sequence $\alpha^{(j)} = \langle \alpha_n^{(j)} \in (0,1) \rangle_{n \in \mathbb{N}}$, $\forall\ j = 1, 2, ..., r$, and studied the degree of approximation via deferred weighted A-statistical convergence method and power series method.

1.2 Bivariate Lagrange–Hermite Operators

Let

$$\alpha^{(i)} = \left\{\alpha_{n_1}^{(i)}\right\}_{n_1 \in \mathbb{N}}, \quad (i = 1, 2, \ldots r_1)$$
$$\beta^{(j)} = \left\{\beta_{n_2}^{(j)}\right\}_{n_2 \in \mathbb{N}}, \quad (j = 1, 2, \ldots r_2)$$

be sequences such that $0 < \alpha_{n_1}^{(i)} < 1$, $0 < \beta_{n_2}^{(j)} < 1$; $(i = 1, 2, \ldots, r_1,\ j = 1, 2, \ldots, r_2,\ r_1, r_2 \in \mathbb{N})$. Further, let $\{q_{n_i}\}$; $q_{n_i} \in (0, 1)$ be a sequence such that $q_{n_i} \to 1$ and $q_{n_i}^{n_i} \to a_i$; $a_i \in [0, 1)$, as $n_i \to \infty$ for $i = 1, 2$.

Suppose that $B(I^2) := \left\{\varphi : \varphi\ is\ \text{bounded}\ on\ I^2\right\}$, with the uniform norm defined as

$$\|\varphi\| = \sup_{(x,y)\in I^2} |\varphi(x, y)|,$$

and $C(I^2) := \left\{\varphi \in B(I^2) : \varphi\ is\ \text{continuous}\ on\ I^2\right\}$.

For $\varphi \in C(I^2)$, the bivariate tensor product of (1.6) is given by

$$\mathfrak{R}_{n_1,q_{n_1},\alpha^{(1)},\ldots,\alpha^{(r_1)}}^{n_2,q_{n_2},\beta^{(1)},\ldots,\beta^{(r_2)}}(\varphi; x, y) = \left\{\prod_{i=1}^{r_1}\prod_{j=1}^{r_2} (\alpha_i x^i; q_{n_1})_{n_1} (\beta_i y^i; q_{n_2})_{n_2}\right\} \sum_{p=0}^{\infty}\sum_{s=0}^{\infty}\Bigg\{ \sum_{l_1+2l_2+\ldots+r_1 l_{r_1}=p}\ \sum_{h_1+2h_2+\ldots+r_2 h_{r_2}=s}$$
$$\left\{\prod_{k=1}^{r_1}\prod_{j=1}^{r_2} (q_{n_1}^{n_1}, q_{n_1})_{l_k} \frac{(\alpha_{n_1}^{(k)})^{l_k}}{(q_{n_1}, q_{n_1})_{l_k}} (q_{n_2}^{n_2}, q_{n_2})_{h_j} \frac{(\beta_{n_2}^{(j)})^{h_j}}{(q_{n_2}, q_{n_2})_{h_j}}\right\}$$
$$\varphi\left(\frac{[l_1]_{q_{n_1}}}{[n_1 + l_1 - 1]_{q_{n_1}}}, \frac{[h_1]_{q_{n_2}}}{[n_2 + h_1 - 1]_{q_{n_2}}}\right)\Bigg\} x^p y^s. \tag{1.7}$$

Let us denote $e_i^j(x, y) = x^i y^j$, $i, j = \overline{0, 2}$, and define

$$\vartheta_{n_1,q_{n_1},r}^{\alpha^{(1)},\alpha^{(2)}\ldots\alpha^{(r_1)}}(x) = \mathfrak{R}_{n_1,q_{n_1},\alpha^{(1)},\ldots,\alpha^{(r_1)}}^{n_2,q_{n_2},\beta^{(1)},\ldots,\beta^{(r_2)}}((t_1 - x)^r; x, y);$$
$$\vartheta_{n_2,q_{n_2},r}^{\beta^{(1)},\beta^{(2)}\ldots\beta^{(r_2)}}(y) = \mathfrak{R}_{n_1,q_{n_1},\alpha^{(1)},\ldots,\alpha^{(r_1)}}^{n_2,q_{n_2},\beta^{(1)},\ldots,\beta^{(r_2)}}((t_2 - y)^r; x, y),\ r \in \mathbb{N} \cup \{0\}$$

Further, let $\Pi_{n_1,q_{n_1},r}^{\alpha^{(1)},\alpha^{(2)}\ldots\alpha^{(r_1)}} = \sup_{x\in I} \vartheta_{n_1,q_{n_1},r}^{\alpha^{(1)},\alpha^{(2)}\ldots\alpha^{(r_1)}}(x)$, and $\Pi_{n_2,q_{n_2},r}^{\beta^{(1)},\beta^{(2)}\ldots\beta^{(r_2)}} = \sup_{y\in I} \vartheta_{n_2,q_{n_2},r}^{\beta^{(1)},\beta^{(2)}\ldots\beta^{(r_2)}}(y)$.

The proof of the other inequalities follow similarly hence the details are omitted.

We begin with the estimation of some raw moments of the operators (1.7).

Lemma 1.1 *The operators* $\mathfrak{R}^{n_2,q_{n_2},\beta^{(1)},\dots,\beta^{(r_2)}}_{n_1,q_{n_1},\alpha^{(1)},\dots,\alpha^{(r_1)}}(.;x,y)$ *satisfy*

$$\mathfrak{R}^{n_2,q_{n_2},\beta^{(1)},\dots,\beta^{(r_2)}}_{n_1,q_{n_1},\alpha^{(1)},\dots,\alpha^{(r_1)}}(1;x,y)=1,\quad \textit{for all } x,y\in I,$$

and

$$\left\|\mathfrak{R}^{n_2,q_{n_2},\beta^{(1)},\dots,\beta^{(r_2)}}_{n_1,q_{n_1},\alpha^{(1)},\dots,\alpha^{(r_1)}}(e_0^0)-e_0^0\right\|=0;$$

Proof Using (1.5), the proof easily follows. □

Lemma 1.2 *For the operators* $\mathfrak{R}^{n_2,q_{n_2},\beta^{(1)},\dots,\beta^{(r_2)}}_{n_1,q_{n_1},\alpha^{(1)},\dots,\alpha^{(r_1)}}(.;x,y)$*, we have*

$$\mathfrak{R}^{n_2,q_{n_2},\beta^{(1)},\dots,\beta^{(r_2)}}_{n_1,q_{n_1},\alpha^{(1)},\dots,\alpha^{(r_1)}}(t_1;x,y)=x\alpha_n^{(1)},\quad \textit{and}\quad \mathfrak{R}^{n_2,q_{n_2},\beta^{(1)},\dots,\beta^{(r_2)}}_{n_1,q_{n_1},\alpha^{(1)},\dots,\alpha^{(r_1)}}(t_2;x,y)=y\beta_{n_2}^{(1)},$$

and hence

$$\left\|\mathfrak{R}^{n_2,q_{n_2},\beta^{(1)},\dots,\beta^{(r_2)}}_{n_1,q_{n_1},\alpha^{(1)},\dots,\alpha^{(r_1)}}(e_1^0)-e_1^0\right\|=1-\alpha_{n_1}^{(1)};$$

$$\left\|\mathfrak{R}^{n_2,q_{n_2},\beta^{(1)},\dots,\beta^{(r_2)}}_{n_1,q_{n_1},\alpha^{(1)},\dots,\alpha^{(r_1)}}(e_0^1)-e_0^1\right\|=1-\beta_{n_2}^{(1)};$$

Proof We just show the proof of the first relation as the second one can be established similarly. Using (1.7), one has

$$\begin{aligned}\mathfrak{R}^{n_2,q_{n_2},\beta^{(1)},\dots,\beta^{(r_2)}}_{n_1,q_{n_1},\alpha^{(1)},\dots,\alpha^{(r_1)}}(t_1;x,y)&=\left\{\prod_{i=1}^{r_1}\prod_{j=1}^{r_2}(\alpha_i x^i;q_{n_1})_{n_1}(\beta_i y^i;q_{n_2})_{n_2}\right\}\sum_{p=0}^{\infty}\sum_{s=0}^{\infty}\Bigg\{\sum_{l_1+2l_2+\dots+r_1l_{r_1}=p}\;\sum_{h_1+2h_2+\dots+r_2h_{r_2}=s}\\&\left\{\prod_{k=1}^{r_1}\prod_{j=1}^{r_2}(q_{n_1}^{n_1},q_{n_1})_{l_k}\frac{(\alpha_{n_1}^{(k)})^{l_k}}{(q_{n_1},q_{n_1})_{l_k}}(q_{n_2}^{n_2},q_{n_2})_{h_j}\frac{(\beta_{n_2}^{(j)})^{h_j}}{(q_{n_2},q_{n_2})_{h_j}}\frac{[l_1]_{q_{n_1}}}{[n_1+l_1-1]_{q_{n_1}}}\right\}\Bigg\}x^py^s.\end{aligned}$$

Using some elementary transformations, $\frac{[l_1]_{q_{n_1}}}{(q_{n_1};q_{n_1})_{l_1}}=\frac{1}{(1-q_{n_1})(q_{n_1};q_{n_1})_{l_1-1}}$ and $\frac{(q_{n_1}^{n_1};q_{n_1})_{l_1}}{[n_1+l_1-1]_{q_{n_1}}}$ $=(1-q_{n_1})(q_{n_1}^{n_1};q_{n_1})_{l_1-1}$ and Lemma 1.1, we have

$$\begin{aligned}\mathfrak{R}^{n_2,q_{n_2},\beta^{(1)},\dots,\beta^{(r_2)}}_{n_1,q_{n_1},\alpha^{(1)},\dots,\alpha^{(r_1)}}(t_1;x,y)&=x\alpha_{n_1}^{(1)}\left\{\prod_{i=1}^{r_1}\prod_{j=1}^{r_2}(\alpha_i x^i;q_{n_1})_{n_1}(\beta_i y^i;q_{n_2})_{n_2}\right\}\sum_{p=1}^{\infty}\sum_{s=0}^{\infty}\Bigg\{\sum_{\substack{l_1+2l_2+\dots+r_1l_{r_1}-1=p-1\\ l_1\geq 1}}\\&\sum_{h_1+2h_2+\dots+r_2h_{r_2}=s}\;\prod_{j=1}^{r_2}\Big\{(q_{n_1}^{n_1};q_{n_1})_{l_1-1}\dots(q_{n_1}^{n_1};q_{n_1})_{l_{r_1}}\\&\frac{(\alpha_{n_1}^{(1)})^{l_1-1}\dots(\alpha_{n_1}^{(r_1)})^{l_{r_1}}}{(q_{n_1};q_{n_1})_{l_1-1}\dots(q_{n_1};q_{n_1})_{l_{r_1}}}\frac{\left(q_{n_2}^{n_2},q_{n_2}\right)_{h_j}(\beta_{n_2}^{(j)})^{h_j}}{(q_{n_2},q_{n_2})_{h_j}}\Big\}\Bigg\}x^{p-1}y^s\\&=x\alpha_{n_1}^{(1)}\left\{\prod_{i=1}^{r_1}(\alpha_i x^i;q_{n_1})_{n_1}\right\}\sum_{p=1}^{\infty}h_{p-1,q_{n_1}}^{(n_1,\dots,n)}(\alpha_{n_1}^{(1)},\alpha_{n_1}^{(2)},\dots,\alpha_{n_1}^{(r_1)})x^{p-1}\\&=x\alpha_{n_1}^{(1)}\left\{\prod_{i=1}^{r_1}(\alpha_i x^i;q_{n_1})_{n_1}\right\}\sum_{p=0}^{\infty}h_{p,q_{n_1}}^{(n_1,\dots,n_1)}(\alpha_{n_1}^{(1)},\alpha_{n_1}^{(2)},\dots,\alpha_{n_1}^{(r_1)})x^{p}\\&=x\alpha_{n_1}^{(1)},\quad \text{in view of (1.5).}\end{aligned}$$

The other equalities easily follow on taking the supremum over I. □

Lemma 1.3 *The operators* $\mathfrak{R}^{n_2,q_{n_2},\beta^{(1)},\dots,\beta^{(r_2)}}_{n_1,q_{n_1},\alpha^{(1)},\dots,\alpha^{(r_1)}}(.;x,y)$ *satisfy*

$$\mathfrak{R}^{n_2,q_{n_2},\beta^{(1)},\dots,\beta^{(r_2)}}_{n_1,q_{n_1},\alpha^{(1)},\dots,\alpha^{(r_1)}}(t_1^2;x,y) \le q_{n_1}(x\alpha^{(1)}_{n_1})^2 + \frac{x\alpha^{(1)}_{n_1}}{[n_1]_{q_{n_1}}};$$

$$\mathfrak{R}^{n_2,q_{n_2},\beta^{(1)},\dots,\beta^{(r_2)}}_{n_1,q_{n_1},\alpha^{(1)},\dots,\alpha^{(r_1)}}(t_2^2;x,y) \le q_{n_2}(y\beta^{(1)}_{n_2})^2 + \frac{y\beta^{(1)}_{n_2}}{[n_2]_{q_{n_2}}};$$

and

$$\left|\mathfrak{R}^{n_2,q_{n_2},\beta^{(1)},\dots,\beta^{(r_2)}}_{n_1,q_{n_1},\alpha^{(1)},\dots,\alpha^{(r_1)}}(t_1^2-x^2;x,y)\right| \le 2x^2(1-\alpha^{(1)}_{n_1}) + \frac{x\alpha^{(1)}_{n_1}}{[n_1]_{q_{n_1}}};$$

$$\left|\mathfrak{R}^{n_2,q_{n_2},\beta^{(1)},\dots,\beta^{(r_2)}}_{n_1,q_{n_1},\alpha^{(1)},\dots,\alpha^{(r_1)}}(t_2^2-y^2;x,y)\right| \le 2y^2(1-\beta^{(1)}_{n_2}) + \frac{y\beta^{(1)}_{n_2}}{[n_2]_{q_{n_2}}}.$$

Hence,

$$\left\|\mathfrak{R}^{n_2,q_{n_2},\beta^{(1)},\dots,\beta^{(r_2)}}_{n_1,q_{n_1},\alpha^{(1)},\dots,\alpha^{(r_1)}}(e_2^0)-e_2^0\right\| \le 2(1-\alpha^{(1)}_{n_1}) + \frac{\alpha^{(1)}_{n_1}}{[n_1]_{q_{n_1}}};$$

$$\left\|\mathfrak{R}^{n_2,q_{n_2},\beta^{(1)},\dots,\beta^{(r_2)}}_{n_1,q_{n_1},\alpha^{(1)},\dots,\alpha^{(r_1)}}(e_0^2)-e_0^2\right\| \le 2(1-\beta^{(1)}_{n_2}) + \frac{\beta^{(1)}_{n_2}}{[n_2]_{q_{n_2}}}.$$

Proof From the definition of the operator $\mathfrak{R}^{n_2,q_{n_2},\beta^{(1)},\dots,\beta^{(r_2)}}_{n_1,q_{n_1},\alpha^{(1)},\dots,\alpha^{(r_1)}}(.;x,y)$, we have

$$\mathfrak{R}^{n_2,q_{n_2},\beta^{(1)},\dots,\beta^{(r_2)}}_{n_1,q_{n_1},\alpha^{(1)},\dots,\alpha^{(r_1)}}(t_1^2;x,y) =$$
$$\left\{\prod_{i=1}^{r_1}\prod_{j=1}^{r_2}(\alpha_i x^i;q_{n_1})_{n_1}(\beta_i y^i;q_{n_2})_{n_2}\right\}\sum_{p=1}^{\infty}\sum_{s=0}^{\infty}\Bigg\{\sum_{\substack{l_1+2l_2+\dots+r_1l_{r_1}=p\\ l_{r_1}\ge 1}}\ \sum_{h_1+2h_2+\dots+r_2h_{r_2}=s}$$
$$\left\{\prod_{k=1}^{r_1}\prod_{j=1}^{r_2}(q^{n_1}_{n_1},q_{n_1})_{l_k}\frac{(\alpha^{(k)}_{n_1})^{l_k}}{(q_{n_1},q_{n_1})_{l_k}}(q^{n_2}_{n_2},q_{n_2})_{h_j}\frac{(\beta^{(j)}_{n_2})^{h_j}}{(q_{n_2},q_{n_2})_{h_j}}\left(\frac{[l_1]_{q_{n_1}}}{[n_1+l_1-1]_{q_{n_1}}}\right)^2\right\}\Bigg\}x^py^s.$$

By using Lemma 1.1, we obtain,

$$\mathfrak{R}^{n_2,q_{n_2},\beta^{(1)},\dots,\beta^{(r_2)}}_{n_1,q_{n_1},\alpha^{(1)},\dots,\alpha^{(r_1)}}(t_1^2;x,y)$$
$$= x\alpha^{(1)}_{n_1}\left\{\prod_{i=1}^{r_1}(\alpha_i x^i;q)_{n_1}\right\}\sum_{p=1}^{\infty}\Bigg\{\sum_{\substack{l_1+2l_2+\dots+r_1l_{r_1}-1=p-1\\ l_1\ge 1}}(q^{n_1}_{n_1};q_{n_1})_{l_1-1}\dots(q^{n_1}_{n_1};q_{n_1})_{l_{r_1}}$$

$$\frac{[l_1]_{q_{n_1}}}{[n_1+l_1-1]_{q_{n_1}}}\frac{(\alpha_{n_1}^{(1)})^{l_1-1}\dots(\alpha_{n_1}^{(r_1)})^{l_{r_1}}}{(q_{n_1};q_{n_1})_{l_1-1}\dots(q_{n_1};q_{n_1})_{l_{r_1}}}\Bigg\}x^{p-1}$$

$$=x\alpha_{n_1}^{(1)}\left\{\prod_{i=1}^{r_1}(\alpha_i x^i;q_{n_1})_{n_1}\right\}\sum_{p=1}^{\infty}\Bigg\{\sum_{\substack{l_1+2l_2+\ldots+r_1l_{r_1}-1=p-1\\ l_1\geq 1}}(q_{n_1}^{n_1};q_{n_1})_{l_1-1}\dots(q_{n_1}^{n_1};q_{n_1})_{l_{r_1}}$$

$$\left(\frac{1+q_{n_1}[l_1-1]_{q_{n_1}}}{[n_1+l_1-1]_{q_{n_1}}}\right)\frac{(\alpha_{n_1}^{(1)})^{l_1-1}\dots(\alpha_{n_1}^{(r_1)})^{l_{r_1}}}{(q_{n_1};q_{n_1})_{l_1-1}\dots(q_{n_1};q_{n_1})_{l_{r_1}}}\Bigg\}x^{p-1}=\sum_1+\sum_2,\quad\text{say.}\tag{1.8}$$

Here,

$$\sum_1=x\alpha_{n_1}^{(1)}\left\{\prod_{i=1}^{r_1}(\alpha_i x^i;q_{n_1})_{n_1}\right\}\sum_{p=1}^{\infty}\Bigg\{\sum_{\substack{l_1+2l_2+\ldots+r_1l_{r_1}-1=p-1\\ l_1\geq 1}}(q_{n_1}^{n_1};q_{n_1})_{l_1-1}\dots(q_{n_1}^{n_1};q_{n_1})_{l_{r_1}}$$

$$\left(\frac{1}{[n_1+l_1-1]_{q_{n_1}}}\right)\frac{(\alpha_{n_1}^{(1)})^{l_1-1}\dots(\alpha_{n_1}^{(r_1)})^{l_{r_1}}}{(q_{n_1};q_{n_1})_{l_1-1}\dots(q_{n_1};q_{n_1})_{l_{r_1}}}\Bigg\}x^{p-1},$$

since $\frac{1}{[n_1+l_1-1]_{q_{n_1}}}\leq\frac{1}{[n_1]_{q_{n_1}}}$, then using (1.5), we have

$$\sum_1\leq\frac{x\alpha_{n_1}^{(1)}}{[n_1]_{q_{n_1}}}.\tag{1.9}$$

Now,

$$\sum_2=q_{n_1}x\alpha_{n_1}^{(1)}\left\{\prod_{i=1}^{r_1}(\alpha_i x^i;q_{n_1})_{n_1}\right\}\sum_{p=1}^{\infty}\Bigg\{\sum_{\substack{l_1+2l_2+\ldots+r_1l_{r_1}-1=p-1\\ l_1\geq 1}}(q_{n_1}^{n_1};q_{n_1})_{l_1-1}\dots(q_{n_1}^{n_1};q_{n_1})_{l_{r_1}}$$

$$\left(\frac{[l_1-1]_{q_{n_1}}}{[n_1+l_1-1]_{q_{n_1}}}\right)\frac{(\alpha_{n_1}^{(1)})^{l_1-1}\dots(\alpha_{n_1}^{(r_1)})^{l_{r_1}}}{(q_{n_1};q_{n_1})_{l_1-1}\dots(q_{n_1};q_{n_1})_{l_{r_1}}}\Bigg\}x^{p-1}$$

$$=q_{n_1}(x\alpha_{n_1}^{(1)})^2\left\{\prod_{i=1}^{r_1}(\alpha_i x^i;q_{n_1})_{n_1}\right\}\sum_{p=2}^{\infty}\Bigg\{\sum_{\substack{l_1+2l_2+\ldots+r_1l_{r_1}-2=p-2\\ l_1\geq 2}}(q_{n_1}^{n_1};q_{n_1})_{l_1-2}\dots(q_{n_1}^{n_1};q_{n_1})_{l_{rn_1}}$$

$$\left(\frac{(1-q_{n_1}^{l_1-1})(1-q_{n_1}^{n_1+l_1-2})}{(1-q_{n_1}^{l_1-1})(1-q_{n_1})[n_1+l_1-1]_{q_{n_1}}}\right)\frac{(\alpha_{n_1}^{(1)})^{l_1-2}\dots(\alpha_{n_1}^{(r_1)})^{l_{r_1}}}{(q_{n_1};q_{n_1})_{l_1-2}\dots(q_{n_1};q_{n_1})_{l_{r_1}}}\Bigg\}x^{p-2}$$

$$=q_{n_1}(x\alpha_{n_1}^{(1)})^2\left\{\prod_{i=1}^{r_1}(\alpha_i x^i;q_{n_1})_{n_1}\right\}\sum_{p=2}^{\infty}\Bigg\{\sum_{\substack{l_1+2l_2+\ldots+r_1l_{r_1}-2=p-2\\ l_1\geq 2}}(q_{n_1}^{n_1};q_{n_1})_{l_1-2}\dots(q_{n_1}^{n_1};q_{n_1})_{l_{r_1}}$$

$$\left(\frac{[n_1+l_1-2]_{q_{n_1}}}{[n_1+l_1-1]_{q_{n_1}}}\right)\frac{(\alpha_{n_1}^{(1)})^{l_1-2}\dots(\alpha_{n_1}^{(r_1)})^{l_{r_1}}}{(q_{n_1};q_{n_1})_{l_1-2}\dots(q_{n_1};q_{n_1})_{l_{r_1}}}\Bigg\}x^{p-2},$$

since $\frac{[n_1+l_1-2]_{q_{n_1}}}{[n_1+l_1-1]_{q_{n_1}}}<1$, in view of (1.5) we get

$$\sum_{2} \leq q_{n_1}(x\alpha_{n_1}^{(1)})^2. \tag{1.10}$$

Finally, using (1.9) and (1.10) in (1.8), we obtain

$$\mathfrak{R}_{n_1,q_{n_1},\alpha^{(1)},\cdots,\alpha^{(r_1)}}^{n_2,q_{n_2},\beta^{(1)},\ldots,\beta^{(r_2)}}(t_1^2;x,y) \leq q_{n_1}(x\alpha_{n_1}^{(1)})^2 + \frac{x\alpha_{n_1}^{(1)}}{[n_1]_{q_{n_1}}}. \tag{1.11}$$

Hence, we can write

$$\mathfrak{R}_{n_1,q_{n_1},\alpha^{(1)},\ldots,\alpha^{(r_1)}}^{n_2,q_{n_2},\beta^{(1)},\ldots,\beta^{(r_2)}}(t_1^2;x,y) - x^2 \leq x^2(q_{n_1}(\alpha_{n_1}^{(1)})^2 - 1) + \frac{x\alpha_{n_1}^{(1)}}{[n_1]_{q_{n_1}}}$$
$$= -x^2(1 - q_{n_1}(\alpha_{n_1}^{(1)})^2) + \frac{x\alpha_{n_1}^{(1)}}{[n_1]_{q_{n_1}}}.$$

As $q_{n_1}, \alpha_{n_1}^{(1)} \in (0,1)$, we have

$$\mathfrak{R}_{n_1,q_{n_1},\alpha^{(1)},\ldots,\alpha^{(r_1)}}^{n_2,q_{n_2},\beta^{(1)},\ldots,\beta^{(r_2)}}(t_1^2;x,y) - x^2 \leq \frac{x\alpha_{n_1}^{(1)}}{[n_1]_{q_{n_1}}}. \tag{1.12}$$

Observe that Lemmas 1.1 and 1.2, allow us to write

$$0 \leq \mathfrak{R}_{n_1,q_{n_1},\alpha^{(1)},\ldots,\alpha^{(r_1)}}^{n_2,q_{n_2},\beta^{(1)},\ldots,\beta^{(r_2)}}((t_1-x)^2;x,y) = \mathfrak{R}_{n_1,q_{n_1},\alpha^{(1)},\ldots,\alpha^{(r_1)}}^{n_2,q_{n_2},\beta^{(1)},\ldots,\beta^{(r_2)}}(t_1^2;x,y) - 2x\mathfrak{R}_{n_1,q_{n_1},\alpha^{(1)},\ldots,\alpha^{(r_1)}}^{n_2,q_{n_2},\beta^{(1)},\ldots,\beta^{(r_2)}}(t_1;x,y) + x^2;$$
$$\implies \quad -2x^2(1-\alpha_{n_1}^{(1)}) \leq \mathfrak{R}_{n_1,q_{n_1},\alpha^{(1)},\ldots,\alpha^{(r_1)}}^{n_2,q_{n_2},\beta^{(1)},\ldots,\beta^{(r_2)}}(t_1^2;x,y) - x^2$$
$$\text{or,} \quad -2x^2(1-\alpha_{n_1}^{(1)}) - \frac{x\alpha_{n_1}^{(1)}}{[n_1]_{q_{n_1}}} \leq \mathfrak{R}_{n_1,q_{n_1},\alpha^{(1)},\ldots,\alpha^{(r_1)}}^{n_2,q_{n_2},\beta^{(1)},\ldots,\beta^{(r_2)}}(t_1^2;x,y) - x^2.$$

Hence, in view of (1.12), we obtain

$$\left|\mathfrak{R}_{n_1,q_{n_1},\alpha^{(1)},\ldots,\alpha^{(r_1)}}^{n_2,q_{n_2},\beta^{(1)},\ldots,\beta^{(r_2)}}(t_1^2;x,y) - x^2\right| \leq 2x^2(1-\alpha_{n_1}^{(1)}) + \frac{x\alpha_{n_1}^{(1)}}{[n_1]_{q_{n_1}}}.$$

□

We begin with the following Korovkin type result for the operators $\mathfrak{R}_{n_1,q_{n_1},\alpha^{(1)},\ldots,\alpha^{(r_1)}}^{n_2,q_{n_2},\beta^{(1)},\ldots,\beta^{(r_2)}}$.

Theorem 1.1 *For all* $\varphi \in C(I^2)$, *and* $n_1, n_2 \in \mathbb{N}$, *we have*

$$\lim_{n_1,n_2\to\infty} \left\|\mathfrak{R}_{n_1,q_{n_1},\alpha^{(1)},\ldots,\alpha^{(r_1)}}^{n_2,q_{n_2},\beta^{(1)},\ldots,\beta^{(r_2)}}(\varphi) - \varphi\right\| = 0$$

if and only if

$$\lim_{n_1\to\infty} \alpha_{n_1}^{(1)} = 1 \ \text{and} \ \lim_{n_2\to\infty} \beta_{n_2}^{(1)} = 1.$$

Proof Suppose $\lim_{n_1\to\infty} \alpha_{n_1}^{(1)} = 1$ and $\lim_{n_2\to\infty} \beta_{n_2}^{(1)} = 1$, then applying Lemmas 1.1–1.3, it follows that

$$\lim_{n_1,n_2\to\infty} \|\mathfrak{R}_{n_1,q_{n_1},\alpha^{(1)},\ldots,\alpha^{(r_1)}}^{n_2,q_{n_2},\beta^{(1)},\ldots,\beta^{(r_2)}}(e_i^j) - e_i^j\| = 0, \quad \text{for } (i,j) \in \{(0,0),(1,0),(0,1)\}$$

$$\text{and} \lim_{n_1,n_2\to\infty} \|\mathfrak{R}_{n_1,q_{n_1},\alpha^{(1)},\ldots,\alpha^{(r_1)}}^{n_2,q_{n_2},\beta^{(1)},\ldots,\beta^{(r_2)}}(e_2^0 + e_0^2) - (e_2^0 + e_0^2)\| = 0.$$

Now using the Volkov theorem [80], we have

$$\lim_{n_1,n_2\to\infty} \left\|\mathfrak{R}_{n_1,q_{n_1},\alpha^{(1)},\ldots,\alpha^{(r_1)}}^{n_2,q_{n_2},\beta^{(1)},\ldots,\beta^{(r_2)}}(\varphi) - \varphi\right\| = 0.$$

Conversely, assume that

$$\lim_{n_1,n_2\to\infty} \left\|\mathfrak{R}_{n_1,q_{n_1},\alpha^{(1)},\ldots,\alpha^{(r_1)}}^{n_2,q_{n_2},\beta^{(1)},\ldots,\beta^{(r_2)}}(\varphi) - \varphi\right\| = 0$$

Using Lemma 1.1, we obtain

$$\left\|\mathfrak{R}_{n_1,q_{n_1},\alpha^{(1)},\ldots,\alpha^{(r_1)}}^{n_2,q_{n_2},\beta^{(1)},\ldots,\beta^{(r_2)}}(e_1^0; x, y) - x\right\| = 1 - \alpha_{n_1}^{(1)};$$
$$\left\|\mathfrak{R}_{n_1,q_{n_1},\alpha^{(1)},\ldots,\alpha^{(r_1)}}^{n_2,q_{n_2},\beta^{(1)},\ldots,\beta^{(r_2)}}(e_0^1; x, y) - y\right\| = 1 - \beta_{n_2}^{(1)};$$

which imply that $\lim_{n_1\to\infty} \alpha_{n_1}^{(1)} = 1$ and $\lim_{n_2\to\infty} \beta_{n_2}^{(1)} = 1$. □

1.3 Rate of Convergence

First, we give some definitions which will be required to examine the convergence rate of (1.7).

Definition 1.1 Let $\varphi \in C(I^2)$ and $\delta > 0$ be arbitrary. The total modulus of continuity and the partial moduli of continuity of φ are defined as

$$\omega^{(c)}(\varphi;\delta) = \sup_{\sqrt{(s_1-t_1)^2+(s_2-t_2)^2}\le\delta} \{|\varphi(s_1,s_2) - \varphi(t_1,t_2)| : (s_1,s_2),(t_1,t_2) \in I^2\} \tag{1.13}$$

and

$$\omega^{(1)}(\varphi;\delta) = \sup_{|s_1-s_2|\le\delta} \sup_{y\in I}\{|\varphi(s_1,y) - \varphi(s_2,y)|\}$$
$$\omega^{(2)}(\varphi;\delta) = \sup_{x\in I} \sup_{|t_1-t_2|\le\delta} \{|\varphi(x,t_1) - \varphi(x,t_2)|\}, \tag{1.14}$$

respectively. Clearly, $\omega^{(c)}(\varphi;\delta)$ is a nondecreasing function of δ and tends to 0, as $\delta \to 0$. Further, it is known that

$$|\varphi(s_1,s_2)-\varphi(t_1,t_2)| \le \omega^{(c)}(\varphi;\delta)\left(1+\frac{\sqrt{(s_1-t_1)^2+(s_2-t_2)^2}}{\delta}\right). \quad (1.15)$$

Evidently, the partial modulus of continuity behaves like the usual modulus of continuity, the details can be seen in [29].

Theorem 1.2 *If $n_1, n_2 \in \mathbb{N}$ and $\varphi \in C(I^2)$, then the following inequality holds:*

$$\left\| \mathfrak{R}^{n_2,q_{n_2},\beta^{(1)},\ldots,\beta^{(r_2)}}_{n_1,q_{n_1},\alpha^{(1)},\ldots,\alpha^{(r_1)}}(\varphi)-\varphi \right\| \le 2\ \omega^{(c)}\left(\varphi;\delta_{n_1,n_2}\right), \quad (1.16)$$

where

$$\delta_{n_1,n_2}=\sqrt{4(1-\alpha^{(1)}_{n_1})+\frac{\alpha^{(1)}_{n_1}}{[n_1]_{q_{n_1}}}+4(1-\beta^{(1)}_{n_2})+\frac{\beta^{(1)}_{n_2}}{[n_2]_{q_{n_2}}}}. \quad (1.17)$$

Proof In view of the property (1.15), Lemma 1.1 and the Cauchy–Schwarz inequality, we obtain

$$\begin{aligned}|\mathfrak{R}^{n_2,q_{n_2},\beta^{(1)},\ldots,\beta^{(r_2)}}_{n_1,q_{n_1},\alpha^{(1)},\ldots,\alpha^{(r_1)}}(\varphi;x,y)-\varphi(x,y)| &\le \mathfrak{R}^{n_2,q_{n_2},\beta^{(1)},\ldots,\beta^{(r_2)}}_{n_1,q_{n_1},\alpha^{(1)},\ldots,\alpha^{(r_1)}}(|\varphi(t_1,t_2)-\varphi(x,y)|;x,y)\\ &\le \omega^{(c)}(\varphi;\delta)\left(1+\frac{\sqrt{\mathfrak{R}^{n_2,q_{n_2},\beta^{(1)},\ldots,\beta^{(r_2)}}_{n_1,q_{n_1},\alpha^{(1)},\ldots,\alpha^{(r_1)}}((t_1-x)^2+(t_2-y)^2;x,y)}}{\delta}\right). \quad (1.18)\end{aligned}$$

Since,

$$\begin{aligned}&\mathfrak{R}^{n_2,q_{n_2},\beta^{(1)},\ldots,\beta^{(r_2)}}_{n_1,q_{n_1},\alpha^{(1)},\ldots,\alpha^{(r_1)}}\left((t_1-x)^2+(t_2-y)^2;x,y\right) \le |\mathfrak{R}^{n_2,q_{n_2},\beta^{(1)},\ldots,\beta^{(r_2)}}_{n_1,q_{n_1},\alpha^{(1)},\ldots,\alpha^{(r_1)}}\left(e^0_2;x,y\right)-e^0_2(x,y)|\\ &+|\mathfrak{R}^{n_2,q_{n_2},\beta^{(1)},\ldots,\beta^{(r_2)}}_{n_1,q_{n_1},\alpha^{(1)},\ldots,\alpha^{(r_1)}}\left(e^2_0;x,y\right)-e^2_0(x,y)|+2x|\mathfrak{R}^{n_2,q_{n_2},\beta^{(1)},\ldots,\beta^{(r_2)}}_{n_1,q_{n_1},\alpha^{(1)},\ldots,\alpha^{(r_1)}}\left(e^0_1;x,y\right)-e^0_1(x,y)|\\ &+2y|\mathfrak{R}^{n_2,q_{n_2},\beta^{(1)},\ldots,\beta^{(r_2)}}_{n_1,q_{n_1},\alpha^{(1)},\ldots,\alpha^{(r_1)}}\left(e^1_0;x,y\right)-e^1_0(x,y)|, \quad (1.19)\end{aligned}$$

applying Lemmas 1.3 and 1.2, we have

$$\begin{aligned}\left\|\mathfrak{R}^{n_2,q_{n_2},\beta^{(1)},\ldots,\beta^{(r_2)}}_{n_1,q_{n_1},\alpha^{(1)},\ldots,\alpha^{(r_1)}}\left((t_1-x)^2+(t_2-y)^2\right)\right\| &\le 4(1-\alpha^{(1)}_{n_1})+\frac{\alpha^{(1)}_{n_1}}{[n_1]_{q_{n_1}}}\\ &+4(1-\beta^{(1)}_{n_2})+\frac{\beta^{(1)}_{n_2}}{[n_2]_{q_{n_2}}}. \quad (1.20)\end{aligned}$$

Therefore, considering relation (1.18) and (1.20), we get

$$|\mathfrak{R}^{n_2,q_{n_2},\beta^{(1)},...,\beta^{(r_2)}}_{n_1,q_{n_1},\alpha^{(1)},...,\alpha^{(r_1)}}(\varphi; x, y) - \varphi(x, y)|$$
$$\leq \omega^{(c)}(\varphi; \delta)\left(1 + \frac{\sqrt{4(1-\alpha^{(1)}_{n_1}) + \frac{\alpha^{(1)}_{n_1}}{[n_1]_{q_{n_1}}} + 4(1-\beta^{(1)}_{n_2}) + \frac{\beta^{(1)}_{n_2}}{[n_2]_{q_{n_2}}}}}{\delta}\right).$$

Now choosing $\delta = \delta_{n_1,n_2} = \sqrt{4(1-\alpha^{(1)}_{n_1}) + \frac{\alpha^{(1)}_{n_1}}{[n_1]_{q_{n_1}}} + 4(1-\beta^{(1)}_{n_2}) + \frac{\beta^{(1)}_{n_2}}{[n_2]_{q_{n_2}}}}$, we obtain the desired result. □

In our further consideration, let us define $\chi_{n_1} = 4(1-\alpha^{(1)}_{n_1}) + \frac{\alpha^{(1)}_{n_1}}{[n_1]_{q_{n_1}}}$ and $\chi_{n_2} = 4(1-\beta^{(2)}_{n_2}) + \frac{\beta^{(2)}_{n_2}}{[n_2]_{q_{n_2}}}$. The next theorem provides the convergence rate of (1.7) via partial moduli of continuity for the functions in $C(I^2)$.

Theorem 1.3 *For all $n_1, n_2 \in \mathbb{N}$ and $\varphi \in C(I^2)$, we have*

$$\left\|\mathfrak{R}^{n_2,q_{n_2},\beta^{(1)},...,\beta^{(r_2)}}_{n_1,q_{n_1},\alpha^{(1)},...,\alpha^{(r_1)}}(\varphi) - \varphi\right\| \leq 2\left\{\omega^{(1)}\left(\varphi; \sqrt{\chi_{n_1}}\right) + \omega^{(2)}\left(\varphi; \sqrt{\chi_{n_2}}\right)\right\}.$$

Proof Using the definition of partial moduli of continuity with respect to x and y, we have

$$|\mathfrak{R}^{n_2,q_{n_2},\beta^{(1)},...,\beta^{(r_2)}}_{n_1,q_{n_1},\alpha^{(1)},...,\alpha^{(r_1)}}(\varphi(t_1,t_2); x, y) - \varphi(x, y)| \leq \{\mathfrak{R}^{n_2,q_{n_2},\beta^{(1)},...,\beta^{(r_2)}}_{n_1,q_{n_1},\alpha^{(1)},...,\alpha^{(r_1)}}(\omega^{(2)}(\varphi; |t_2 - y|); x, y)$$
$$+ \mathfrak{R}^{n_2,q_{n_2},\beta^{(1)},...,\beta^{(r_2)}}_{n_1,q_{n_1},\alpha^{(1)},...,\alpha^{(r_1)}}(\omega^{(1)}(\varphi; |t_1 - x|); x, y)\}.$$

Using Cauchy–Schwarz inequality with $\delta_1 = \sqrt{\chi_{n_1}}$ and $\delta_2 = \sqrt{\chi_{n_2}}$, we obtain the desired result. □

Now, we propose to discuss the convergence behavior of the operators $\mathfrak{R}^{n_2,q_{n_2},\beta^{(1)},...,\beta^{(r_2)}}_{n_1,q_{n_1},\alpha^{(1)},...,\alpha^{(r_1)}}(.; x, y)$ for functions in the Lipschitz class $Lip_M(\gamma_1, \gamma_2)$, for $0 < \gamma_1, \gamma_2 \leq 1$. We know that a function φ belongs to $Lip_M(\gamma_1, \gamma_2)$ $(M > 0 \text{ and } 0 < \gamma_1, \gamma_2 \leq 1)$, provided the following inequality holds:

$$|\varphi(s_1, s_2) - \varphi(t_1, t_2)| \leq M\,|s_1 - t_1|^{\gamma_1}\,|s_2 - t_2|^{\gamma_2}, \ (s_1, s_2), (t_1, t_2) \in I^2. \tag{1.21}$$

Theorem 1.4 *For all $n_1, n_2 \in \mathbb{N}$ and $\varphi \in Lip_M(\gamma_1, \gamma_2)$, there holds the inequality*

$$\left\|\mathfrak{R}^{n_2,q_{n_2},\beta^{(1)},...,\beta^{(r_2)}}_{n_1,q_{n_1},\alpha^{(1)},...,\alpha^{(r_1)}}(\varphi) - \varphi\right\| \leq M(\chi_{n_1})^{\gamma_1/2}(\chi_{n_2})^{\gamma_2/2}. \tag{1.22}$$

Proof Let $\varphi \in Lip_M(\gamma_1, \gamma_2)$, $0 < \gamma_1, \gamma_2 \leq 1$ and $(t_1, t_2), (x, y) \in I^2$. By linearity and monotonicity of $\mathfrak{R}^{n_2,q_{n_2},\beta^{(1)},\ldots,\beta^{(r_2)}}_{n_1,q_{n_1},\alpha^{(1)},\ldots,\alpha^{(r_1)}}$, and using (1.21) we reach to the following

$$\begin{aligned}|\mathfrak{R}^{n_2,q_{n_2},\beta^{(1)},\ldots,\beta^{(r_2)}}_{n_1,q_{n_1},\alpha^{(1)},\ldots,\alpha^{(r_1)}}(\varphi; x, y) - \varphi(x, y)| &\leq \mathfrak{R}^{n_2,q_{n_2},\beta^{(1)},\ldots,\beta^{(r_2)}}_{n_1,q_{n_1},\alpha^{(1)},\ldots,\alpha^{(r_1)}}(|\varphi(t_1, t_2) - \varphi(x, y)|; x, y)\\ &\leq M\mathfrak{R}^{n_2,q_{n_2},\beta^{(1)},\ldots,\beta^{(r_2)}}_{n_1,q_{n_1},\alpha^{(1)},\ldots,\alpha^{(r_1)}}(|t_1 - x|^{\gamma_1}\, |t_2 - y|^{\gamma_2}\, ; x, y)\\ &= M\mathfrak{R}^{n_2,q_{n_2},\beta^{(1)},\ldots,\beta^{(r_2)}}_{n_1,q_{n_1},\alpha^{(1)},\ldots,\alpha^{(r_1)}}(|t_1 - x|^{\gamma_1}\, ; x, y)\\ &\times \mathfrak{R}^{n_2,q_{n_2},\beta^{(1)},\ldots,\beta^{(r_2)}}_{n_1,q_{n_1},\alpha^{(1)},\ldots,\alpha^{(r_1)}}(|t_2 - y|^{\gamma_2}\, ; x, y). \qquad (1.23)\end{aligned}$$

Applying the Hölder's inequality for $u_1 = \frac{2}{\gamma_1}$ and $v_1 = \frac{2}{2-\gamma_1}$, and $u_2 = \frac{2}{\gamma_2}$ and $v_2 = \frac{2}{2-\gamma_2}$, in view of Lemma 1.3 and (1.20), we get

$$\begin{aligned}\left\|\mathfrak{R}^{n_2,q_{n_2},\beta^{(1)},\ldots,\beta^{(r_2)}}_{n_1,q_{n_1},\alpha^{(1)},\ldots,\alpha^{(r_1)}}(\varphi) - \varphi\right\| &\leq M\big[\mathfrak{R}^{n_2,q_{n_2},\beta^{(1)},\ldots,\beta^{(r_2)}}_{n_1,q_{n_1},\alpha^{(1)},\ldots,\alpha^{(r_1)}}((t_1 - x)^2\, ; x, y)\big]^{\frac{\gamma_1}{2}}\\ &\times \big[\mathfrak{R}^{n_2,q_{n_2},\beta^{(1)},\ldots,\beta^{(r_2)}}_{n_1,q_{n_1},\alpha^{(1)},\ldots,\alpha^{(r_1)}}((t_2 - y)^2\, ; x, y)\big]^{\frac{\gamma_2}{2}}\\ &\leq M(\chi_{n_1})^{\gamma_1/2}(\chi_{n_2})^{\gamma_2/2},\end{aligned}$$

which proves the required result. □

For $s \in \mathbb{N} \cup \{0\}$, let $C^s(I^2) := \left\{\varphi : \frac{\partial^{i+j}\varphi}{\partial x^i \partial y^j} \in C(I^2),\ \forall\, 0 \leq i + j \leq s,\ i, j \in \mathbb{N} \cup \{0\}\right\}$.

Our next theorem provides the rate of convergence by the operators $\mathfrak{R}^{n_2,q_{n_2},\beta^{(1)},\ldots,\beta^{(r_2)}}_{n_1,q_{n_1},\alpha^{(1)},\ldots,\alpha^{(r_1)}}$ for functions in $C^1(I^2)$.

Theorem 1.5 *For $\varphi \in C^1(I^2)$, we have*

$$\left\|\mathfrak{R}^{n_2,q_{n_2},\beta^{(1)},\ldots,\beta^{(r_2)}}_{n_1,q_{n_1},\alpha^{(1)},\ldots,\alpha^{(r_1)}}(\varphi) - \varphi\right\| \leq \|\varphi'_x\|\, \sqrt{\chi_{n_1}} + \|\varphi'_y\|\, \sqrt{\chi_{n_2}}.$$

Proof Since $\varphi \in C^1(I^2)$, we can write as follows:

$$\varphi(t_1, t_2) - \varphi(x, y) = \int_x^{t_1} \varphi'_w(w, t_2)dw + \int_y^{t_2} \varphi'_u(x, u)du.$$

Consequently, $\mathfrak{R}^{n_2,q_{n_2},\beta^{(1)},\dots,\beta^{(r_2)}}_{n_1,q_{n_1},\alpha^{(1)},\dots,\alpha^{(r_1)}}(.;\ x,y)$, we are led to

$$\left|\mathfrak{R}^{n_2,q_{n_2},\beta^{(1)},\dots,\beta^{(r_2)}}_{n_1,q_{n_1},\alpha^{(1)},\dots,\alpha^{(r_1)}}(\varphi;x,y)-\varphi(x,y)\right| \leq \mathfrak{R}^{n_2,q_{n_2},\beta^{(1)},\dots,\beta^{(r_2)}}_{n_1,q_{n_1},\alpha^{(1)},\dots,\alpha^{(r_1)}}\left(\left|\int_x^{t_1}|\varphi'_w(w,t_2)|dw\right|;x,y\right)$$
$$+\mathfrak{R}^{n_2,q_{n_2},\beta^{(1)},\dots,\beta^{(r_2)}}_{n_1,q_{n_1},\alpha^{(1)},\dots,\alpha^{(r_1)}}\left(\left|\int_y^{t_2}|\varphi'_u(x,u)|du\right|;x,y\right).$$

Since

$$\left|\int_x^{t_1}|\varphi'_w(w,t_2)|dw\right| \leq \|\varphi'_x\|\ |t_1-x| \ \text{ and } \ \left|\int_y^{t_2}|\varphi'_u(x,u)|du\right| \leq \|\varphi'_y\|\ |t_2-y|,$$

we easily reach

$$\left|\mathfrak{R}^{n_2,q_{n_2},\beta^{(1)},\dots,\beta^{(r_2)}}_{n_1,q_{n_1},\alpha^{(1)},\dots,\alpha^{(r_1)}}(\varphi;x,y)-\varphi(x,y)\right| \leq \|\varphi'_x\|\ \mathfrak{R}^{n_2,q_{n_2},\beta^{(1)},\dots,\beta^{(r_2)}}_{n_1,q_{n_1},\alpha^{(1)},\dots,\alpha^{(r_1)}}(|t_1-x|;x,y)$$
$$+\|\varphi'_y\|\ \mathfrak{R}^{n_2,q_{n_2},\beta^{(1)},\dots,\beta^{(r_2)}}_{n_1,q_{n_1},\alpha^{(1)},\dots,\alpha^{(r_1)}}(|t_2-y|;x,y).$$

Applying the Cauchy-Schwarz inequality, Lemmas 1.1 and 1.3, we get

$$\left\|\mathfrak{R}^{n_2,q_{n_2},\beta^{(1)},\dots,\beta^{(r_2)}}_{n_1,q_{n_1},\alpha^{(1)},\dots,\alpha^{(r_1)}}(\varphi)-\varphi\right\| \leq \|\varphi'_x\|\left(\Pi^{\alpha^{(1)},\alpha^{(2)}\dots\alpha^{(r_1)}}_{n_1,q_{n_1},2}\right)^{\frac{1}{2}}\left(\mathfrak{R}^{n_2,q_{n_2},\beta^{(1)},\dots,\beta^{(r_2)}}_{n_1,q_{n_1},\alpha^{(1)},\dots,\alpha^{(r_1)}}(e_{00})\right)^{\frac{1}{2}}$$
$$+\|\varphi'_y\|\left(\Pi^{\beta^{(1)},\beta^{(2)}\dots\beta^{(r_2)}}_{n_2,q_{n_2},2}\right)^{\frac{1}{2}}\left(\mathfrak{R}^{n_2,q_{n_2},\beta^{(1)},\dots,\beta^{(r_2)}}_{n_1,q_{n_1},\alpha^{(1)},\dots,\alpha^{(r_1)}}(e_{00})\right)^{\frac{1}{2}}$$
$$\leq \|\varphi'_y\|\left\{4(1-\alpha^{(r_1)}_{n_1})+\frac{\alpha^{(r_1)}_{n_1}}{[n_1]_{q_{n_1}}}\right\}^{\frac{1}{2}}+\|\varphi'_y\|\left\{4(1-\beta^{(r_2)}_{n_2})+\frac{\beta^{(r_2)}_{n_2}}{[n_2]_{q_{n_2}}}\right\}^{\frac{1}{2}}$$
$$\leq \|\varphi'_x\|\ \sqrt{\chi_{n_1}}+\|\varphi'_y\|\ \sqrt{\chi_{n_2}}.$$

□

Our next result presents the order of convergence of the operators (1.7) for functions in $C(I^2)$, via the approach of Peetre's $\mathcal{K}$-functional. For any $\psi \in C^2(I^2)$, let us consider the norm on the space $C^2(I^2)$ defined as

$$\|\psi\|_{C^2(I^2)} = \|\psi\| + \sum_{i=1}^{2}\left(\left\|\frac{\partial^i\psi}{\partial x^i}\right\| + \left\|\frac{\partial^i\psi}{\partial y^i}\right\|\right) + \left\|\frac{\partial^2\psi}{\partial x\partial y}\right\|.$$

For $\varphi \in C(I^2)$ and any $\delta > 0$, the Peetre's $\mathcal{K}$-functional and the second-order modulus of continuity are given by

$$\mathcal{K}(\varphi;\delta) = \inf_{h\in C^2(I^2)} \left\{\|\varphi - h\| + \delta\|h\|_{C^2(I^2)}\right\}, \tag{1.24}$$

and

$$\omega^2(\varphi;\delta) = \sup_{\sqrt{u^2+v^2}\le\delta} \left\|\Delta^2_{u,v}\varphi(x,y)\right\|,$$

respectively, where $\Delta^2_{u,v}\varphi(x,y) = \sum_{j=0}^{2}(-1)^{2-j}\binom{2}{j}\varphi(s+ju, t+jv)$.

From Devore and Lorentz [29], there exists a constant $C > 0$ such that

$$\mathcal{K}(\varphi;\delta) \le C\omega^2\left(\varphi;\sqrt{\delta}\right). \tag{1.25}$$

Theorem 1.6 *For every $n_1, n_2 \in \mathbb{N}$, for all $\varphi \in C(I^2)$, the following inequalities hold:*

$$\left\|\mathfrak{R}^{n_2,q_{n_2},\beta^{(1)},\ldots,\beta^{(r_2)}}_{n_1,q_{n_1},\alpha^{(1)},\ldots,\alpha^{(r_1)}}(\varphi) - \varphi\right\| \le 2C\omega^2\left(\varphi;\sqrt{\frac{\delta^{n_2}_{n_1}}{2}}\right),$$

where C is a positive constant, $\delta^{n_2}_{n_1} = \max\left\{(1-\alpha^{(1)}_{n_1}), (1-\beta^{(1)}_{n_2})\right.$ and $\left.\frac{1}{2}\chi_{n_1}, \frac{1}{2}\chi_{n_2}, \sqrt{\chi_{n_1}\chi_{n_2}}\right\}$.

Proof For any $h \in C^2(I^2)$, by the Taylor formula, we can write

$$\begin{aligned} h(t_1,t_2) - h(x,y) &= h'_x(x,y)(t_1-x) + \int_x^{t_1}(t_1-\theta)h''_{\theta\theta}(\theta,y)d\theta + h'_y(x,y)(t_2-y) \\ &+ \int_y^{t_2}(t_2-\phi)h''_{\phi\phi}(x,\phi)d\phi + \int_x^{t_1}\int_y^{t_2} h''_{\theta\phi}(\theta,\phi)d\phi d\theta \\ &= h'_x(x,y)(t_1-x) + \int_0^{t_1-x}(t_1-x-\theta)h''_{\theta\theta}(\theta+x,y)d\theta \\ &+ h'_y(x,y)(t_2-y) + \int_0^{t_2-y}(t_2-y-\phi)h''_{\phi\phi}(x,\phi+y)d\phi + \int_x^{t_1}\int_y^{t_2} h''_{\theta\phi}(\theta,\phi)d\phi d\theta. \end{aligned}$$

Then using Lemma 1.1, it follows that

$$\mathfrak{R}^{n_2,q_{n_2},\beta^{(1)},\ldots,\beta^{(r_2)}}_{n_1,q_{n_1},\alpha^{(1)},\ldots,\alpha^{(r_1)}}(t_1-x;x,y) = x(\alpha^{(1)}_{n_1}-1) \text{ and } \mathfrak{R}^{n_2,q_{n_2},\beta^{(1)},\ldots,\beta^{(r_2)}}_{n_1,q_{n_1},\alpha^{(1)},\ldots,\alpha^{(r_1)}}(t_2-y;x,y) = y(\beta^{(1)}_{n_2}-1). \tag{1.26}$$

Hence, using (1.26), we obtain

$$\begin{aligned}
\left|\mathfrak{R}^{n_2,q_{n_2},\beta^{(1)},\ldots,\beta^{(r_2)}}_{n_1,q_{n_1},\alpha^{(1)},\ldots,\alpha^{(r_1)}}(h;x,y)-h(x,y)\right| \le& \left|h'_x(x,y)\right|\left|\mathfrak{R}^{n_2,q_{n_2},\beta^{(1)},\ldots,\beta^{(r_2)}}_{n_1,q_{n_1},\alpha^{(1)},\ldots,\alpha^{(r_1)}}(t_1-x;x,y)\right| \\
&+\left|\mathfrak{R}^{n_2,q_{n_2},\beta^{(1)},\ldots,\beta^{(r_2)}}_{n_1,q_{n_1},\alpha^{(1)},\ldots,\alpha^{(r_1)}}\left(\int_0^{t_1-x}(t_1-x-\theta)h''_{\theta\theta}(\theta+x,y)d\theta;x,y\right)\right| \\
&+\left|h'_y(x,y)\right|\left|\mathfrak{R}^{n_2,q_{n_2},\beta^{(1)},\ldots,\beta^{(r_2)}}_{n_1,q_{n_1},\alpha^{(1)},\ldots,\alpha^{(r_1)}}(t_2-y;x,y)\right| \\
&+\left|\mathfrak{R}^{n_2,q_{n_2},\beta^{(1)},\ldots,\beta^{(r_2)}}_{n_1,q_{n_1},\alpha^{(1)},\ldots,\alpha^{(r_1)}}\left(\int_0^{t_2-y}(t_2-y-\phi)h''_{\phi\phi}(x,\phi+y)d\phi;x,y\right)\right| \\
&+\left|\mathfrak{R}^{n_2,q_{n_2},\beta^{(1)},\ldots,\beta^{(r_2)}}_{n_1,q_{n_1},\alpha^{(1)},\ldots,\alpha^{(r_1)}}\left(\int_x^{t_1}\int_y^{t_2}h''_{\theta\phi}(\theta,\phi)d\phi d\theta;x,y\right)\right| \\
\le&(1-\alpha^{(1)}_{n_1})\|h'_x\|_{C(I^2)}+(1-\beta^{(1)}_{n_2})\|h'_y\|_{C(I^2)} \\
&+\frac{1}{2}\|h''_{xx}\|\left|\mathfrak{R}^{n_2,q_{n_2},\beta^{(1)},\ldots,\beta^{(r_2)}}_{n_1,q_{n_1},\alpha^{(1)},\ldots,\alpha^{(r_1)}}\left((t_1-x)^2;x,y\right)\right| \\
&+\frac{1}{2}\|h''_{yy}\|\left|\mathfrak{R}^{n_2,q_{n_2},\beta^{(1)},\ldots,\beta^{(r_2)}}_{n_1,q_{n_1},\alpha^{(1)},\ldots,\alpha^{(r_1)}}\left((t_2-y)^2;x,y\right)\right| \\
&+\|h''_{xy}\|\mathfrak{R}^{n_2,q_{n_2},\beta^{(1)},\ldots,\beta^{(r_2)}}_{n_1,q_{n_1},\alpha^{(1)},\ldots,\alpha^{(r_1)}}(|t_1-x||t_2-y|;x,y) \\
\le&(1-\alpha^{(1)}_{n_1})\|h'_x\|_{C(I^2)}+(1-\beta^{(1)}_{n_2})\|h'_y\|_{C(I^2)} \\
&+\frac{1}{2}\|h''_{xx}\|\left(4(1-\alpha^{(1)}_{n_1})+\frac{\alpha^{(1)}_{n_1}}{[n_1]_{q_{n_1}}}\right)+\frac{1}{2}\|h''_{yy}\|\left(4(1-\beta^{(1)}_{n_2})+\frac{\beta^{(1)}_{n_2}}{[n_2]_{q_{n_2}}}\right) \\
&+\|h''_{xy}\|(4(1-\alpha^{(1)}_{n_1})+\frac{\alpha^{(1)}_{n_1}}{[n_1]_{q_{n_1}}})^{\frac{1}{2}}(4(1-\beta^{(1)}_{n_2})+\frac{\beta^{(1)}_{n_2}}{[n_2]_{q_{n_2}}})^{\frac{1}{2}} \\
\le&\delta^{n_2}_{n_1}\|h\|_{C^2(I^2)}.
\end{aligned}$$

Hence, for $\varphi\in C\left(I^2\right)$ and any $h\in C^2\left(I^2\right)$, from (1.26), we find that

$$\begin{aligned}
\left|\mathfrak{R}^{n_2,q_{n_2},\beta^{(1)},\ldots,\beta^{(r_2)}}_{n_1,q_{n_1},\alpha^{(1)},\ldots,\alpha^{(r_1)}}(\varphi;x,y)-\varphi(x,y)\right| \le& \left\|\mathfrak{R}^{n_2,q_{n_2},\beta^{(1)},\ldots,\beta^{(r_2)}}_{n_1,q_{n_1},\alpha^{(1)},\ldots,\alpha^{(r_1)}}(\varphi-h)\right\|+\|\varphi-h\| \\
&+\left\|\mathfrak{R}^{n_2,q_{n_2},\beta^{(1)},\ldots,\beta^{(r_2)}}_{n_1,q_{n_1},\alpha^{(1)},\ldots,\alpha^{(r_1)}}(h)-h\right\| \\
\le&2\left(\|\varphi-h\|+\frac{1}{2}\delta^{n_2}_{n_1}\|h\|_{C^2(I^2)}\right).
\end{aligned}$$

Hence,

$$\begin{aligned}
\left|\mathfrak{R}^{n_2,q_{n_2},\beta^{(1)},\ldots,\beta^{(r_2)}}_{n_1,q_{n_1},\alpha^{(1)},\ldots,\alpha^{(r_1)}}(\varphi;x,y)-\varphi(x,y)\right| \le& 2\inf_{h\in C^2(I^2)}\left(\|\varphi-h\|+\frac{1}{2}\delta^{n_2}_{n_1}\|h\|_{C^2(I^2)}\right) \\
=&2\,\mathcal{K}(\varphi;\frac{1}{2}\delta^{n_2}_{n_1}).
\end{aligned}$$

Taking into account (1.25), we reach the desired conclusion. □

1.4 Generalization of $\mathfrak{R}^{n_2,q_{n_2},\beta^{(1)},\dots,\beta^{(r_2)}}_{n_1,q_{n_1},\alpha^{(1)},\dots,\alpha^{(r_1)}}$

We know that the rate of convergence of a linear positive operator improves with the smoothness of a function but there is a limit to it. Voronovskaya [81] and Korovkin [54] showed that the rate of approximation by positive linear operators can not be improved beyond $O(\frac{1}{n^2})$, however smooth the function may be. To overcome this problem, Kirov and Popova [53] generalized a linear positive operator by using the sth order Taylor polynomial. Subsequently, significant contributions were made by researchers on this topic (see, for instance, [11, 52, 57, 68, 73, 83]) and the references therein. Inspired by these studies, we consider such a generalization of our operator $\mathfrak{R}^{n_2,q_{n_2},\beta^{(1)},\dots,\beta^{(r_2)}}_{n_1,q_{n_1},\alpha^{(1)},\dots,\alpha^{(r_1)}}$, as follows:

For $\varphi \in C^s(I^2)$, the sth order generalization of the operators $\mathfrak{R}^{n_2,q_{n_2},\beta^{(1)},\dots,\beta^{(r_2)}}_{n_1,q_{n_1},\alpha^{(1)},\dots,\alpha^{(r_1)}}(.;x,y)$ is defined by

$$\widehat{\mathfrak{R}}^{n_2,q_{n_2},\beta^{(1)},\dots,\beta^{(r_2)}}_{s,n_1,q_{n_1},\alpha^{(1)},\dots,\alpha^{(r_1)}}(\varphi;x,y)=\left\{\prod_{i=1}^{r_1}\prod_{j=1}^{r_2}(\alpha_i x^i;q_{n_1})_{n_1}(\beta_i y^i;q_{n_2})_{n_2}\right\}\sum_{p=0}^{\infty}\sum_{s=0}^{\infty}\Bigg\{\sum_{l_1+2l_2+\dots+r_1 l_{r_1}=p}\;\sum_{h_1+2h_2+\dots+r_2 h_{r_2}=s}$$
$$\left\{\prod_{k=1}^{r_1}\prod_{j=1}^{r_2}(q_{n_1}^{n_1},q_{n_1})_{l_k}\frac{(\alpha_{n_1}^{(k)})^{l_k}}{(q_{n_1},q_{n_1})_{l_k}}(q_{n_2}^{n_2},q_{n_2})_{h_j}\frac{(\beta_{n_2}^{(j)})^{h_j}}{(q_{n_2},q_{n_2})_{h_j}}\right\}$$
$$\bigoplus_{\frac{[l_1]_{q_{n_1}}}{[n_1+l_1-1]_{q_{n_1}}}}^{\frac{[h_1]_{q_{n_2}}}{[n_2+h_1-1]_{q_{n_2}}}}\left(x-\frac{[l_1]_{q_{n_1}}}{[n_1+l_1-1]_{q_{n_1}}},\,y-\frac{[h_1]_{q_{n_2}}}{[n_2+h_1-1]_{q_{n_2}}}\right)\Bigg\}x^p y^s, \tag{1.27}$$

where

$$\bigoplus_{\frac{[l_1]_{q_{n_1}}}{[n_1+l_1-1]_{q_{n_1}}}}^{\frac{[h_1]_{q_{n_2}}}{[n_2+h_1-1]_{q_{n_2}}}}\left(x-\frac{[l_1]_{q_{n_1}}}{[n_1+l_1-1]_{q_{n_1}}},\,y-\frac{[h_1]_{q_{n_2}}}{[n_2+h_1-1]_{q_{n_2}}}\right)=\sum_{w=0}^{s}\sum_{i_1+i_2=w}\frac{1}{w!}\binom{w}{i_2}\times$$
$$\varphi_{x^{i_1}}^{y^{i_2}}\left(\frac{[l_1]_{q_{n_1}}}{[n_1+l_1-1]_{q_{n_1}}},\frac{[h_1]_{q_{n_2}}}{[n_2+h_1-1]_{q_{n_2}}}\right)\left(x-\frac{[l_1]_{q_{n_1}}}{[n_1+l_1-1]_{q_{n_1}}}\right)^{i_1}\left(y-\frac{[h_1]_{q_{n_2}}}{[n_2+h_1-1]_{q_{n_2}}}\right)^{i_2}, \tag{1.28}$$

and $\varphi_{x^i}^{y^j}=\dfrac{\partial^s\varphi(x,y)}{\partial x^i\partial y^j}$. Now let us write

$$\left(x-\frac{[l_1]_{q_{n_1}}}{[n_1+l_1-1]_{q_{n_1}}},\,y-\frac{[h_1]_{q_{n_2}}}{[n_2+h_1-1]_{q_{n_2}}}\right)=z.(\chi_1,\chi_2), \tag{1.29}$$

where $z=\left|x-\frac{[l_1]_{q_{n_1}}}{[n_1+l_1-1]_{q_{n_1}}},\,y-\frac{[h_1]_{q_{n_2}}}{[n_2+h_1-1]_{q_{n_2}}}\right|$ and (χ_1,χ_2) is a unit vector. Then $x=\frac{[l_1]_{q_{n_1}}}{[n_1+l_1-1]_{q_{n_1}}}+z\chi_1$ and $y=\frac{[h_1]_{q_{n_2}}}{[n_2+h_1-1]_{q_{n_2}}}+z\chi_2$. Now let us assume that

$$\begin{aligned}\Delta(z) &= \varphi\left(\frac{[l_1]_{q_{n_1}}}{[n_1+l_1-1]_{q_{n_1}}}+z\chi_1, \frac{[h_1]_{q_{n_2}}}{[n_2+h_1-1]_{q_{n_2}}}+z\chi_2\right)\\ &= \varphi\left(\frac{[l_1]_{q_{n_1}}}{[n_1+l_1-1]_{q_{n_1}}}+\left(x-\frac{[l_1]_{q_{n_1}}}{[n_1+l_1-1]_{q_{n_1}}}\right), \frac{[h_1]_{q_{n_2}}}{[n_2+h_1-1]_{q_{n_2}}}+\left(y-\frac{[h_1]_{q_{n_2}}}{[n_2+h_1-1]_{q_{n_2}}}\right)\right)\\ &= \varphi(x, y).\end{aligned} \tag{1.30}$$

Observe that, if we take $s = 0$, in (1.27), then we immediately get (1.7), i.e.,

$$\widehat{\mathfrak{R}}^{n_2,q_{n_2},\beta^{(1)},\dots,\beta^{(r_2)}}_{0,n_1,q_{n_1},\alpha^{(1)},\dots,\alpha^{(r_1)}}(\varphi; x, y) = \mathfrak{R}^{n_2,q_{n_2},\beta^{(1)},\dots,\beta^{(r_2)}}_{n_1,q_{n_1},\alpha^{(1)},\dots,\alpha^{(r_1)}}(\varphi; x, y).$$

Remark 1.1 *(see Chap. 3 of* [26]) We have

$$\Delta^{(s)}(z) = \sum_{i+j=s}\binom{s}{j}\varphi^{y^j}_{x^i}\left(\frac{[l_1]_{q_{n_1}}}{[n_1+l_1-1]_{q_{n_1}}}+z\chi_1, \frac{[h_1]_{q_{n_2}}}{[n_2+h_1-1]_{q_{n_2}}}+z\chi_2\right)\chi_1^i\chi_2^j, \quad s \in \mathbb{N}.$$

Remark 1.2 By Taylor's theorem, we obtain

$$\Delta(z) = \Delta(0) + \Delta'(0)z + \frac{\Delta''(0)}{2!}z^2 + \dots + \frac{\Delta^{(s-1)}(0)}{(s-1)!}z^{s-1} + \frac{z^s}{(s-1)!}\int_0^1 \Delta^{(s)}(tz)(1-t)^{s-1}dt.$$

Using Remark 1.1 and (1.29), we may write

$$\begin{aligned}\Delta(z) &= \sum_{n_2=0}^{s-1}\sum_{i+j=n_2}\frac{1}{n_2!}\binom{n_2}{j}\varphi^{y^j}_{x^i}\left(\frac{[l_1]_{q_{n_1}}}{[n_1+l_1-1]_{q_{n_1}}}, \frac{[h_1]_{q_{n_2}}}{[n_2+h_1-1]_{q_{n_2}}}\right)\\ &\times\left(x-\frac{[l_1]_{q_{n_1}}}{[n_1+l_1-1]_{q_{n_1}}}\right)^i\left(y-\frac{[h_1]_{q_{n_2}}}{[n_2+h_1-1]_{q_{n_2}}}\right)^j\\ &+\frac{1}{(s-1)!}\int_0^1(1-t)^{s-1}\sum_{i+j=s}\binom{s}{j}\left[x-\frac{[l_1]_{q_{n_1}}}{[n_1+l_1-1]_{q_{n_1}}}\right]^i\left[y-\frac{[h_1]_{q_{n_2}}}{[n_2+h_1-1]_{q_{n_2}}}\right]^j\\ &\times\varphi^{y^j}_{x^i}\left(\frac{[l_1]_{q_{n_1}}}{[n_1+l_1-1]_{q_{n_1}}}+t\left(x-\frac{[l_1]_{q_{n_1}}}{[n_1+l_1-1]_{q_{n_1}}}\right), \frac{[h_1]_{q_{n_2}}}{[n_2+h_1-1]_{q_{n_2}}}+t\left(y-\frac{[h_1]_{q_{n_2}}}{[n_2+h_1-1]_{q_{n_2}}}\right)\right)dt.\end{aligned}$$

For a given $M > 0$ *and* $0 < \gamma \leq 1$, the Lipschitz class is defined as

$$Lip_M(\gamma) := \left\{\varphi : |\varphi(s_1, s_2) - \varphi(t_1, t_2)| \leq M[(s_1-t_1)^2 + (s_2-t_2)^2]^{\frac{\gamma}{2}}, \ (s_1, s_2), (t_1, t_2) \in I^2\right\}.$$

Now, we discuss the approximation of $\varphi \in C^s\left(I^2\right)$ such that $\Delta^{(s)}(z) \in Lip_M(\gamma)$, $s \in \mathbb{N}\cup\{0\}$, by using the Lagrange- Hermite-Taylor operators $\widehat{\mathfrak{R}}^{n_2,q_{n_2},\beta^{(1)},\dots,\beta^{(r_2)}}_{s,n_1,q_{n_1},\alpha^{(1)},\dots,\alpha^{(r_1)}}$ $(\varphi; x, y)$.

Theorem 1.7 *Assume* $n_1, n_2 \in \mathbb{N}$ *and* $s \in \mathbb{N}$. *Then for all* $\varphi \in C^s\left(I^2\right)$ *such that* $\Delta^{(s)}(z) \in Lip_M(\gamma)$, $0 < \gamma \leq 1$, *we have*

$$\left\| \varphi - \widehat{\mathfrak{R}}^{n_2,q_{n_2},\beta^{(1)},\ldots,\beta^{(r_2)}}_{s,n_1,q_{n_1},\alpha^{(1)},\ldots,\alpha^{(r_1)}}(\varphi) \right\|_{C(I^2)} \leq \frac{M}{(s-1)!}\frac{\gamma}{\gamma+s} B(\gamma,s) \times \left\| \widehat{\mathfrak{R}}^{n_2,q_{n_2},\beta^{(1)},\ldots,\beta^{(r_2)}}_{s,n_1,q_{n_1},\alpha^{(1)},\ldots,\alpha^{(r_1)}}\left(|(x,y)-(u,v)|^{s+\gamma} \right) \right\|_{C(I^2)}, \tag{1.31}$$

where $B(\gamma, s)$ is the usual Beta function.

Proof Let $(x, y) \in I^2$ be arbitrary. From the definition (1.27), we get

$$\begin{aligned}
&\varphi(x,y) - \widehat{\mathfrak{R}}^{n_2,q_{n_2},\beta^{(1)},\ldots,\beta^{(r_2)}}_{s,n_1,q_{n_1},\alpha^{(1)},\ldots,\alpha^{(r_1)}}(\varphi;x,y) \\
&= \left\{ \prod_{i=1}^{r_1}\prod_{j=1}^{r_2} \left(\alpha_i x^i; q_{n_1}\right)_{n_1} \left(\beta_i y^i; q_{n_2}\right)_{n_2} \right\} \\
&\sum_{p=0}^{\infty}\sum_{s=0}^{\infty}\Bigg\{ \sum_{l_1+2l_2+\ldots+r_1 l_{r_1}=p} \; \sum_{h_1+2h_2+\ldots+r_2 h_{r_2}=s} \\
&\left\{ \prod_{k=1}^{r_1}\prod_{j=1}^{r_2} \left(q_{n_1}^{n_1}, q_{n_1}\right)_{l_k} \frac{(\alpha_{n_1}^{(k)})^{l_k}}{\left(q_{n_1}, q_{n_1}\right)_{l_k}} \left(q_{n_2}^{n_2}, q_{n_2}\right)_{h_j} \frac{(\beta_{n_2}^{(j)})^{h_j}}{\left(q_{n_2}, q_{n_2}\right)_{h_j}} \right\} \\
&\varphi(x,y) - \bigoplus_{\frac{[l_1]_{q_{n_1}}}{[n_1+l_1-1]_{q_{n_1}}}}^{\frac{[h_1]_{q_{n_2}}}{[n_2+h_1-1]_{q_{n_2}}}} \left(x - \frac{[l_1]_{q_{n_1}}}{[n_1+l_1-1]_{q_{n_1}}}, y - \frac{[h_1]_{q_{n_2}}}{[n_2+h_1-1]_{q_{n_2}}} \right) \Bigg\} x^p y^s.
\end{aligned} \tag{1.32}$$

By Taylor's formula for $\varphi(x, y)$ at $\left(\frac{[l_1]_{q_{n_1}}}{[n_1+l_1-1]_{q_{n_1}}}, \frac{[h_1]_{q_{n_2}}}{[n_2+h_1-1]_{q_{n_2}}} \right)$ and (1.28), we arrive at

$$\begin{aligned}
&\varphi(x,y) - \bigoplus_{\frac{[l_1]_{q_{n_1}}}{[n_1+l_1-1]_{q_{n_1}}}}^{\frac{[h_1]_{q_{n_2}}}{[n_2+h_1-1]_{q_{n_2}}}} \left(x - \frac{[l_1]_{q_{n_1}}}{[n_1+l_1-1]_{q_{n_1}}}, y - \frac{[h_1]_{q_{n_2}}}{[n_2+h_1-1]_{q_{n_2}}} \right) \\
&= \frac{1}{(s-1)!}\int_0^1 (1-\nu)^{s-1} \sum_{i+j=s} \binom{s}{j} \left[x - \frac{[l_1]_{q_{n_1}}}{[n_1+l_1-1]_{q_{n_1}}} \right]^i \left[y - \frac{[h_1]_{q_{n_2}}}{[n_2+h_1-1]_{q_{n_2}}} \right]^j \\
&\times \varphi_{x^i}^{y^j}\left(\frac{[l_1]_{q_{n_1}}}{[n_1+l_1-1]_{q_{n_1}}} + \nu\left(x - \frac{[l_1]_{q_{n_1}}}{[n_1+l_1-1]_{q_{n_1}}} \right), \frac{[h_1]_{q_{n_2}}}{[n_2+h_1-1]_{q_{n_2}}} + \nu\left(y - \frac{[h_1]_{q_{n_2}}}{[n_2+h_1-1]_{q_{n_2}}} \right) \right) d\nu.
\end{aligned} \tag{1.33}$$

Using Remark 1.2, Eq. (1.33) yields us:

$$\Delta(z) - \sum_{n_2=0}^{s} \frac{1}{n_2!}\Delta^{(n_2)}(0)z^{n_2} = \frac{z^s}{(s-1)!}\int_0^1 \left[\Delta^{(s)}(\nu z) - \Delta^{(s)}(0) \right](1-\nu)^{s-1} dt. \tag{1.34}$$

Since $\Delta^{(s)}(z) \in Lip_M(\gamma)$, from (1.34) it follows that

$$\left|\varphi(x,y) - \bigoplus_{\frac{[l_1]_{q_{n_1}}}{[n_1+l_1-1]_{q_{n_1}}}}^{\frac{[h_1]_{q_{n_2}}}{[n_2+h_1-1]_{q_{n_2}}}} \left(x - \frac{[l_1]_{q_{n_1}}}{[n_1+l_1-1]_{q_{n_1}}}, y - \frac{[h_1]_{q_{n_2}}}{[n_2+h_1-1]_{q_{n_2}}}\right)\right|$$

$$= \left|\Delta(z) - \sum_{n_2=0}^{s} \frac{1}{n_2!}\Delta^{(n_2)}(0)z^{n_2}\right| \leq \frac{|z|^s}{(s-1)!}\int_0^1 \left|\Delta^{(s)}(\nu z) - \Delta^{(s)}(0)\right|(1-\nu)^{s-1}d\nu. \tag{1.35}$$

On the other hand, using the property of Beta function, (1.35) yields us

$$\left|\varphi(x,y) - \bigoplus_{\frac{[l_1]_{q_{n_1}}}{[n_1+l_1-1]_{q_{n_1}}}}^{\frac{[h_1]_{q_{n_2}}}{[n_2+h_1-1]_{q_{n_2}}}} \left(x - \frac{[l_1]_{q_{n_1}}}{[n_1+l_1-1]_{q_{n_1}}}, y - \frac{[h_1]_{q_{n_2}}}{[n_2+h_1-1]_{q_{n_2}}}\right)\right| \leq \frac{|u|^{s+\gamma}}{(s-1)!}\frac{\gamma M}{\gamma+s}B(\gamma,s)$$

$$\leq \frac{M}{(s-1)!}\frac{\gamma}{\gamma+s}B(\gamma,s)\left|x - \frac{[l_1]_{q_{n_1}}}{[n_1+l_1-1]_{q_{n_1}}}, y - \frac{[h_1]_{q_{n_2}}}{[n_2+h_1-1]_{q_{n_2}}}\right|^{s+\gamma}. \tag{1.36}$$

Finally, considering (1.36) in (1.32), we easily obtain (1.31). □

For $g \in C(I^2)$, consider

$$g(t,s) = |(t,s) - (x,y)|^{s+\gamma}. \tag{1.37}$$

Clearly $g(x,y) = 0$, hence using Theorem 1.1, we obtain

$$\left\|\widehat{\mathfrak{R}}^{n_2,q_{n_2},\beta^{(1)},\dots,\beta^{(r_2)}}_{s,n_1,q_{n_1},\alpha^{(1)},\dots,\alpha^{(r_1)}}(g)\right\|_{C(I^2)} \to 0, \text{ as } \quad n_1, n_2 \to \infty.$$

Thus from Theorem 1.7, it follows that

$$\left\|\widehat{\mathfrak{R}}^{n_2,q_{n_2},\beta^{(1)},\dots,\beta^{(r_2)}}_{s,n_1,q_{n_1},\alpha^{(1)},\dots,\alpha^{(r_1)}}(\varphi)\right\|_{C(I^2)} \to 0, \text{ as } \quad n_1, n_2 \to \infty.$$

Now, we obtain the rates of convergence of the bivariate Hermite–Lagrange–Taylor operators defined by (1.27) by means of the elements of the Lipschitz class $Lip_M(\gamma)$.

Taking into consideration Theorem 2 and $\Delta^{(s)}(z) \in Lip_M(\gamma)$, from Theorem 1.7, we deduce:

Corollary 1.1 *Under the assumptions of Theorem 1.7, we have*

$$\left\| \varphi - \widehat{\mathfrak{R}}^{n_2,q_{n_2},\beta^{(1)},\ldots,\beta^{(r_2)}}_{s,n_1,q_{n_1},\alpha^{(1)},\ldots,\alpha^{(r_1)}}(\varphi) \right\|_{C(I^2)} \leq \frac{2M}{(s-1)!}\frac{\gamma}{\gamma+s}B(\gamma,s)\,\omega^{(c)}(g;\delta_{n_1,n_2}),$$

where $\Delta^{(s)}(z)$, δ_{n_1,n_2} and g are the same as in Remark 1.1, Theorem 1.2 and (1.37), respectively.

Corollary 1.2 *Let $\varphi \in C^s\left(I^2\right)$ such that $\Delta^{(s)}(z) \in Lip_M(\gamma)$, and assuming that $g \in Lip_{2^{\frac{s}{2}}}(\gamma)$ in Theorem 1.2, we have*

$$\left\| \varphi - \widehat{\mathfrak{R}}^{n_2,q_{n_2},\beta^{(1)},\ldots,\beta^{(r_2)}}_{s,n_1,q_{n_1},\alpha^{(1)},\ldots,\alpha^{(r_1)}}(\varphi) \right\|_{C(I^2)} \leq \frac{2^{\frac{s}{2}+1}M}{(s-1)!}\frac{\gamma}{\gamma+s}B(\gamma,s)\delta^{\gamma}_{n_1,n_2},$$

where $\Delta^{(s)}(z)$ and δ_{n_1,n_2} are the same as in Remark 1.1, and Theorem 1.2, respectively.

1.5 GBS (Generalized Boolean Sum) Operators

Bögel [24, 25] introduced the concepts of B-continuous (Bögel continuous) and B-differentiable (Bögel differentiable) functions. Dobrescu and Matei [31] showed that a B-continuous function on a bounded interval can be uniformly approximated by the GBS of bivariate Bernstein polynomials. Subsequently, Badea and Cottin [20] gave theorems of Korovkin type for GBS operators. In [22], Baxhaku et al. defined the GBS operators associated with the bivariate operators based on q-Lagrange polynomials and examined the approximation degree of these operators in the Bögel space by utilizing mixed modulus of smoothness. For more details, see, for instance, [7, 21, 43, 66, 67]. The purpose of this section is to define the GBS operators related with $\mathfrak{R}^{n_2,q_{n_2},\beta^{(1)},\cdots,\beta^{(r_2)}}_{n_1,q_{n_1},\alpha^{(1)},\cdots,\alpha^{(r_1)}}(.;x,y)$ defined in (1.7), and discuss the approximation of B-continuous and B-differentiable functions. In the sequel, we shall need the following definitions.

A function $\varphi : I^2 \to \mathbb{R}$ is called B-bounded (Bögel bounded) if we can find a constant $M > 0$, such that

$$|\Delta[\varphi(t_1,t_2);(x,y)]| \leq M \quad \text{for all } (x,y),(t_1,t_2) \in I^2,$$

where $\Delta[\varphi(t_1,t_2);(x,y)] = \varphi(x,y) - \varphi(x,t_2) - \varphi(t_1,y) + \varphi(t_1,t_2)$ is the mixed difference of φ. We denote $B_b(I^2) := \left\{\varphi : \varphi \; is \; B - bounded \; on \; I^2\right\}$. The function φ is said to be B-continuous on I^2 if for every $(x,y) \in I^2$

$$\lim_{(t_1,t_2)\to(x,y)} \Delta[\varphi(t_1,t_2);(x,y)] = 0.$$

Let $C_b(I^2) := \{\varphi : \varphi \; is \; B - continuous \; on \; I^2\}$. We call a function φ to be B-differentiable on I^2 if for each point $(x, y) \in I^2$, the following limit

$$D_B(\varphi; x, y) = \lim_{(t_1,t_2)\to(x,y)} \frac{\Delta[\varphi(t_1, t_2); (x, y)]}{(t_1 - x)(t_2 - y)},$$

exists and is finite. Assume $D_b(I^2) := \{\varphi : \varphi \; is \; B - \text{differentiable} \; on \; I^2\}$.

For $\varphi \in C_b(I^2)$, the GBS operator associated with (1.7) is given by:

$$\begin{aligned}
&\mathfrak{H}^{n_2,q_{n_2},\beta^{(1)},\ldots,\beta^{(r_2)}}_{n_1,q_{n_1},\alpha^{(1)},\ldots,\alpha^{(r_1)}}(\varphi; x, y) \\
&= \left\{\prod_{i=1}^{r_1}\prod_{j=1}^{r_2}(\alpha_i x^i; q_{n_1})_{n_1}(\beta_i y^i; q_{n_2})_{n_2}\right\}\sum_{p=0}^{\infty}\sum_{s=0}^{\infty}\Bigg\{\sum_{l_1+2l_2+\ldots+r_1 l_{r_1}=p}\;\sum_{h_1+2h_2+\ldots+r_2 h_{r_2}=s} \\
&\left\{\prod_{k=1}^{r_1}\prod_{j=1}^{r_2}(q_{n_1}^{n_1}, q_{n_1})_{l_k}\frac{(\alpha_{n_1}^{(k)})^{l_k}}{(q_{n_1}, q_{n_1})_{l_k}}(q_{n_2}^{n_2}, q_{n_2})_{h_j}\frac{(\beta_{n_2}^{(j)})^{h_j}}{(q_{n_2}, q_{n_2})_{h_j}}\right\}\varphi\left(x, \frac{[h_1]_{q_{n_2}}}{[n_2 + h_1 - 1]_{q_{n_2}}}\right) \\
&+\varphi\left(\frac{[l_1]_{q_{n_1}}}{[n_1 + l_1 - 1]_{q_{n_1}}}, y\right) - \varphi\left(\frac{[l_1]_{q_{n_1}}}{[n_1 + l_1 - 1]_{q_{n_1}}}, \frac{[h_1]_{q_{n_2}}}{[n_2 + h_1 - 1]_{q_{n_2}}}\right)\Bigg\}x^p y^s. \qquad (1.38)
\end{aligned}$$

Evidently, the operator (1.38) is a linear operator from $C_b(I^2)$ into $C(I^2)$.

First, we discuss the uniform convergence theorem of Korovkin type for the operator given by (1.38).

Theorem 1.8 *For $\varphi \in C_b(I^2)$, the GBS operators involving multivariable q-Lagrange–Hermite polynomials $\{\mathfrak{H}^{n_2,q_{n_2},\beta^{(1)},\ldots,\beta^{(r_2)}}_{n_1,q_{n_1},\alpha^{(1)},\ldots,\alpha^{(r_1)}}(\varphi; x, y)\}_{n_1,n_2}$ converge uniformly to φ in $(x, y) \in I^2$, as $n_1, n_2 \to \infty$.*

Proof Applying the test function theorem given in [18] and Lemmas 1.1–1.3, we easily reach the required result. □

For any $\varphi \in C_b(I^2)$ and $\delta_1, \delta_2 > 0$, the mixed modulus of smoothness is defined by

$$\omega_B(\varphi; \delta_1, \delta_2) = \sup_{(t_1,t_2),(x,y)\in I^2}\{|\Delta[\varphi(t_1, t_2); (x, y)]| : |t_1 - x| < \delta_1, |t_2 - y| < \delta_2\},$$

where $(x, y), (t_1, t_2) \in \mathbb{R}^2$, with $\omega_B : (0, \infty) \times (0, \infty) \to \mathbb{R}$. The properties of ω_B can be found in (cf. [19, 20]). Since

$$\omega_B(\varphi; a_1\delta_1, a_2\delta_2) \le (1 + a_1)(1 + a_2)\omega_B(\varphi; \delta_1, \delta_2), \;\; for \; a_1, a_2 > 0,$$

we have,

$$|\Delta\varphi[(t_1, t_2); (x, y)]| \leq \left(1 + \frac{|t_1 - x|}{\delta_1}\right)\left(1 + \frac{|t_2 - y|}{\delta_2}\right)\omega_B(\varphi; \delta_1, \delta_2), \quad (1.39)$$

for every $(t_1, t_2), (x, y) \in I^2$.

The following result yields a convergence estimate of φ by $\mathfrak{H}^{n_2,q_{n_2},\beta^{(1)},\ldots,\beta^{(r_2)}}_{n_1,q_{n_1},\alpha^{(1)},\ldots,\alpha^{(r_1)}}$ via $\omega_B(\varphi; \delta_1, \delta_2)$ (cf. [43], p. 242).

Theorem 1.9 *Assume $\varphi \in C_b(I^2)$. Then, we have*

$$\left\|\mathfrak{H}^{n_2,q_{n_2},\beta^{(1)},\ldots,\beta^{(r_2)}}_{n_1,q_{n_1},\alpha^{(1)},\ldots,\alpha^{(r_1)}}(\varphi) - \varphi\right\| \leq 4\,\omega_B\left(\varphi; \sqrt{\chi_{n_1}}, \sqrt{\chi_{n_2}}\right).$$

Proof Using the definition of the operators $\mathfrak{H}^{n_2,q_{n_2},\beta^{(1)},\ldots,\beta^{(r_2)}}_{n_1,q_{n_1},\alpha^{(1)},\ldots,\alpha^{(r_1)}}(\varphi; x, y)$ we can write

$$\begin{aligned}
\mathfrak{H}^{n_2,q_{n_2},\beta^{(1)},\ldots,\beta^{(r_2)}}_{n_1,q_{n_1},\alpha^{(1)},\ldots,\alpha^{(r_1)}}(\varphi(t_1, t_2); x, y) &= \mathfrak{R}^{n_2,q_{n_2},\beta^{(1)},\ldots,\beta^{(r_2)}}_{n_1,q_{n_1},\alpha^{(1)},\ldots,\alpha^{(r_1)}}(\varphi(t_1, y) + \varphi(x, t_2) - \varphi(t_1, t_2); x, y) \\
&= \varphi(x, y)\mathfrak{R}^{n_2,q_{n_2},\beta^{(1)},\ldots,\beta^{(r_2)}}_{n_1,q_{n_1},\alpha^{(1)},\ldots,\alpha^{(r_1)}}(1; x, y) - \mathfrak{R}^{n_2,q_{n_2},\beta^{(1)},\ldots,\beta^{(r_2)}}_{n_1,q_{n_1},\alpha^{(1)},\ldots,\alpha^{(r_1)}}(|\Delta\varphi[(t_1, t_2); (x, y)]|; x, y).
\end{aligned}$$

By considering the inequality (1.39), for any $\delta_1, \delta_2 > 0$, we have

$$\begin{aligned}
|\mathfrak{H}^{n_2,q_{n_2},\beta^{(1)},\ldots,\beta^{(r_2)}}_{n_1,q_{n_1},\alpha^{(1)},\ldots,\alpha^{(r_1)}}(\varphi; x, y) - \varphi(x, y)| &\leq \mathfrak{R}^{n_2,q_{n_2},\beta^{(1)},\ldots,\beta^{(r_2)}}_{n_1,q_{n_1},\alpha^{(1)},\ldots,\alpha^{(r_1)}}(|\Delta\varphi[(t_1, t_2); (x, y)]|; x, y) \\
&\leq \omega_B\left(\varphi; \delta_1, \delta_2\right)\left[\mathfrak{R}^{n_2,q_{n_2},\beta^{(1)},\ldots,\beta^{(r_2)}}_{n_1,q_{n_1},\alpha^{(1)},\ldots,\alpha^{(r_1)}}(e_0^0; x, y)\right. \\
&+ \frac{1}{\delta_1\,\delta_2}\mathfrak{R}^{n_2,q_{n_2},\beta^{(1)},\ldots,\beta^{(r_2)}}_{n_1,q_{n_1},\alpha^{(1)},\ldots,\alpha^{(r_1)}}(|t_1 - x|; x, y) \\
&\quad \mathfrak{R}^{n_2,q_{n_2},\beta^{(1)},\ldots,\beta^{(r_2)}}_{n_1,q_{n_1},\alpha^{(1)},\ldots,\alpha^{(r_1)}}(|t_2 - y|; x, y) \\
&+ \frac{1}{\delta_2}\mathfrak{R}^{n_2,q_{n_2},\beta^{(1)},\ldots,\beta^{(r_2)}}_{n_1,q_{n_1},\alpha^{(1)},\ldots,\alpha^{(r_1)}}(|t_2 - y|; x, y) \\
&\left.+ \frac{1}{\delta_1}\mathfrak{R}^{n_2,q_{n_2},\beta^{(1)},\ldots,\beta^{(r_2)}}_{n_1,q_{n_1},\alpha^{(1)},\ldots,\alpha^{(r_1)}}(|t_1 - x|; x, y)\right].
\end{aligned}$$

Considering Cauchy–Schwarz inequality and Lemma 1.3, one obtains

$$\begin{aligned}
\left\|\mathfrak{H}^{n_2,q_{n_2},\beta^{(1)},\ldots,\beta^{(r_2)}}_{n_1,q_{n_1},\alpha^{(1)},\ldots,\alpha^{(r_1)}}(\varphi) - \varphi\right\| &\leq \omega_B(\varphi; \delta_1, \delta_2)\left[1 + \frac{1}{\delta_1}\sqrt{\Pi^{\alpha^{(1)},\alpha^{(2)}\ldots\,\alpha^{(r_1)}}_{n_1,q_{n_1},2}}\right. \\
&\left.+ \frac{1}{\delta_1\,\delta_2}\sqrt{\Pi^{\alpha^{(1)},\alpha^{(2)}\ldots\,\alpha^{(r_1)}}_{n_1,q_{n_1},2}\;\Pi^{\beta^{(1)},\beta^{(2)}\ldots\,\beta^{(r_2)}}_{n_2,q_{n_2},2}} + \frac{1}{\delta_2}\sqrt{\Pi^{\beta^{(1)},\beta^{(2)}\ldots\,\beta^{(r_2)}}_{n_2,q_{n_2},2}}\right] \\
&\leq \omega_B(f; \delta_1, \delta_2)\left[1 + \frac{1}{\delta_1}\sqrt{4(1 - \alpha^{(1)}_{n_1}) + \frac{\alpha^{(1)}_{n_1}}{[n_1]_{q_{n_1}}}}\right.
\end{aligned}$$

$$+\frac{1}{\delta_1\,\delta_2}\sqrt{4(1-\alpha_{n_1}^{(1)})+\frac{\alpha_{n_1}^{(1)}}{[n_1]_{q_{n_1}}}}\sqrt{4(1-\beta_{n_2}^{(1)})+\frac{\beta_{n_2}^{(1)}}{[n_2]_{q_{n_2}}}}$$
$$+\frac{1}{\delta_2}\sqrt{4(1-\beta_{n_2}^{(1)})+\frac{\beta_{n_2}^{(1)}}{[n_2]_{q_{n_2}}}}\Bigg].$$

Now, choose $\delta_1=\sqrt{\chi_{n_1}}$ and $\delta_2=\sqrt{\chi_{n_2}}$ to complete the proof. □

Next, we obtain the rate of convergence of the operator $\mathfrak{H}^{n_2,q_{n_2},\beta^{(1)},\ldots,\beta^{(r_2)}}_{n_1,q_{n_1},\alpha^{(1)},\ldots,\alpha^{(r_1)}}(\varphi;x,y)$ for functions belonging to $Lip^{(\lambda)}_{M,b}(I^2)$, $0<\lambda\le 1$, the Lipschitz class of B- continuous functions defined as:

$$Lip^{(\lambda)}_{M,b}(I^2):=\{\varphi\in C_b(I^2):|\Delta\varphi[(t_1,t_2);(x,y)]|\le M\|u-v\|_2^{\lambda},\ \forall\,(t_1,t_2),(x,y)\in I^2\},$$

where $u=(t_1,t_2),\ v=(x,y)\in I^2$, $\|u-v\|_2=\{(t_1-x)^2+(t_2-y)^2\}^{\frac{1}{2}}$ and $M>0$, is a constant.

Theorem 1.10 *If $\varphi\in Lip^{(\lambda)}_{M,b}(I^2)$, then for every $(x,y)\in I^2$ and for all $n_1,n_2\in\mathbb{N}$ we have*

$$\left\|\mathfrak{H}^{n_2,q_{n_2},\beta^{(1)},\ldots,\beta^{(r_2)}}_{n_1,q_{n_1},\alpha^{(1)},\ldots,\alpha^{(r_1)}}(\varphi)-\varphi\right\|\le M(\chi_{n_1}+\chi_{n_2})^{\gamma/2},$$

for $M>0$, and $\gamma\in(0,1]$.

Proof From the hypothesis,

$$\left|\mathfrak{H}^{n_2,q_{n_2},\beta^{(1)},\ldots,\beta^{(r_2)}}_{n_1,q_{n_1},\alpha^{(1)},\ldots,\alpha^{(r_1)}}(\varphi;x,y)-\varphi(x,y)\right|\le\mathfrak{R}^{n_2,q_{n_2},\beta^{(1)},\ldots,\beta^{(r_2)}}_{n_1,q_{n_1},\alpha^{(1)},\ldots,\alpha^{(r_1)}}\left(\left|\Delta\varphi[(t_1,t_2);(x,y)]\right|;x,y\right)$$
$$\le M\,\mathfrak{R}^{n_2,q_{n_2},\beta^{(1)},\ldots,\beta^{(r_2)}}_{n_1,q_{n_1},\alpha^{(1)},\ldots,\alpha^{(r_1)}}\left(\|u-v\|^{\lambda};x,y\right).$$

Using Hölder inequality with $u_1=\frac{2}{\lambda}$, $v_1=\frac{2}{2-\lambda}$ and Lemma 1.3, the required result easily follows. Hence, we omit the details. □

Finally, we derive an error estimate in approximating a function in $D_B\varphi\in B(I^2)$ by the operators (1.38).

Theorem 1.11 *If $\varphi\in D_b(I^2)$ and $D_B\varphi\in B(I^2)$, then for $\delta_1,\ \delta_2>0$, the operator $\mathfrak{H}^{n_2,q_{n_2},\beta^{(1)},\ldots,\beta^{(r_2)}}_{n_1,q_{n_1},\alpha^{(1)},\ldots,\alpha^{(r_1)}}$ verifies the following inequality:*

$$\|\mathfrak{H}^{n_2,q_{n_2},\beta^{(1)},\ldots,\beta^{(r_2)}}_{n_1,q_{n_1},\alpha^{(1)},\ldots,\alpha^{(r_1)}}(\varphi)-\varphi\|\le 3\|D_B\varphi\|\sqrt{\Pi^{\alpha^{(1)},\alpha^{(2)}\ldots\,\alpha^{(r_1)}}_{n_1,q_{n_1},2}\ \Pi^{\beta^{(1)},\beta^{(2)}\ldots\,\beta^{(r_2)}}_{n_2,q_{n_2},2}}$$
$$+\Bigg[\sqrt{\Pi^{\alpha^{(1)},\alpha^{(2)}\ldots\,\alpha^{(r_1)}}_{n_1,q_{n_1},2}\ \Pi^{\beta^{(1)},\beta^{(2)}\ldots\,\beta^{(r_2)}}_{n_2,q_{n_2},2}}$$

$$+\delta_1^{-1}\sqrt{\Pi_{n_1,q_{n_1},4}^{\alpha^{(1)},\alpha^{(2)}\dots\alpha^{(r_1)}}\;\Pi_{n_2,q_{n_2},2}^{\beta^{(1)},\beta^{(2)}\dots\beta^{(r_2)}}}$$

$$+\delta_2^{-1}\sqrt{\Pi_{n_1,q_{n_1},2}^{\alpha^{(1)},\alpha^{(2)}\dots\alpha^{(r_1)}}\;\Pi_{n_2,q_{n_2},4}^{\beta^{(1)},\beta^{(2)}\dots\beta^{(r_2)}}}$$

$$+\delta_1^{-1}\delta_2^{-1}\Pi_{n_1,q_{n_1},2}^{\alpha^{(1)},\alpha^{(2)}\dots\alpha^{(r_1)}}\;\Pi_{n_2,q_{n_2},2}^{\beta^{(1)},\beta^{(2)}\dots\beta^{(r_2)}}\Bigg]\omega_B(D_B\varphi;\delta_1,\delta_2).$$

Proof By our assumption

$$\Delta[\varphi(t_1,t_2);(x,y)]=(t_1-x)(t_2-y)D_B\varphi(a,b),\quad x<a<t_1;\ y<b<t_2. \quad (1.40)$$

Since $D_B\varphi\in B(I^2)$, $|D_B\varphi(x,y)|\le\|D_B\varphi\|$, for every $(x,y)\in I^2$. From the definition of mixed difference, we have

$$D_B\varphi(a,b)=\Delta[D_B\varphi(a,b);x,y]+D_B\varphi(a,y)+D_B\varphi(x,b)-D_B\varphi(x,y).$$

Hence, in view of (1.40),

$$\begin{aligned}
&\left|\mathfrak{R}_{n_1,q_{n_1},\alpha^{(1)},\dots,\alpha^{(r_1)}}^{n_2,q_{n_2},\beta^{(1)},\dots,\beta^{(r_2)}}(\Delta[\varphi(t_1,t_2);(x,y)];x,y)\right|\\
&=\left|\mathfrak{R}_{n_1,q_{n_1},\alpha^{(1)},\dots,\alpha^{(r_1)}}^{n_2,q_{n_2},\beta^{(1)},\dots,\beta^{(r_2)}}\big((t_1-x)(t_2-y)D_B\varphi(a,b);x,y\big)\right|\\
&\le\Big\{\mathfrak{R}_{n_1,q_{n_1},\alpha^{(1)},\dots,\alpha^{(r_1)}}^{n_2,q_{n_2},\beta^{(1)},\dots,\beta^{(r_2)}}\big(|t_1-x||t_2-y||\,\Delta\,[D_B\varphi(a,b);x,y]|;x,y\big)\\
&+\mathfrak{R}_{n_1,q_{n_1},\alpha^{(1)},\dots,\alpha^{(r_1)}}^{n_2,q_{n_2},\beta^{(1)},\dots,\beta^{(r_2)}}\big(|t_1-x||t_2-y|(|D_B\varphi(a,y)|+|D_B\varphi(x,b)|\\
&+|D_B\varphi(x,y)|);x,y\big)\Big\}.
\end{aligned}$$

Consequently, we obtain

$$\begin{aligned}
\left|\mathfrak{R}_{n_1,q_{n_1},\alpha^{(1)},\dots,\alpha^{(r_1)}}^{n_2,q_{n_2},\beta^{(1)},\dots,\beta^{(r_2)}}(\Delta[\varphi(t_1,t_2);(x,y)];x,y)\right| &\le\Big\{\mathfrak{R}_{n_1,q_{n_1},\alpha^{(1)},\dots,\alpha^{(r_1)}}^{n_2,q_{n_2},\beta^{(1)},\dots,\beta^{(r_2)}}\Big(|t_1-x||t_2-y|\\
&\quad\omega_B(D_B\varphi;|a-x|,|b-y|);x,y\Big)\\
&\quad+3\|D_B\varphi\|_\infty\mathfrak{R}_{n_1,q_{n_1},\alpha^{(1)},\dots,\alpha^{(r_1)}}^{n_2,q_{n_2},\beta^{(1)},\dots,\beta^{(r_2)}}(|t_1-x||t_2-y|;x,y)\Big\}.
\end{aligned}$$

Now, using (1.39), the Cauchy–Schwarz inequality and Lemmas 1.1–1.3, for $\delta_1,\delta_2>0$, we have

$$\left|\mathfrak{H}^{n_2,q_{n_2},\varsigma^{(1)},\varsigma^{(2)}\ldots\varsigma^{(r_2)}}_{n_1,q_{n_1},\eta^{(1)},\eta^{(2)}\ldots\eta^{(r_1)}}(\varphi;x,y)-\varphi(x,y)\right| \le 3\|D_B\varphi\|\sqrt{\vartheta^{\alpha^{(1)},\alpha^{(2)}\ldots\alpha^{(r_1)}}_{n_1,q_{n_1},2}(x)\,\vartheta^{\beta^{(1)},\beta^{(2)}\ldots\beta^{(r_2)}}_{n_2,q_{n_2},2}(y)}$$
$$+\Bigg[\sqrt{\vartheta^{\alpha^{(1)},\alpha^{(2)}\ldots\alpha^{(r_1)}}_{n_1,q_{n_1},2}(x)\,\vartheta^{\beta^{(1)},\beta^{(2)}\ldots\beta^{(r_2)}}_{n_2,q_{n_2},2}(y)}$$
$$+\delta_1^{-1}\sqrt{\vartheta^{\alpha^{(1)},\alpha^{(2)}\ldots\alpha^{(r_1)}}_{n_1,q_{n_1},4}(x)\,\vartheta^{\beta^{(1)},\beta^{(2)}\ldots\beta^{(r_2)}}_{n_2,q_{n_2},2}(y)}$$
$$+\delta_2^{-1}\sqrt{\vartheta^{\alpha^{(1)},\alpha^{(2)}\ldots\alpha^{(r_1)}}_{n_1,q_{n_1},2}(x)\,\vartheta^{\beta^{(1)},\beta^{(2)}\ldots\beta^{(r_2)}}_{n_2,q_{n_2},4}(y)}$$
$$+\delta_1^{-1}\delta_2^{-1}\vartheta^{\alpha^{(1)},\alpha^{(2)}\ldots\alpha^{(r_1)}}_{n_1,q_{n_1},2}(x)\,\vartheta^{\beta^{(1)},\beta^{(2)}\ldots\beta^{(r_2)}}_{n_2,q_{n_2},2}(y)\Bigg]\omega_B(D_B\varphi;\delta_1,\delta_2),$$

for all $(x,y)\in I^2$, leading us to the required result. □

1.6 Future Scope of Work

We define a Kantorovich version of the bivariate Lagrange–Hermite operators defined by (1.7) via Taylor polynomial and GBS operators.

1.6.1 Kantorovich Type Operators Involving Multivariable Q-Lagrange–Hermite Polynomials

For a function $\varphi\in C(I^2)$, the bivariate Kantorovich type operators involving multivariable q-Lagrange–Hermite polynomials are defined a

$$\mathfrak{W}^{n_2,q_{n_2},\beta^{(1)},\ldots,\beta^{(r_2)}}_{n_1,q_{n_1},\alpha^{(1)},\ldots,\alpha^{(r_1)}}(\varphi;x,y)$$
$$=\left\{\prod_{i=1}^{r_1}\prod_{j=1}^{r_2}\left(\alpha_i x^i;q_{n_1}\right)_{n_1}\left(\beta_i y^i;q_{n_2}\right)_{n_2}\right\}\sum_{p=0}^{\infty}\sum_{s=0}^{\infty}\Bigg\{\sum_{l_1+2l_2+\ldots+r_1l_{r_1}=p}\ \sum_{h_1+2h_2+\ldots+r_2h_{r_2}=s}$$
$$\left\{\prod_{k=1}^{r_1}\prod_{j=1}^{r_2}\left(q_{n_1}^{n_1},q_{n_1}\right)_{l_k}\frac{(\alpha_{n_1}^{(k)})^{l_k}}{\left(q_{n_1},q_{n_1}\right)_{l_k}}\left(q_{n_2}^{n_2},q_{n_2}\right)_{h_j}\frac{(\beta_{n_2}^{(j)})^{h_j}}{\left(q_{n_2},q_{n_2}\right)_{h_j}}\right\}A(n_1,n_2)\Bigg\}x^p y^s,\tag{1.41}$$

where

$$A(n_1,n_2)=[n_1+l_1-1]_{q_{n_1}}[n_2+h_1-1]_{q_{n_2}}\int_{\frac{[l_1]q_{n_1}}{[n_1+l_1-1]q_{n_1}}}^{\frac{[l_1+1]q_{n_1}}{[n_1+l_1-1]q_{n_1}}}\int_{\frac{[h_1]q_{n_2}}{[n_2+h_1-1]q_{n_2}}}^{\frac{[h_1+1]q_{n_2}}{[n_2+h_1-1]q_{n_2}}}\varphi(t,s)dtds.$$

1.6.2 Kantorovich–Taylor Type Operators Involving Multivariable Q-Lagrange–Hermite Polynomials

We define the rth order generalization $\mathfrak{X}^{n_2,q_{n_2},\beta^{(1)},\ldots,\beta^{(r_2)}}_{n_1,q_{n_1},\alpha^{(1)},\ldots,\alpha^{(r_1)}}$ of the above linear positive operators (1.41):

$$\mathfrak{X}^{n_2,q_{n_2},\beta^{(1)},\ldots,\beta^{(r_2)}}_{s,n_1,q_{n_1},\alpha^{(1)},\ldots,\alpha^{(r_1)}}(\varphi;x,y)$$
$$=\left\{\prod_{i=1}^{r_1}\prod_{j=1}^{r_2}\left(\alpha_i x^i;q_{n_1}\right)_{n_1}\left(\beta_i y^i;q_{n_2}\right)_{n_2}\right\}\sum_{p=0}^{\infty}\sum_{s=0}^{\infty}\Bigg\{\sum_{l_1+2l_2+\ldots+r_1l_{r_1}=p}\ \sum_{h_1+2h_2+\ldots+r_2h_{r_2}=s}$$
$$\left\{\prod_{k=1}^{r_1}\prod_{j=1}^{r_2}\left(q_{n_1}^{n_1},q_{n_1}\right)_{l_k}\frac{(\alpha_{n_1}^{(k)})^{l_k}}{\left(q_{n_1},q_{n_1}\right)_{l_k}}\left(q_{n_2}^{n_2},q_{n_2}\right)_{h_j}\frac{(\beta_{n_2}^{(j)})^{h_j}}{\left(q_{n_2},q_{n_2}\right)_{h_j}}\right\}TA(n_1,n_2)\Bigg\}x^p y^s, \tag{1.42}$$

where

$$TA(n_1,n_2)=[n_1+l_1-1]_{q_{n_1}}[n_2+h_1-1]_{q_{n_2}}$$
$$\int_{\frac{[l_1]_{q_{n_1}}}{[n_1+l_1-1]_{q_{n_1}}}}^{\frac{[l_1+1]_{q_{n_1}}}{[n_1+l_1-1]_{q_{n_1}}}}\int_{\frac{[h_1]_{q_{n_2}}}{[n_2+h_1-1]_{q_{n_2}}}}^{\frac{[h_1+1]_{q_{n_2}}}{[n_2+h_1-1]_{q_{n_2}}}}\left\{\varphi(x,y)-\sum_{l=0}^{r}\frac{d^l\varphi(t,s)}{l!}\right\}dtds$$

and

$$d^l\varphi(t,s)=\sum_{i=0}^{l}\binom{l}{i}\frac{\partial^l\varphi(t,s)}{\partial x^{l-i}\partial y^i}(x-t)^{l-i}(y-s)^i.$$

For $r=0$, from (1.42) and (1.41), there follows the equality $\mathfrak{X}^{n_2,q_{n_2},\beta^{(1)},\ldots,\beta^{(r_2)}}_{s,n_1,q_{n_1},\alpha^{(1)},\ldots,\alpha^{(r_1)}}(\varphi;x,y)$ $=\mathfrak{W}^{n_2,q_{n_2},\beta^{(1)},\ldots,\beta^{(r_2)}}_{n_1,q_{n_1},\alpha^{(1)},\ldots,\alpha^{(r_1)}}(\varphi;x,y)$.

1.6.3 GBS-Kantorovich Type Operators Involving Multivariable Q-Lagrange–Hermite Polynomials

We propose the GBS case of the Kantorovich operators involving multivariable q-Lagrange–Hermite polynomials as follows:

$$\mathfrak{S}^{n_2,q_{n_2},\beta^{(1)},\ldots,\beta^{(r_2)}}_{n_1,q_{n_1},\alpha^{(1)},\ldots,\alpha^{(r_1)}}(\varphi;x,y)$$

$$= \left\{ \prod_{i=1}^{r_1} \prod_{j=1}^{r_2} \left(\alpha_i x^i; q_{n_1}\right)_{n_1} \left(\beta_i y^i; q_{n_2}\right)_{n_2} \right\} \sum_{p=0}^{\infty} \sum_{s=0}^{\infty} \Bigg\{ \sum_{l_1+2l_2+\ldots+r_1 l_{r_1}=p} \; \sum_{h_1+2h_2+\ldots+r_2 h_{r_2}=s}$$

$$\left\{ \prod_{k=1}^{r_1} \prod_{j=1}^{r_2} \left(q_{n_1}^{n_1}, q_{n_1}\right)_{l_k} \frac{(\alpha_{n_1}^{(k)})^{l_k}}{\left(q_{n_1}, q_{n_1}\right)_{l_k}} \left(q_{n_2}^{n_2}, q_{n_2}\right)_{h_j} \frac{(\beta_{n_2}^{(j)})^{h_j}}{\left(q_{n_2}, q_{n_2}\right)_{h_j}} \right\} G(n_1, n_2) \Bigg\} x^p y^s, \tag{1.43}$$

where

$$G(n_1, n_2) = [n_1 + l_1 - 1]_{q_{n_1}} [n_2 + h_1 - 1]_{q_{n_2}} \int_{\frac{[l_1]_{q_{n_1}}}{[n_1+l_1-1]_{q_{n_1}}}}^{\frac{[l_1+1]_{q_{n_1}}}{[n_1+l_1-1]_{q_{n_1}}}} \int_{\frac{[h_1]_{q_{n_2}}}{[n_2+h_1-1]_{q_{n_2}}}}^{\frac{[h_1+1]_{q_{n_2}}}{[n_2+h_1-1]_{q_{n_2}}}} (\varphi(t_1, y) + \varphi(x, t_2) - \varphi(t_1, t_2)) dt_1 dt_2.$$

We remark here that all the results obtained in this chapter could be extended for the operators introduced in this section.

Acknowledgements The authors are extremely grateful to the learned reviewers for their valuable comments leading to a considerable improvement of the paper.

References

1. Abel, U., Ivan, M.: Asymptotic approximation of functions and their derivatives by Müller's gamma operators. Results Math. **43**, 1–12 (2003)
2. Acar, T., Mursaleen, M., Deveci, S.N.: Gamma operators reproducing exponential functions. Adv. Differ. Eqn. **423** (2020). https://doi.org/10.1186/s13662-020-02880-x
3. Agrawal, P.N., Gupta, P.: Durrmeyer variant of q-Favard-Szász operators based on Appell polynomials. Creat. Math. Inform. **26**(1), 9–17 (2017)
4. Agrawal, P.N., Ispir, N.: Degree of approximation for bivariate Chlodowsky-Szász-Charlier type operators. Results Math. **69**(3–4), 369–385 (2016)
5. Agrawal, P.N., Baxhaku, B., Singh, J.K.: A q-Erkuş-Srivastava polynomials operator. Math. Methods Appl. Sci. **47**(8), 7079–7102 (2024). https://doi.org/10.1002/mma.9958
6. Agrawal, P.N., Shukla, R., Baxhaku, B.: On some summability methods for a q-analogue of an integral type operator based on multivariate q-Lagrange polynomials. Numer. Funct. Anal. Optim. **43**(7), 796–815 (2022). https://doi.org/10.1080/01630563.2022.2056199
7. Agrawal, P.N., Baxhaku, B., Chauhan, R.: The approximation of bivariate Chlodowsky-Szász-Kantorovich-Charlier type operators. J. Inequal. Appl. **195**, 23 (2017)
8. Agrawal, P.N., Shukla, R., Baxhaku, B.: Characterization of Deferred type statistical convergence and P-summability method for operators: Applications to q-Lagrange-Hermite operator. Math. Methods Appl. Sci. **46**(4), 4449–4465 (2023). https://doi.org/10.1002/mma.8770

9. Aktaş, R., Duman, E.E.: The Laguerre polynomials in the several variables. Math. Slovaca **63**(3), 531–544 (2013)
10. Altin, A., Aktaş, R., Duman, E.E.: On a multivariate extension for the extended Jacobi polynomials. J. Math. Anal. Appl. **353**, 121–133 (2009)
11. Altin, A., Doğru, O., Taşdelen, F.: The generalization of Meyer-König and Zeller operators by generating functions. J. Math. Anal. Appl. **312**, 181–194 (2005)
12. Altin, A., Erkuş, E.: On a multivariate extension of the Lagrange-Hermite polynomials. Integral Transforms Spec. Funct. **1717**(4), 239–244 (2006)
13. Altin, A., Erkuş, E., Taşdelen, F.: The q-Lagrange polynomilas in several variables. Taiwan. J. Math. **5**, 1131–1137 (2006)
14. Andrews, G.E., Askey, R., Roy, R.: Special Functions. Cambridge University Press (1999)
15. Ansari, K.J., Mursaleen, M., Rahman, S.: Approximation by Jakimovski-Leviatan operators of Durrmeyer type involving multiple Appell polynomials. RACSAM **113**, 1007–1024 (2019)
16. Aral, A., Gupta, V., Agarwal, R.P.: Application of q-Calculus in Operator Theory. Springer, New York (2013)
17. Atakut, Ç., Büyükyazici, I.: Approximation by Kantorovich-Szász type operators based on Brenke type polynomials. Numer. Funct. Anal. Optim. **37**(12), 1488–1502 (2016)
18. Badea, C., Badea, I., Gonska, H.H.: A test function theorem and approximation by pseudopolynomials. Bull. Aust. Math. Soc. **34**(1), 53–64 (1986)
19. Badea, C., Badea, I., Gonska, H.H.: Notes on the degree of approximation of B-continuous and B-differentiable functions. J. Approx. Theory Appl. **4**, 95–108 (1988)
20. Badea, C., Cottin, C.: Korovkin-type theorems for Generalized Boolean Sum operators. Colloq. Math. Soc. János Bolyai, Approx. Theory, Kecskemét (Hungary) **58**, 51–67 (1990)
21. Baxhaku, B., Agrawal, P.N.: Degree of approximation for bivariate extension of Chlodowsky-type q-Bernstein-Stancu-Kantorovich operators. Appl. Math. Comput. **306**, 56–72 (2017)
22. Baxhaku, B., Agrawal, P.N., Shukla, R.: Bivariate positive linear operators constructed by means of q-Lagrange polynomials. J. Math. Anal. Appl. **491**(2), 124337 (2020). https://doi.org/10.1016/j.jmaa.2020.124337
23. Betus, Ö., Usta, F.: Approximation of functions by a new type of Gamma operators. Numer. Methods Partial Differ. Eqn. 1–12 (2020). https://doi.org/10.1002/num.22660
24. Bögel, K.: Mehrdimensionale differentiation von funktionen mehrerer reeller veränderlichen. J. Reine. Angew. Math. **170**, 197–217 (1934)
25. Bögel, K.: Über die mehrdimensionale differentiation, integration and bescharänkte variation. J. Reine. Angew. Math. **173**, 5–30 (1935)
26. Callahan, J.: Advanced Calculus. Lecture Notes. Smith College, USA
27. Chan, W.C.C., Chyan, C.J., Srivastava, H.M.: The Lagrange polynomials in several variables. Integral Transforms Spec. Funct. **12**, 139–148 (2001)
28. Ciupa, A.: Positive linear operators obtained by means of Appell polynomials. In: Approximation and Optimization, vol. II, pp. 63–68. Cluj-Napoca, Transilvania, Cluj (1996)
29. DeVore, R.A., Lorentz, G.G.: Constructive Approximation. Springer, Berlin (1993)
30. Dhamija, M., Deo, N.: Jain-Durrmeyer operators associated with the inverse Pólya-Eggenberger distribution. Appl. Math. Comput. **286**, 15–22 (2016)
31. Dobrescu, E., Matei, I.: The approximation by Bernstein type polynomials of bidimensionally continuous functions. Univ. Timisoara Ser. Sti. Mat. Fiz **4**, 85–90 (1966)
32. Erkuş-Duman, E.: A q-extension of the Erkuş-Srivastava polynomials in several variables. Taiwanese J. Math. **12**(2), 539–543 (2008)
33. Erkuş- Duman, E.: Statistical approximation by means of operators constructed by q-Lagrange polynomials. J. Comput. Anal. Appl. **14**(1), 67–77 (2012)
34. Erkuş, E.: The multivariable q-Lagrange-Hermite polynomials. Int. J. Comput Numer. Anal. Appl. **6**, 143–151 (2004)
35. Erkuş, E., Duman, O.: Integral-type generalizations of operators obtained from certain multivariate polynomials. Calcolo **45**, 53–67 (2008)
36. Erkuş, E., Duman, O., Srivastava, H.M.: Statistical approximation of certain positive linear operators constructed by means of the Chan-Chyan-Srivastava polynomials. Appl. Math. Comput. **182**, 213–222 (2006)

37. Erkuş, E., Srivastava, H.M.: A unified presentation of some families ofmultivariable Lagrange polynomials. Integral Transforms Spec. Funct. **17**, 267–273 (2006)
38. Garg, T., Agrawal, P.N., Araci, S.: Rate of convergence by Kantorovich-Szász type operators based on Brenke type polynomials. J. Inequal. Appl. **2017**, 156 (2017)
39. Gupta, V.: Convergence estimates for Gamma operator. Bull. Malays. Math. Soc. **43**(3), 2065–2075 (2020)
40. Gupta, P., Agrawal, P.N.: Jakimovski-Leviatan operators of Durrmeyer type involving Appell polynomials. Turk. J. Math. **42**, 1457–1470 (2018)
41. Gupta, P., Agrawal, P.N.: Quantitative Voronovskaja and Grüss Voronovskaja-type theorems for operators of Kantorovich type involving multiple Appell polynomials. Iran. J. Sci. Technol. Trans. Sci. **43**, 1679–1687 (2019)
42. Gupta, V., Greubel, G.C.: Moment estimations of new Szász- Mirakian- Durrmeyer operators. Appl. Math. Comput. **271**, 540–547 (2015)
43. Gupta, V., Rassias, MTh., Agrawal, P.N., Acu, A.M.: Recent Advances in Constructive Approximation Theory. Springer Nature, Cham (2018)
44. İçöz, G., Varma, S., Sucu, S.: Approximation by operators including generalized Appell polynomials. Filomat **30**(2), 429–440 (2016)
45. Ismail, M.E.H.: On a generalization of Szász operators. Mathematica (Cluj) **39**, 259–267 (1974)
46. Jakimovski, A., Leviatan, D.: Generalized Szász operators for the approximation in the infinite interval. Mathematica (Cluj) **34**, 97–103 (1969)
47. Kajla, A., Agrawal, P.N.: Szász-Durrmeyer type operators based on Charlier polynomials. Appl. Math. Comput. **268**, 1001–1014 (2015)
48. Kajla, A., Agrawal, P.N.: Approximation properties of Szász type operators based on Charlier polynomials. Turk. J. Math. **39**, 990–1003 (2015)
49. Kajla, A., Agrawal, P.N.: Szász-Kantorovich type operators based on Charlier polynomials. Kyungpook Math. J. **56**(3), 877–897 (2016)
50. Karaisa, A.: Approximation by Durrmeyer type Jakimovski-Leviatan operators. Math. Meth. Appl. Sci. **39**(9), 2401–2410 (2016)
51. Karsli, H.: Rate of convergence of new gamma type operators for functions with derivatives of bounded variation. Math. Comput. Model. **45**(5–6), 617–624 (2007)
52. Kirov, G.: A generalization of S. N. Bernstein's polynomials. Math Balkanica **6**, 147–154 (1992)
53. Kirov, G., Popova, I.: A generalization of linear positive operators. Math Balkanica **7**, 149–162 (1993)
54. Korovkin, P.P.: Linear Operators and the Theory of Approximations. Fizmatgiz, Moscow (1959)
55. Lupas, A.: Die Folge der Beta Operatoren, Dissertation, Universität Stuttgart (1972)
56. Lupas, A., Müller, M.: Approximations eigenschaften der Gamma operatoren. Math. Z. **98**(3), 208–226 (1967)
57. Mahmudov, N.I.: Higher order limit q-Bernstein operators. Math. Meth. Appl. Sci. **34**, 1618–1626 (2011)
58. Miheşan, V: The beta approximating operators of first kind. Studia Univ. Babes Bolyai Mathematica **49**(2), 65–77 (2004)
59. Miheşan, V.: The beta approximating operators of second kind. Studia Univ. Babes Bolyai Mathematica **49**(2), 79–87 (2004)
60. Milovanović, G.V., Rassias, M.Th. (eds.): Analytic Number Theory. Approximation Theory and Special Functions. Springer (2014)
61. Mursaleen, M., Al-Abied, A.H., Acu, A.M.: Approximation by Chlodowsky type of Szász operators based on Boas-Buck type polynomials. Turk. J. Math. **42**, 2243–2259 (2018)
62. Mursaleen, M., Ansari, K.J.: On Chlodowsky variant of Szász operators by Brenke type polynomials. Appl. Math. Comput. **271**, 991–1003 (2015)
63. Mursaleen, M., Khan, A., Srivastava, H.M., Nisar, K.S.: Operators constructed by means of q-Lagrange polynomials and A-statistical approximation. Appl. Math. Comput. **219**, 6911–6918 (2013)

64. Neer, T., Acu, A.M., Agrawal, P.N.: Baskakov-Durrmeyer type operators involving generalized Appell polynomials. Math. Methods Appl. Sci. 1–13 (2019). http://doi.org/10.1002/mma.6089
65. Öksüzer, Ö., Karsli, H., Taşdelen, F.: Approximation by Kantorovich variant of Szász operators based on Brenke type polynomials. Mediterr. J. Math. **13**, 3327–3340 (2016)
66. Pop, O.T., Farcas, M.D.: Approximation of B-continuous and B-differentiable functions by GBS operators of Bernstein bivariate polynomials. J. Inequal. Pure Appl. Math. **7**(3), 9 pp. (electronic) (2006)
67. Ruchi, R., Baxhaku, B., Agrawal, P.N.: GBS operators of bivariate Bernstein-Durrmeyer type on a triangle. Math. Methods Appl. Sci. **41**(7), 2673–2683 (2018)
68. Sharma, H.: Approximation properties of rth order generalized Bernstein polynomials based on q-calculus. Anal. Theory Appl. **27**(1), 40–50 (2001)
69. Sidharth, M., Agrawal, P.N., Araci, S.: Szász-Durrmeyer operators involving Boas-Buck polynomials of blending type. J. Inequal. Appl. **127** (2017)
70. Singh, J.K., Agrawal, P.N., Kajla, A.: Approximation by modified q-Gamma type operators via A-statistical convergence and power series method. Linear Multilinear Algebra (2021). https://doi.org/10.1080/03081087.2021.1960260
71. Singh, J.K., Agrawal, P.N., Kajla, A.: Approximation by modified q-Gamma type operators in a polynomial weighted space. Math. Methods Appl. Sci. (2022). https://doi.org/10.1002/mma.8687
72. Sucu, S., İçöz, G., Varma, S.: On some extension of Szász operators including Boas-Buck type polynomials. Absr. Appl. Anal. **2012**. Hindawi Publishing Corporation (2012)
73. Taşdelen, F., Olgun, A., Başcanbaz-Tunca, G.: Approximation of functions of two variables by certain linear positive operators. Proc. Math. Sci. **117**(3), 387–399 (2007)
74. Taşdelen, F., Aktaş, R., Altin, A.: A Kantorovich type of Szász operators including Brenke type polynomials . Abs. Appl. Anal. Article Number 867203 (2012). https://doi.org/10.1155/2012/867203
75. Upreti, R.: Approximation properties of beta operators. J. Approx. Theory **45**(2), 85–89 (1985)
76. Usta, F., Betus, Ö.: A new modification of Gamma operators with a better error estimation. Linear Multilinear Algebra 2198–2209 (2020)
77. Varma, S.: On a generalization of Szász operators by multiple Appell polynomials, Studia. Universitatis Babes-Bolyai, Mathematica **58**(3), 361–369 (2013)
78. Varma, S., Sucu, S., İçöz, G.: Generalization of Szász operators involving Brenke type polynomials. Comput. Math. Appl. **64**(2), 121–127 (2012)
79. Varma, S., Taşdelen, F.: Szász type operators involving Charlier polynomials. Math. Comput. Model. **56**, 118–122 (2012)
80. Volkov, V.I.: On the convergence of sequence of linear positive operators in the space of continuous functions of two variable. Dokl. Akad. Nauk. SSSR (N.S) **115**, 17–19 (1957)
81. Voronovskaya, E.V.: Determining the asymptotic kind of the approximations of functions by S. N. Bernstein's polynomials. DAN SSSR A **4**, 74–85 (1934)
82. Wafi, A., Rao, N., Deepmala.: On Kantorovich form of generalized Szász type operators using Charlier polynomials. Korean J. Math. **25**(1), 99–116 (2017)
83. Walczak, Z.: Bernstein-Durrmeyer type operators. Acta Math. Univ. Ostrav. **12**(1), 65–72 (2004)
84. Yilik, Ö. Ö., Garg, T., Agrawal, P.N.: Convergence rate of Szász operators involving Boas-Buck type polynomials. Proc. Natl. Acad. Sci., India, Sect. A Phys. Sci. **92**(2) (2024). https://doi.org/10.1007/s40010-020-00663-3
85. Zeng, X.M.: Approximation properties of Gamma operators. J. Math. Anal. Appl. **311**(2), 389–401 (2005)

Chapter 2
Approximation Process of the Fuzzy Meyer-König and Zeller Operators

Esma Yıldız Özkan

Abstract We construct fuzzy Meyer-König and Zeller operators and fuzzy post-quantum Meyer-König and Zeller operators. We investigate their fuzzy Korovkin-type approximation results and inequalities estimating their approximations.

Keywords Meyer König-Zeller operators · Fuzzy Korovkin-type approximation · Rate of convergence

2.1 Introduction

In 1960, Meyer-König and Zeller [19] defined the following operators:

$$B_n(f;x) = \sum_{k=0}^{\infty} f\left(\frac{k}{n+k}\right)\binom{n+k-1}{k} x^k(1-x)^n, x \in [0, 1), \qquad (2.1)$$

where f is such a function that makes the above series convergent. These operators are known as Bernstein power series, which do not reproduce linear functionals. Therefore, Cheney and Sharma [5] defined with a little modification the following operator:

$$M_n(f;x) = \sum_{k=0}^{\infty} f\left(\frac{k}{n+k}\right)\binom{n+k}{k} x^k(1-x)^{n+1}, x \in [0, 1), \qquad (2.2)$$

for each $n \in \mathbb{N}$ and for any continuous function f defined on $[0, 1]$. If $f(1)$ exists, and f is continuous to the left at 1, then the limit

$$\lim_{x\to 1^-} M_n(f;x) = f(1)$$

E. Y. Özkan (✉)
Faculty of Sciences, Department of Mathematics, Gazi University, 06500 Ankara, Türkiye
e-mail: esmayildiz@gazi.edu.tr

B. C. Tripathy et al. (eds.), *Operators, Inequalities and Approximation*, Industrial and Applied Mathematics, https://doi.org/10.1007/978-981-97-3238-8_2

exists. Thus, $M_n(f; 1)$ is defined by as follows:

$$M_n(f; 1) := f(1). \tag{2.3}$$

These modified operators are called Meyer-König and Zeller operators. For the Meyer-König and Zeller operators, we have

$$M_n(1; x) = 1 \tag{2.4}$$

$$M_n(t; x) = x. \tag{2.5}$$

Many authors [1, 4, 18, 21, 29, 35] were dealt with estimations of the second moment; however, an explicit expression for $M_n\left(t^2; x\right)$ does not occur. We consider in this chapter the following expression given by Müller [21, p. 61] in 1967 because of being closest estimation of the second moment:

$$0 \leq M_n\left((t-x)^2 f; x\right) \leq \frac{x(1-x)}{n+1}, x \in [0, 1). \tag{2.6}$$

Müller [22] and Totik [35] investigated the approximation properties of the integral-type Meyer-König and Zeller operators in the L_p space. In [6], Doğru obtained a generalization of the Meyer-König and Zeller operators. Later, Agratini [2] gave another generalization of Meyer-König and Zeller operators including the operators generalized by Doğru. Doğru and Özalp [8] introduced the Kantorovich-type generalization of the Meyer-König and Zeller operators. Liu and Zhang gave in [17] the strong Voronovskaja-type asymptotic expansion of Meyer-König and Zeller operators.

Fuzzy sets were introduced by Zareh [41]. The fuzzy real numbers were defined by Wu and Ming [38, 39]. Some significant results of the approximation theory have recently been extended to the fuzzy function space. For instance, Gal [9, 10], Wu, and Danghang [37] gave the Weierstrass approximation theorem in a fuzzy sense. Anastassiou [3, pp. 102–104] presented the corresponding fuzzy Shisha-Mond inequality of Shisha-Mond inequality in [30] and the corresponding fuzzy Korovkin-type theorem of the well-known Korovkin theorem [16], and he applied the results for the varied fuzzy operators [3]. For more results and applications of fuzzy operators, readers can see [24–28]. Moreover, some recent studies such as [31–34] have the potential of extending to the fuzzy function space.

2.2 Some Auxilary Results

Firstly, we recall some basic definitions and results of fuzzy mathematics. The details can be found in the references [3, 11, 15, 20, 38, 39].

Let $z : \mathbb{R} \to [0, 1]$ be a function satisfying the following properties:

(1) There exists at least an element $\rho_0 \in \mathbb{R}$ satisfying $z(\rho_0) = 1$,
(2) For each $\lambda \in [0, 1]$ and $\rho, \sigma \in \mathbb{R}$

$$z(\lambda\rho + (1 - \lambda)\sigma) \geq min\{z(\rho), z(\sigma)\},$$

equivalently, for all $\eta \in [0, 1]$, the set $\{\rho \in \mathbb{R} : z(\rho) \geq \eta\}$ is convex.
(3) There exists at least a neighborhood $V(\rho_0)$ satisfying $z(\rho) \leq z(\rho_0) + \varepsilon$ for all $\rho \in V(\rho_0)$,
(4) The set $\overline{\{\rho \in \mathbb{R} : z(\rho) > 0\}}$ is compact, where $\overline{A}$ is the closure of any set A,

then z is called a fuzzy real number.

Let by $\mathbb{R}_{\mathcal{F}}$ be denoted the set of all the fuzzy real numbers. Let by $[z]^r$ be denoted by $[z]^r = \{\rho \in \mathbb{R} : z(\rho) \geq r\}$, which is represented with a bounded and closed interval of $\mathbb{R}$ for each $r \in [0, 1]$, i.e., there exist $z_-^r, z_+^r \in \mathbb{R}$ such that $[z]^r = \left[z_-^r, z_+^r\right]$. Let $r_1, r_2 \in [0, 1]$ such that $r_1 \leq r_2$ then $[z]^{r_1} \subseteq [z]^{r_2}$.

The addition denoted by $w \oplus z$ and the scalar product denoted by $k \odot z$ on $\mathbb{R}_{\mathcal{F}}$ are defined by as follows:

$$[w \oplus z]^r = [w]^r + [z]^r = \left[z_-^r, z_+^r\right] + \left[w_-^r, w_+^r\right] = \left[z_-^r + w_-^r, z_+^r + w_+^r\right],$$

$$[k \odot z]^r = k[z]^r = k\left[z_-^r, z_+^r\right] = \left[kz_-^r, kz_+^r\right],$$

uniquely for $w, z \in \mathbb{R}_{\mathcal{F}}$, $k \in \mathbb{R}$ and $r \in [0, 1]$.

Let $D : \mathbb{R}_{\mathcal{F}} \times \mathbb{R}_{\mathcal{F}} \to [0, \infty)$ be defined by

$$D(w, v) = \sup_{r \in [0,1]} \max\left\{\left|w_-^r - z_-^r\right|, \left|w_+^r - z_+^r\right|\right\},$$

where $[z]^r = \left[z_-^r, z_+^r\right]$, $[w]^r = \left[w_-^r, w_+^r\right] \subset \mathbb{R}$. D is a metric on $\mathbb{R}_{\mathcal{F}}$. $(\mathbb{R}_{\mathcal{F}}, D)$ is a complete metric space, which has very close properties of a metric derived by the norm space.

A partial order on $\mathbb{R}_{\mathcal{F}}$ denoted by "$\precsim$" is defined as the partial order of the bounded and closed intervals $[w]^r, [z]^r$ for $w, z \in \mathbb{R}_{\mathcal{F}}$ such that $[w]^r \precsim [z]^r$ for each $r \in [0, 1]$ if and only if $w_-^r \leq z_-^r$, $w_+^r \leq z_+^r$ for each $r \in [0, 1]$, where $w_-^r, w_+^r, z_-^r, z_+^r \in \mathbb{R}$.

Let $k, l \in \mathbb{R}$, $w, z \in \mathbb{R}_{\mathcal{F}}$ and $\tilde{o} = \chi_{\{0\}}$ be the characteristic function of $\{0\}$. Then the addition and the scalar product on $\mathbb{R}_{\mathcal{F}}$ has the following properties:

(a) A neutral element with respect to $\oplus$ is $\tilde{o} \in \mathbb{R}_{\mathcal{F}}$,
(b) For each $z \neq \tilde{o}$, $z \in \mathbb{R}_{\mathcal{F}}$ has not an opposite in $\mathbb{R}_{\mathcal{F}}$,

(c) If $k, l > 0$ or $k, l \leq 0$, we have

$$(k + l) \odot z = k \odot z \oplus l \odot z,$$

(The above property does not hold for general $k, l \in \mathbb{R}$.)

(d)

$$k \odot (w \oplus z) = k \odot w \oplus k \odot z,$$

(e)

$$k \odot (l \odot z) = (k \cdot l) \odot z.$$

Let $(z_k)_{k \in \mathbb{N}}$ be any sequence in $\mathbb{R}_{\mathcal{F}}$. The fuzzy infinite series denoted by $\sum_{k=1}^{\infty} *z_k$ is defined by the following pair:

$$\left(z_k, \sum_{k=1}^{n} *z_k \right),$$

where the sum $\sum_{k=1}^{n} *$ denotes the finite fuzzy summation. The series $\sum_{k=1}^{\infty} *z_k$ converges to $z \in \mathbb{R}_{\mathcal{F}}$ if and only if the limit:

$$\lim_{n \to \infty} D\left(z_k, \sum_{k=1}^{n} *z_k \right) = 0,$$

is exist, and valid. The convergence of the series $\sum_{k=1}^{\infty} *z_k$ to z is denoted by $\sum_{k=1}^{\infty} *z_k = z$, which is equivalent to the convergence of the following series:

$$\left(\sum_{k=1}^{\infty} *z_k \right)_{\pm}^{r} = z_{\pm}^{r} = \sum_{k=1}^{\infty} (z_k)_{\pm}^{r},$$

for each $r \in [0, 1]$ respect to $\pm$, respectively.

For any two convergent sequences $(z_k)_{k \in \mathbb{N}}$ and $(w_k)_{k \in \mathbb{N}}$ in $\mathbb{R}_{\mathcal{F}}$ and $\alpha \in \mathbb{R}$, one has the following equality:

$$\sum_{k=1}^{\infty} *(z_k \oplus w_k) = z \oplus w = \sum_{k=1}^{\infty} *z_k \oplus \sum_{k=1}^{\infty} *w_k.$$

For more details see [3, pp. 126–127].

Let (X, d) be a metric space and $S \subset X$ be a subset. A function $\varphi : S \to \mathbb{R}_{\mathcal{F}}$ is called a fuzzy real number valued function on S or shortly a fuzzy function on S, which is represented with the following bounded and closed interval:

$$[\varphi(x)]^r = \left[(\varphi(x))_-^r, (\varphi(x))_+^r\right],$$

for each $x \in S$, $r \in [0, 1]$, where $(\varphi(x))_-^r$ and $(\varphi(x))_+^r$ are left and right points of the bounded and closed interval $[\varphi(x)]^r$, respectively.

Let "$\preccurlyeq$" denote the partial order relation of two fuzzy functions φ, ϕ on S such that $\varphi \preccurlyeq \phi$ if and only if $[\varphi(x)]^r \lesssim [\phi(x)]^r$ for each $x \in S$ if and only if $\varphi_+^r(x) \leq \phi_+^r(x)$ and $\varphi_-^r(x) \leq \phi_-^r(x)$ for each $x \in S$, $r \in [0, 1]$, where "$\leq$" is the partial order relation on $\mathbb{R}$.

Any function $\varphi : S \to \mathbb{R}_{\mathcal{F}}$ is called a fuzzy continuous at $x_0 \in S$ if and only if φ is sequential continuous at $x_0 \in S$. If φ is continuous for each $x_0 \in S$, then φ is called a fuzzy continuous function on S.

Any function $\varphi : S \to \mathbb{R}_{\mathcal{F}}$ is called fuzzy bounded function on S if and only if there exists $M > 0$ such that $D(f(x), \tilde{o}) \leq M$ for all $x \in S$.

The space of all fuzzy continuous functions is denoted by $C_{\mathcal{F}}(S)$. The addition "+" and the product with scalar "." on $C_{\mathcal{F}}(S)$ are defined by

$$(\varphi + \gamma)(x) = \varphi(x) \oplus \gamma(x),$$

$$(\alpha.\varphi)(x) = \alpha \odot \varphi(x),$$

for $\alpha \in \mathbb{R}$, $\gamma, \varphi \in C_{\mathcal{F}}(S)$. If $\varphi \in C_{\mathcal{F}}(S)$, then there exist $\varphi_-^r, \varphi_+^r \in C(S)$ for each $r \in [0, 1]$ such that the below representation is used

$$\varphi(x) = \left[\varphi_-^r(x), \varphi_+^r(x)\right],$$

for each $r \in [0, 1]$, $x \in S$.

Let S be a compact subset of real numbers. Let $D^* : C_{\mathcal{F}}(S) \times C_{\mathcal{F}}(S) \to [0, \infty)$ be defined by

$$D^*(\gamma, \varphi) = \sup_{x \in S} \sup_{r \in [0,1]} \max\left\{\left|\gamma_-^r(x) - \varphi_-^r(x)\right|, \left|\gamma_+^r(x) - \varphi_+^r(x)\right|\right\},$$

where $\gamma, \varphi \in C_{\mathcal{F}}(S)$ such that $\gamma_-^r, \gamma_+^r, \varphi_-^r, \varphi_+^r \in C(S)$ D^* is a metric on $C_{\mathcal{F}}(S)$. $(C_{\mathcal{F}}(S), D^*)$ is a complete metric space, which has very close properties of a metric derived by the norm space.

Let $L_{\mathcal{F}}$ be any operator from $C_{\mathcal{F}}(S)$ into itself, such that the following equality is valid:

$$L_{\mathcal{F}}(c_1\varphi + c_2\phi) = c_1 L_{\mathcal{F}}(\varphi) + c_2 L_{\mathcal{F}}(\phi),$$

for all $c_1, c_2 \in \mathbb{R}$ and $\varphi, \phi \in C_{\mathcal{F}}(S)$, then $L_{\mathcal{F}}$ is called a fuzzy linear operator on the space $C_{\mathcal{F}}(S)$, which is represented by the following bounded and closed interval:

$$[L_{\mathcal{F}}(\varphi)(x)]^r = \left[(L_{\mathcal{F}}(\varphi)(x))_-^r, (L_{\mathcal{F}}(\varphi)(x))_+^r\right], \tag{2.7}$$

for all $x \in S$, $r \in [0, 1]$.

Let $L_{\mathcal{F}}$ be a fuzzy linear operator from $C_{\mathcal{F}}(S)$ into itself. $L_{\mathcal{F}}$ is called positive if and only if whenever $\varphi, \phi \in C_{\mathcal{F}}(S)$ such that $\varphi \preccurlyeq \phi$ then $L_{\mathcal{F}}(\varphi) \preccurlyeq L_{\mathcal{F}}(\phi)$ if and only if $(L_{\mathcal{F}}(\varphi)(x))_+^r \leq (L_{\mathcal{F}}(\phi)(x))_+^r$ and $(L_{\mathcal{F}}(\varphi)(x))_-^r \leq (L_{\mathcal{F}}(\phi))_-^r(x)$ for each $x \in S$, $r \in [0, 1]$. Here, "$\preccurlyeq$" and "$\leq$" are the partial relation order on $C_{\mathcal{F}}(S)$ and $\mathbb{R}$, respectively.

For example, the fuzzy Bernstein operator is a fuzzy linear positive operator [10].

Let S be a compact subset of any metric space (X, d).

Theorem 2.1 ([3, p.136]) *Let L be a linear positive operator from $C(S)$ into itself. Then there exists a unique fuzzy linear positive operator $L_{\mathcal{F}}$ on $C_{\mathcal{F}}(S)$ produced by L satisfying the following:*

$$(L_{\mathcal{F}}(\varphi))_\pm^r = L(\varphi_\pm^r), \tag{2.8}$$

respectively in $\pm$, for all $r \in [0, 1]$, $\varphi \in C_{\mathcal{F}}(S)$ such that $\varphi_-^r, \varphi_+^r \in C(S)$. Additionally, if $L(\varphi_\pm^r)$ are equi-continuous regard to $r \in [0, 1]$ respectively in $\pm$, then $L_{\mathcal{F}}(\varphi) \in C_{\mathcal{F}}(S)$ whenever $\varphi \in C_{\mathcal{F}}(S)$.

Assumption 2.2 ([3, p.134]) *Let $L_{\mathcal{F}}$ be a fuzzy linear positive operator from $C_{\mathcal{F}}(S)$ into itself. Assume that there exists a linear positive operator L from $C(S)$ into itself satisfying the condition given by (8), respectively in $\pm$, for each $r \in [0, 1]$ and $\varphi \in C_{\mathcal{F}}(S)$.*

For example, the classical Bernstein operators and the corresponding fuzzy Bernstein operators fulfill the condition given by (2.8) (see Example 9.7 in [3]).

2.3 Some Fuzzy Korovkin-Type Approximation Results

In this part, we construct fuzzy Meyer-König and Zeller operators and fuzzy post-quantum Meyer-König and Zeller operators. We investigate their fuzzy Korovkin-type approximation process.

The fuzzy Meyer-König and Zeller operators are defined for the first time by Yang and Zhang, and they established its fuzzy asymptotic expansion of Voronovskaja [40].

We consider the following fuzzy Meyer-König and Zeller operators:

$$M_n^{\mathcal{F}}(\varphi; x) = \sum_{k=1}^{\infty} * \binom{n+k}{k} x^k (1-x)^{n+1} \odot \varphi\left(\frac{k}{n+k}\right), x \in [0, 1),$$

$$M_n^{\mathcal{F}}(\varphi; 1) := \varphi(1), \tag{2.9}$$

where $n \in \mathbb{N}, \varphi \in C_{\mathcal{F}}[0, 1]$, $\sum_{k=1}^{\infty} *$ denotes the fuzzy series and $\odot$ is the fuzzy scalar product on $\mathbb{R}_{\mathcal{F}}$. $M_n^{\mathcal{F}}$ is a small modification of the operator defined by Yang and Zang in [40]. By the definition of addition, product with scalar and the fuzzy series, since $\varphi \in C_{\mathcal{F}}[0, 1], M_n^{\mathcal{F}}(\varphi; .) \in C_{\mathcal{F}}[0, 1]$, that is, each $M_n^{\mathcal{F}}$ is a fuzzy operator mapping $C_{\mathcal{F}}[0, 1]$ into itself.

Let α and β be real scalars and $\gamma, \varphi \in C_{\mathcal{F}}[0, 1]$. Then, we demonstrate the linearity of the fuzzy operator $M_n^{\mathcal{F}}$ as follows:

$$\begin{aligned} M_n^{\mathcal{F}}(\alpha\varphi + \beta\gamma; x) &= \lim_{n\to\infty} \sum_{k=0}^{n} *\nu_{k,n}(x) \odot (\alpha\varphi + \beta\gamma)\left(\frac{k}{n+k}\right) \\ &= \lim_{n\to\infty} \sum_{k=0}^{n} *\nu_{k,n}(x) \odot \alpha \odot \varphi\left(\frac{k}{n+k}\right) \\ &+ \lim_{n\to\infty} \sum_{k=0}^{n} *\nu_{k,n}(x) \odot \beta \odot \gamma\left(\frac{k}{n+k}\right) \\ &= \alpha \odot \lim_{n\to\infty} \sum_{k=0}^{n} *\nu_{k,n}(x) \odot \varphi\left(\frac{k}{n+k}\right) \\ &+ \beta \odot \lim_{n\to\infty} \sum_{k=0}^{n} *\nu_{k,n}(x) \odot \gamma\left(\frac{k}{n+k}\right) \\ &= \alpha \odot M_n^{\mathcal{F}}(\varphi; x) \oplus \beta \odot M_n^{\mathcal{F}}(\gamma; x), \end{aligned}$$

which implies

$$M_n^{\mathcal{F}}(\alpha\varphi + \beta\gamma; .) = \alpha M_n^{\mathcal{F}}(\varphi; .) + \beta M_n^{\mathcal{F}}(\gamma; .).$$

Then $M_n^{\mathcal{F}}$ is a fuzzy linear operator on the space $C_{\mathcal{F}}[0, 1]$ represented with

$$\left[M_n^{\mathcal{F}}(\varphi; x)\right]^r = \left[\left(M_n^{\mathcal{F}}(\varphi; x)\right)_-^r, \left(M_n^{\mathcal{F}}(\varphi; x)\right)_+^r\right],$$

for each $x, r \in [0, 1]$.

Considering the definition of the fuzzy series, for each $x, r \in [0, 1]$ and $\varphi \in C_{\mathcal{F}}[0, 1], \varphi_-^r, \varphi_+^r \in C[0, 1]$, we directly prove the following equality:

$$\begin{aligned} \left(M_n^{\mathcal{F}}(\varphi; x)\right)_-^r &= \left(\lim_{n\to\infty} \sum_{k=0}^{n} *\nu_{k,n}(x) \odot \varphi\left(\frac{k}{n+k}\right)\right)_-^r \\ &= \lim_{n\to\infty} \sum_{k=0}^{n} \nu_{k,n}(x) \varphi_-^r\left(\frac{k}{n+k}\right) \end{aligned}$$

$$= M_n(\varphi_-^r; x).$$

Similarly

$$\left(M_n^{\mathcal{F}}(\varphi; x)\right)_+^r = M_n(\varphi_+^r; x).$$

Thus, for each $n \in \mathbb{N}$, $M_n^{\mathcal{F}}$ fulfills the condition (2.8).

Let $\gamma, \varphi \in C_{\mathcal{F}}[0, 1]$ such that $\varphi \preccurlyeq \gamma$ then $\varphi_\pm^r, \gamma_\pm^r \in C[0, 1]$ such that $\varphi_\pm^r(x) \le \gamma_\pm^r(x)$ for each $x, r \in [0, 1]$ respect to $+$ and $-$, respectively. Since $v_{k,n}(x) \ge 0$, we have

$$\sum_{k=0}^{n} v_{k,n}(x)\varphi_-^r\left(\frac{k}{n+k}\right) \le \sum_{k=0}^{n} v_{k,n}(x)\gamma_-^r\left(\frac{k}{n+k}\right),$$

which implies

$$M_n(\varphi_\pm^r; x) \le M_n(\gamma_\pm^r; x)$$

for each $x, r \in [0, 1]$; therefore, by the definition of the positivity, we get

$$\left[M_n^{\mathcal{F}}(\varphi; x)\right]^r \precsim \left[M_n^{\mathcal{F}}(\gamma; x)\right]^r,$$

That is

$$M_n^{\mathcal{F}}(\varphi; .) \preccurlyeq M_n^{\mathcal{F}}(\gamma; .).$$

Hence, $M_n^{\mathcal{F}}$ is a positive linear fuzzy operator.

Now, we investigate fuzzy Korovkin-type results for the fuzzy Meyer-König and Zeller operators.

We initially recall the following fuzzy Korovkin-type theorem.

Theorem 3 ([3]) *Assume that there exists a corresponding sequence of linear positive operators $\left(\tilde{L}_n\right)_{n\in\mathbb{N}}$ from $C[a, b]$ into itself for any sequence of fuzzy linear positive operators $(L_n)_{n\in\mathbb{N}}$ from $C_{\mathcal{F}}[a, b]$ into itself for $[a, b] \subset \mathbb{R}$ satisfying the condition (8) for each $r \in [0, 1]$ and $\varphi \in C_{\mathcal{F}}[0, 1]$. Moreover, suppose that $\left(\tilde{L}_n\left(t^i\right)\right)_{n\in\mathbb{N}}$ converge x^i for $i = 0, 1, 2$ uniformly as $n \to \infty$ then $D^*(L_n(\varphi), \varphi)$ converges to zero uniformly as $n \to \infty$ for any $\varphi \in C_{\mathcal{F}}[0, 1]$, i.e., the sequence $(L_n(\varphi))_{n\in\mathbb{N}}$ converges to φ respect to D^* as $n \to \infty$.*

Now, considering the above fuzzy Korovkin-type theorem, we present the following fuzzy Korovkin-type theorem for the fuzzy Meyer-König and Zeller operators.

Theorem 4 *If $\varphi \in C_{\mathcal{F}}[0, 1]$ then $\left(M_n^{\mathcal{F}}(\varphi; x)\right)_{n\in\mathbb{N}}$ converges to φ uniformly in D^*, i.e., $\left(M_n^{\mathcal{F}}\right)_{n\in\mathbb{N}}$ converges to the unit operator I in the fuzzy sense, as $n \to \infty$.*

Proof For the proof of the theorem, we use Theorem 2.3 (fuzzy Korovkin-type theorem). By the constructive properties of the fuzzy Meyer-König and Zeller operators, we know that.

$$\left(M_n^{\mathcal{F}}(\varphi; x)\right)_+^r = M_n\left(\varphi_+^r; x\right),$$

for each $x, r \in [0, 1]$ and $\varphi \in C_{\mathcal{F}}[0, 1], \varphi_-^r, \varphi_+^r \in C[0, 1]$. By the results of Theorem 2.1 and Assumption 2.2, the real corresponding operator M_n acts on $C[0, 1]$ while the fuzzy operator $M_n^{\mathcal{F}}$ acts on $C_{\mathcal{F}}[0, 1]$. On the other hand, $M_n\left(t^i; x\right)$ converge to x^i uniformly for $i = 0, 1, 2$ [5, p. 242]; therefore, the hypotheses of Theorem 2.3 are verified, which completes the proof.

For $\varphi \in C_{\mathcal{F}}[0, 1]$, the first order fuzzy modulus of continuity of φ is defined by

$$\omega_1^{\mathcal{F}}(\varphi; \delta) := \sup_{\substack{x, y \in S \\ d(x, y) \leq \delta}} D(\varphi(x), \varphi(y)), \tag{2.10}$$

for $0 < \delta < diameter(S)$. If $\delta > diameter(S)$ then

$$\omega_1^{\mathcal{F}}(\varphi; \delta) := \omega_1^{\mathcal{F}}(\varphi; diameter(S)).$$

If $\varphi \in C_{\mathcal{F}}(S)$ then $\lim_{\delta \to 0^+} \omega_1^{\mathcal{F}}(\varphi; \delta)$. For $\delta > 0$, assume that $\omega_1^{\mathcal{F}}(\varphi; \delta), \omega_1\left(\varphi_-^r; \delta\right)$ and $\omega_1\left(\varphi_+^r; \delta\right)$ are finite. Then it holds

$$\omega_1^{\mathcal{F}}(\varphi; \delta) = \sup_{r \in [0,1]} \max\left\{\omega_1\left(\varphi_-^r; \delta\right), \omega_1\left(\varphi_+^r; \delta\right)\right\}, \tag{2.11}$$

where $\omega_1\left(\varphi_\pm^r; \delta\right)$ are the first order modulus of continuity defined by

$$\omega_1\left(\varphi_\pm^r; \delta\right) = \sup_{\substack{t, x \in S \\ |t - x| \leq \delta}} \left|\varphi_\pm^r(t) - \varphi_\pm^r(x)\right|,$$

for each $r \in [0, 1]$, $\varphi_-^r, \varphi_+^r \in C(S)$ satisfying

$$\left|\varphi_\pm^r(t) - \varphi_\pm^r(x)\right| \leq \omega_1\left(\varphi_\pm^r; \delta\right)\left(1 + \frac{|t - x|}{\delta}\right),$$

respectively. For $\varphi \in C_{\mathcal{F}}(S)$, $\omega_1^{\mathcal{F}}(\varphi; \delta)$ has the following properties:

$$\omega_1^{\mathcal{F}}(\varphi; \delta_1 + \delta_2) \leq \omega_1^{\mathcal{F}}(\varphi; \delta_1) + \omega_1^{\mathcal{F}}(\varphi; \delta_2), \delta_1, \delta_2 > 0,$$
$$\omega_1^{\mathcal{F}}(\varphi; n\delta) \leq n\omega_1^{\mathcal{F}}(\varphi; \delta), \delta > 0, n \in \mathbb{N},$$
$$\omega_1^{\mathcal{F}}(\varphi; \kappa\delta) \leq (\kappa + 1)\omega_1^{\mathcal{F}}(\varphi; \delta), \delta > 0, \kappa > 0.$$

The details about the fuzzy modulus of continuity can be found in [3].

We give the following result estimating the approximation of the fuzzy Meyer-König and Zeller operators given the first-order fuzzy modulus of continuity.

Theorem 5 *If* $\varphi \in C_{\mathcal{F}}[0,1]$ *then, for each* $n \in \mathbb{N}$, $x \in [0,1]$, *the following inequality.*

$$D^*\big(M_n^{\mathcal{F}}(\varphi;.),\varphi\big) \le 2\omega_1^{\mathcal{F}}(\varphi;\delta_n)$$

is valid, where $\delta_n = \sqrt{\frac{1}{4(n+1)}}$.

Proof By the definition of the fuzzy function, the fuzzy series and the property of the addition and the scalar product on $\mathbb{R}_{\mathcal{F}}$, since $\sum\limits_{k=0}^{\infty} v_{k,n}(x) = 1$, we can write the below equalities:

$$\begin{aligned}
\varphi(x) &= \left(\sum_{k=0}^{\infty} v_{k,n}(x)\right) \odot \varphi(x) \\
&= \left(\lim_{n\to\infty}\sum_{k=0}^{n} v_{k,n}(x)\right) \odot \varphi(x) \\
&= \lim_{n\to\infty}\sum_{k=0}^{n}{}^{*}\big(v_{k,n}(x) \odot \varphi(x)\big) \\
&= \sum_{k=0}^{\infty}{}^{*}\big(v_{k,n}(x) \odot \varphi(x)\big). \qquad (2.12)
\end{aligned}$$

Considering (2.12), the properties of the addition and the scalar product on $\mathbb{R}_{\mathcal{F}}$ and the properties of the metric D on $\mathbb{R}_{\mathcal{F}}$, we obtain the following inequality:

$$\begin{aligned}
&D\big(M_n^{\mathcal{F}}(\varphi;x),\varphi(x)\big) \\
&= D\left(\lim_{n\to\infty}\sum_{k=0}^{n}{}^{*}\left(v_{k,n}(x)\odot\varphi\left(\frac{k}{n+k}\right)\right), \lim_{n\to\infty}\sum_{k=0}^{n}{}^{*}\big(v_{k,n}(x)\odot\varphi(x)\big)\right) \qquad (2.13)\\
&\le \sum_{k=0}^{\infty} v_{k,n}(x) D\left(\varphi\left(\frac{k}{n+k}\right),\varphi(x)\right).
\end{aligned}$$

By the properties of the fuzzy modulus of continuity, we have

$$\begin{aligned}
D\left(\varphi\left(\frac{k}{n+k}\right),\varphi(x)\right) &\le \omega_1^{\mathcal{F}}\left(\varphi;\left|\frac{k}{n+k}-x\right|\right) \\
&\le \left(1+\frac{1}{\delta}\left|\frac{k}{n+k}-x\right|\right)\omega_1^{\mathcal{F}}(\varphi;\delta), \qquad (2.14)
\end{aligned}$$

for each $n \in \mathbb{N}$, $x \in [0, 1]$, $\varphi \in C_{\mathcal{F}}[0, 1]$ and $\delta > 0$.

Substituting (2.13) and (2.14), we get

$$D\big(M_n^{\mathcal{F}}(\varphi; x), \varphi(x)\big) \le \omega_1^{\mathcal{F}}(\varphi; \delta) + \frac{1}{\delta}\omega_1^{\mathcal{F}}(\varphi; \delta) M_n(|t - x| f; x), \tag{2.15}$$

for each $n \in \mathbb{N}$, $x \in [0, 1]$. Applying the Cauchy–Schwarz inequality to (2.15), we obtain the below inequality:

$$D\big(M_n^{\mathcal{F}}(\varphi; x), \varphi(x)\big) \le \omega_1^{\mathcal{F}}(\varphi; \delta) + \frac{1}{\delta}\omega_1^{\mathcal{F}}(\varphi; \delta)\sqrt{M_n\big((t - x)^2 f; x\big)}. \tag{2.16}$$

On the other hand, for each $n \in \mathbb{N}$, $x \in [0, 1]$, using (2.6), we get

$$M_n\big((t - x)^2 f; x\big) \le \frac{1}{4(n + 1)}. \tag{2.17}$$

Applying (2.17) to (2.16) and selecting $\delta = \delta_n = \sqrt{\frac{1}{4(n+1)}}$, we obtain the desired result.

In 2000, Trif [36] introduced the quantum Meyer-König and Zeller operators by

$$\begin{aligned} M_n^q(f; x) &= \sum_{k=0}^{\infty} f\left(\frac{[k]_q}{[n + k]_q}\right) \begin{bmatrix} n + k \\ k \end{bmatrix}_q x^k \prod_{s=0}^{n} (1 - q^s x), x \in [0, 1), \\ M_n^q(f; 1) &= f(1), \end{aligned} \tag{2.18}$$

For each $n \in \mathbb{N}$ and $f \in C[0, 1]$, where $[k]_q$ denotes the quantum integer of k, which is defined by

$$[k]_q = \frac{1 - q^k}{1 - q}, q > 0,$$

$[k]_q!$ denotes the quantum factorial of k, which is defined by

$$[k]_q! = \prod_{r=1}^{k} [r]_q, [0]_q! = 1.$$

For $n, k \in \mathbb{N}$ such that $0 \le k \le n$, $\begin{bmatrix} n \\ k \end{bmatrix}_q$ denotes the quantum binomial coefficients defined by

$$\begin{bmatrix} n \\ k \end{bmatrix}_q = \frac{[n]_q!}{[k]_q! [n - k]_q!}.$$

The details about quantum calculus can be found in [13].

For $q = 1$, the quantum Meyer-König and Zeller operators are reduced to the Meyer-König and Zeller operators defined by (2.2). Trif [36] investigated the monotonicity and convexity of these operators, which was also studied by Heping [12]. But it was not possible to obtain the explicit formula for the second moment for Trif's operator. Therefore, Doğru and Duman [7] gave in 2006 another generalization of the quantum Meyer-König and Zeller operators which is reduced to Cheney and Sharma's operator when $q = 1$. Later, Özarslan and Duman [23] introduced a new generalization of the quantum Meyer-König and Zeller operators which is including the operator defined in 2006. They investigated the statistical approximation properties of this new generalization.

Kadak et al. [14] defined the post-quantum generalization of Meyer-König and Zeller operators for $f \in C[0, 1]$ by

$$\begin{aligned} M_n^{p,q}(f; x) &= \sum_{k=0}^{\infty} f\left(\eta_{k,n}^{p,q}\right) v_{k,n}(x; p, q), x \in [0, 1), \\ M_n^{p,q}(f; 1) &= f(1), \end{aligned} \tag{2.19}$$

where

$$\eta_{k,n}^{p,q} := \frac{p^n [k]_{p,q}}{[n+k]_{p,q}},$$

$$v_{k,n}(x; p, q) := p^{-kn} p^{-\frac{n(n+1)}{2}} \begin{bmatrix} n+k \\ k \end{bmatrix}_{p,q} x^k \prod_{s=0}^{n} (p^s - q^s x).$$

Let $0 < q < p \leq 1$. Here, $[k]_{p,q}$ denotes the quantum integer of non-negative integer k, which is defined by

$$[k]_{p,q} = \frac{p^k - q^k}{p - q},$$

$[k]_{p,q}!$ denotes the quantum factorial of k, which is defined by

$$[k]_{p,q}! = \prod_{r=1}^{k} [r]_{p,q}, \ [0]_{p,q}! = 1.$$

For $n, k \in \mathbb{N}$ such that $0 \leq k \leq n$, $\begin{bmatrix} n \\ k \end{bmatrix}_{p,q}$ denotes the quantum binomial coefficients defined by

$$\begin{bmatrix} n \\ k \end{bmatrix}_{p,q} = \frac{[n]_{p,q}!}{[k]_{p,q}![n-k]_{p,q}!}.$$

By taking $p = 1$ in the definition of the post-quantum generalization of Meyer-König and Zeller operators, these operators are reduced to the operators defined by Trif. Kadak et al. [14] gave for each $x \in [0, 1]$ and $0 < q < p \leq 1$ the following results:

$$M_n^{p,q}(1; x) = 1, \tag{2.20}$$

$$M_n^{p,q}(t; x) = x, \tag{2.21}$$

$$x^2 \leq M_n^{p,q}\left(t^2; x\right) \leq x^2 + \frac{p^n x}{[n+1]_{p,q}}. \tag{2.22}$$

Lastly, let us construct corresponding fuzzy post-quantum Meyer-König and Zeller operators of the post-quantum Meyer-König and Zeller operators defined by (2.19).

We define fuzzy post-quantum Meyer-König and Zeller operators for $\varphi \in C_{\mathcal{F}}[0, 1]$ by

$$\begin{aligned} M_{n,p,q}^{\mathcal{F}}(\varphi; x) &= \sum_{k=0}^{\infty} * \varphi\left(\eta_{k,n}^{p,q}\right) v_{k,n} \odot (x; p, q), x \in [0, 1), \\ M_{n,p,q}^{\mathcal{F}}(\varphi; 1) &= \varphi(1), \end{aligned} \tag{2.23}$$

where $\sum_{k=1}^{\infty} *$ denotes the fuzzy series and $\odot$ is the fuzzy scalar product on $\mathbb{R}_{\mathcal{F}}$,

$$\eta_{k,n}^{p,q} := \frac{p^n [k]_{p,q}}{[n+k]_{p,q}},$$

$$v_{k,n}(x; p, q) := p^{-kn} p^{-\frac{n(n+1)}{2}} \begin{bmatrix} n+k \\ k \end{bmatrix}_{p,q} x^k \prod_{s=0}^{n} (p^s - q^s x).$$

The fuzzy post-quantum Meyer-König and Zeller operators map $C_{\mathcal{F}}[0, 1]$ into itself, and they are fuzzy linear positive operators satisfying the expression:

$$\left(M_{n,p,q}^{\mathcal{F}}(\varphi; x)\right)_{\pm}^{r} = M_n^{p,q}\left(\varphi_{\pm}^{r}; x\right) \tag{2.24}$$

These operators are reduced the corresponding fuzzy operators of Trif's quantum Meyer-König and Zeller operators defined by (2.18) when $p = 1$.

Let (p_n) and (q_n) such two sequences that $q_n \in (0, 1)$ and $p_n \in (q_n, 1]$ such that for each $n \in \mathbb{N}$ fulfilling the following conditions:

$$\lim_{n\to\infty} p_n = 1 = \lim_{n\to\infty} q_n, 0 < \lim_{n\to\infty} p_n^n < 1. \tag{2.25}$$

Such sequences can be found always. For instance, $(p_n) = \left(\frac{n-1}{n}\right)$ and $(q_n) = \left(\frac{n^2-1}{n^2}\right)$ satisfy the conditions (2.21).

Now, we give the following fuzzy Korovkin-type result for the fuzzy post-quantum Meyer-König and Zeller operators.

Theorem 6 *Let (p_n) and (q_n) such sequences that $q_n \in (0, 1)$ and $p_n \in (q_n, 1]$ for each $n \in \mathbb{N}$ fulfilling the following conditions* (2.25). *If $\varphi \in C_{\mathcal{F}}[0, 1]$ then $\left(M^{\mathcal{F}}_{n,p,q}(\varphi; x)\right)_{n\in\mathbb{N}}$ converges to φ uniformly in D^*, i.e., $\left(M^{\mathcal{F}}_{n,p,q}\right)_{n\in\mathbb{N}}$ converges to the unit operator I in the fuzzy sense, as $n \to \infty$.*

Proof Considering (2.20)–(2.22) and (2.24)–(2.25), by Theorem 2.3, the proof of the theorem is obtained similarly to the proof of Theorem 2.4; therefore, we omit the proof.

We have the following inequality estimating the approximation of the fuzzy operator $M^{\mathcal{F}}_{n,p,q}$ using the fuzzy modulus of continuity.

Theorem 7 *If $\varphi \in C_{\mathcal{F}}[0, 1]$ then, for each $n \in \mathbb{N}$, $0 < q < p \leq 1$, $x \in [0, 1]$, the following inequality holds:*

$$D^*\left(M^{\mathcal{F}}_{n,p,q}(\varphi; .), \varphi\right) \leq 2\omega^{\mathcal{F}}_1(\varphi; \delta_n),$$

where

$$\delta_n = \sqrt{\frac{p^n}{[n+1]_{p,q}}}. \tag{2.26}$$

Proof By the definition of the fuzzy function, the fuzzy series and the properties of the addition and the scalar product on $\mathbb{R}_{\mathcal{F}}$, since $\sum\limits_{k=0}^{\infty} v_{k,n}(x) = 1$, we can write that.

$$\varphi(x) = \sum_{k=0}^{\infty} {}^*\left(v_{k,n}(x) \odot \varphi(x)\right). \tag{2.27}$$

Considering (2.27), the properties of the addition and the scalar product on $\mathbb{R}_{\mathcal{F}}$ and the properties of the metric D on $\mathbb{R}_{\mathcal{F}}$, we obtain

$$D\left(M^{\mathcal{F}}_{n,p,q}(\varphi; x), \varphi(x)\right) \leq \sum_{k=0}^{\infty} v_{k,n}(x; p, q) D\left(\varphi\left(\eta^{p,q}_{k,n}\right), \varphi(x)\right). \tag{2.28}$$

By the properties of the fuzzy modulus of continuity, for each $n \in \mathbb{N}$, $x \in [0, 1]$,

$$D\left(\varphi\left(\eta^{p,q}_{k,n}\right), \varphi(x)\right) \leq \left(1 + \frac{1}{\delta}\left|\eta^{p,q}_{k,n} - x\right|\right)\omega^{\mathcal{F}}_1(\varphi; \delta), \tag{2.29}$$

Applying (2.29)–(2.28) and taking the Cauchy–Schwarz inequality into account, we get

$$D\Big(M^{\mathcal{F}}_{n,p,q}(\varphi;x),\varphi(x)\Big) \leq \omega_1^{\mathcal{F}}(\varphi;\delta) + \frac{1}{\delta}\omega_1^{\mathcal{F}}(\varphi;\delta)\sqrt{M_n^{p,q}\big((t-x)^2f;x\big)}. \tag{2.30}$$

On the other hand, for $x \in [0,1]$, by using (2.20)–(2.22), we have

$$0 \leq M_n^{p,q}\big((t-x)^2f;x\big) \leq \frac{p^n}{[n+1]_{p,q}} \tag{2.31}$$

By applying (2.31) to (2.30) and selecting $\delta = \delta_n = \sqrt{\frac{p^n}{[n+1]_{p,q}}}$, the proof of the theorem is completed.

Let $\varphi \in C_{\mathcal{F}}[a,b]$. If there exists $\varphi'(x) \in \mathbb{R}_{\mathcal{F}}$ such that

$$\varphi'(x) = \lim_{n\to\infty} \frac{\varphi(x+t)-\varphi(x)}{t},$$

then φ is differentiable at $x \in [a,b]$. The k-times derivatives $\varphi^{(k)}(x)$ of the fuzzy function φ for each $k \in \mathbb{N}$ is defined similarly.

Let by $C^2_{\mathcal{F}}[0,1]$ be denoted the space of all fuzzy continuous functions defined on $[0,1]$, which are k-times differentiable continuously.

We estimate the fuzzy asymptotic expansion of the fuzzy post-quantum Meyer-König and Zeller operators.

Theorem 8 *If any $\varphi \in C^2_{\mathcal{F}}[0,1]$ then we have.*

$$\begin{aligned}
&D^*\left(M^{\mathcal{F}}_{n,p,q}(\varphi;.),\varphi \oplus \frac{1}{2}M_n^{p,q}\big((t-x)^2f;x\big)\odot\varphi''\right)\\
&\leq D\big(\omega_1^{\mathcal{F}}(\varphi;\delta_n),\tilde{o}\big) + \frac{1}{2}\delta_n^2 D^*(\varphi'',\widetilde{0}),
\end{aligned}$$

where $\tilde{o}$ is the neutral element regard to $\oplus$ in $\mathbb{R}_{\mathcal{F}}$, $\widetilde{0}$ is a fuzzy function such that $\widetilde{0}(x) = \tilde{o}$ for each $x \in [0,1]$ and δ_n is defined as in (2.26).

Proof For $\varphi \in C^2_{\mathcal{F}}[0,1]$, we have the representation $\varphi(x) = \big[\varphi_-^r(x),\varphi_+^r(x)\big]$ such that $\varphi_-^r,\varphi_+^r \in C^2[0,1]$ and $\big(\varphi''\big)^r_-,\big(\varphi''\big)^r_+$ are bounded in $\mathbb{R}$. Considering the definition of the post-quantum operator $M_n^{p,q}$, we can write the below inequality:

$$\begin{aligned}
&\left|M_n^{p,q}(\varphi_\pm^r;x) - \varphi_\pm^r(x) - \frac{1}{2}\big(\varphi''\big)^r_- M_n^{p,q}\big((.-x)^2;x\big)\right|\\
&\leq \left|M_n^{p,q}(\varphi_\pm^r;x) - \varphi_\pm^r(x)\right| + \frac{1}{2}\big(\varphi''\big)^r_-\left|M_n^{p,q}\big((.-x)^2;x\big)\right|,
\end{aligned} \tag{2.32}$$

respect to $-$ and $+$, respectively.

On the other hand, since $\varphi_-^r, \varphi_+^r \in C^2[0,1]$, taking the properties of the first modulus of continuity and applying Cauchy–Schwarz inequality, we have the following inequality:

$$\left|M_n^{p,q}\left(\varphi_\pm^r; x\right) - \varphi_\pm^r(x)\right| \leq \omega_1\left(\varphi_\pm^r; \delta\right)\left(\frac{1}{\delta}\sqrt{M_n^{p,q}\left((. - x)^2; x\right)} + 1\right). \tag{2.33}$$

Replacing (2.33) in (2.32), selecting $\delta = \delta_n = \sqrt{\frac{p^n}{[n+1]_{p,q}}}$, for $x \in [0,1]$, we obtain the below expression:

$$\begin{aligned}
&\left|M_n^{p,q}\left(\varphi_\pm^r; x\right) - \varphi_\pm^r(x) - \frac{1}{2}\left(\varphi''\right)_\pm^r M_n^{p,q}\left((. - x)^2; x\right)\right| \\
&\leq 2\omega_1\left(\varphi_\pm^r; \delta\right) + \frac{1}{2}\left(\varphi''\right)_\pm^r \delta_n^2.
\end{aligned} \tag{2.34}$$

On the other hand, considering the definition of the metric D the representation of the fuzzy function and the fuzzy derivative and the definition of the fuzzy modulus of continuity, we get following statement:

$$\begin{aligned}
&D\left(M_{n,p,q}^{\mathcal{F}}(\varphi; x), \varphi \oplus \frac{1}{2} M_n^{p,q}\left((t-x)^2 f; x\right) \odot \varphi''(x)\right) \\
&= D\Big(\Big[\left(M_{n,p,q}^{\mathcal{F}}(\varphi; x)\right)_-^r, \left(M_{n,p,q}^{\mathcal{F}}(\varphi; x)\right)_+^r\Big], \\
&\left[\varphi_-^r(x), \varphi_+^r(x)\right] + \frac{1}{2} M_n^{p,q}\left((. - x)^2; x\right)\left[\left(\varphi''\right)_-^r(x), \left(\varphi''\right)_+^r(x)\right]\Big) \\
&= \sup_{r\in[0,1]} \max\left\{\left|M_n^{p,q}\left(\varphi_-^r; x\right) - \varphi_-^r(x) - \frac{1}{2}\left(\varphi''\right)_-^r M_n^{p,q}\left((. - x)^2; x\right)\right|\right. \\
&\left.\left|M_n^{p,q}\left(\varphi_+^r; x\right) - \varphi_+^r(x) - \frac{1}{2}\left(\varphi''\right)_+^r M_n^{p,q}\left((. - x)^2; x\right)\right|\right\} \\
&\leq \sup_{r\in[0,1]} \max\left\{2\omega_1\left(\varphi_-^r; \delta_n\right) + \frac{1}{2}\left(\varphi''\right)_-^r \delta_n^2, 2\omega_1\left(\varphi_+^r; \delta_n\right) + \frac{1}{2}\left(\varphi''\right)_+^r \delta_n^2\right\} \\
&\leq \sup_{r\in[0,1]} \max\left\{2\omega_1\left(\varphi_-^r; \delta_n\right), 2\omega_1\left(\varphi_+^r; \delta_n\right)\right\} \\
&+ \sup_{r\in[0,1]} \max\left\{\frac{1}{2}\left(\varphi''\right)_-^r \delta_n^2, \frac{1}{2}\left(\varphi''\right)_+^r \delta_n^2\right\}
\end{aligned} \tag{2.35}$$

Substituting (34) and (35), we complete the proof of the theorem.

2.4 Consequence

In this chapter, we have interested in the approximation process of the fuzzy Meyer-König and Zeller and the fuzzy post-quantum Meyer-König and Zeller operators.

In the results of Theorems 2.6, 2.7, and 2.8, if we choose $p = 1$, then we obtain the approximation results for fuzzy quantum Meyer-König and Zeller operators. Additionally, when $p = 1 = q$, the results are reduced to that for the fuzzy Meyer-König and Zeller operators.

Let us replace (p_n) and (q_n) instead of p and q in the definition of the post-quantum Meyer-König and Zeller operators. Under the condition (2.25), in Theorem 2.7, we have

$$\lim_{n\to\infty} \delta_n = 0,$$

which implies

$$\lim_{n\to\infty} \omega_1^{\mathcal{F}}(\varphi; \delta_n) = 0,$$

i.e.,

$$\lim_{n\to\infty} D\big(\omega_1^{\mathcal{F}}(\varphi; \delta_n), \tilde{o}\big) = 0.$$

Thus, Theorem 2.7 indicates that the sequence of $\left(M^{\mathcal{F}}_{n,p_n,q_n}(\varphi; .)\right)_{n\in\mathbb{N}}$ converges to φ in the metric D^* in fuzzy sense for all $\varphi \in C_{\mathcal{F}}[0, 1]$.

If $0 < \lim\limits_{n\to\infty} \sqrt{p_n^n} < \frac{1}{2}$ under the condition (2.25), then the approximation degree of the fuzzy post-quantum Meyer-König and Zeller operators $M^{\mathcal{F}}_{n,p_n,q_n}$ in Theorem 2.7 is better than of the fuzzy Meyer-König and Zeller operators $M_n^{\mathcal{F}}$ in Theorem 2.5.

Theorem 2.8 demonstrates the approximation to the fuzzy asymptotic expansion of the fuzzy post-quantum Meyer-König and Zeller operators.

References

1. Abel, U.: The complete asymptotic expansion by Meyer-König and Zeller operators. Journal of Mathematical Analysis and Application **280**, 109–119 (1997)
2. Agratini, A.: Korovkin type error estimates for Meyer-König and Zeller operators. Mathematical Inequalities and Applications **4**(2), 119–126 (2001)
3. Anastassiou, G.A.: Fuzzy Mathematics: Approximation Theory. Springer-Verlag, Berlin (2010)
4. Becker, M., Nessel, R.J.: A globe approximation theorem for Meyer-König and Zeller operators. Math. Z. **160**, 195–206 (1978)
5. Cheney, E., Sharma, A.: Bernstein Power Series. Can. J. Math. **16**, 241–252 (1964)
6. Doğru, O.: Approximation order and asymptotic approximation for generalized Meyer-König and Zeller operators. Mathematica Balkanica, New Series, no **3–4**(12), 359–368 (1998)

7. Doğru, O., Duman, O.: Statistical approximation of Meyer-König and Zeller operators based on q-integers. Publicationes Mathematicae Debrecen **68**(1–2), 199–214 (2006)
8. Doğru, D., Özalp, N.: Approximation by Kantorovich type generalization of Meyer-König and Zeller operators. Glasnik Matematicki **36**(56), 311–318 (2001)
9. Gal, S.G.: A fuzzy variant of the Weierstrass' approximation theorem. The Journal of Fuzzy Mathematics **1**(4), 865–872 (1993)
10. Gal, S.G.: Degree of approximation of fuzzy mappings by fuzzy polynomials. The Journal of Fuzzy Mathematics **2**(4), 847–853 (1994)
11. Goetschel, R., Voxman, W.: Elementary fuzzy calculus. Fuzzy Sets Syst. **18**, 31–43 (1986)
12. Heping, W.: Properties of convergence for the q-Meyer-König and Zeller operators. The Journal of Mathematical Analysis and Applications **335**, 1360–1371 (2007)
13. Kac, V., Cheung, P.: Quantum Calculus. Springer-Verlag, New York (2000)
14. Kadak, U., Khan, A., Mursaleen, M.: Approximation by the Meyer-König and Zeller operators based on (p, q)-analogue. Filomat **35**(11), 3767–3779 (2021)
15. Kim, Y.K., Ghil, B.M.: Integrals of fuzzy real number valued functions. Fuzzy Sets Syst. **86**, 213–222 (1997)
16. Korovkin, P.P.: On the convergence of linear positive operators in the space of continuous functions (Russian). Doklady Akademii Nauk SSSR (NS) **90**, 961–964 (1953)
17. Liu, Z.D., Zhang, C.G.: The strong Voronovskaja-type asymptotic expansion of Meyer-König and Zeller operators. Journal of Nanchang University (Natural S, cience) **30**(6), 31–43 (2006)
18. Lupaş, A., Müller, M.W.: Approximation properties of the Mn-operators. Aequationes Math. **5**, 19–37 (1970)
19. Meyer-König, W., Zeller, K.: Bernsteinsche potenzreihen. Stud. Math. **19**, 84–89 (1960)
20. Ming, M.: On embedding problem of fuzzy number space Part IV. Fuzzy Sets Syst. **58**, 185–193 (1993)
21. Müller, M.W. (1967) Die Folge der Gamma Operatoren. Dissertation, Stuttgart.
22. Müller, M.W.: Lp approximation by the method of integral type Meyer-König and Zeller operators. Stud. Math. **69**, 81–88 (1978)
23. Özarslan, M.A., Duman, O.: Approximation theorems by Meyer-König and Zeller type operators. Chaos, Solitans and Fractals **41**(1), 451–456 (2009)
24. Özkan, E.Y., Hazarika, B.: Approximation results by fuzzy-Bernstein type rational functions via interval-valued fuzzy number. Soft. Comput. **27**, 6893–6901 (2023)
25. Özkan, E.Y.: Inequalities for approximation of new defined fuzzy post-quantum Bernstein polynomials via interval-valued fuzzy numbers. Symmetry **14**(4), 696 (2022)
26. Özkan, E.Y.: Approximation by fuzzy (p, q)-Bernstein-Chlodowsky operators. Sahand Communications in Mathematical Analysis **19**(2), 113–132 (2022)
27. Özkan, E.Y. (2020) On fuzzy Bernstein polynomials. In: Academic Studies in Science and Mathematics, Turgay Seçkin (Editor), Gece Publishing, Ankara, Turkey, 1–12.
28. Özkan, E.Y. (2020) Approximation by fuzzy Bernstein type rational functions. In: Academic Studies in Science and Mathematics Sciences, Sema Isisag Ucuncu&Tanja V. Soldatovic (Editors), Livre de Lyon, Lyon, France, 71–85.
29. Sekkime, P.C.: On the asymptotic approximation with operators of Meyer-König and Zeller. Indag. Math. **32**, 428–440 (1970)
30. Shisha, O., Mond, B. (1968) The degree of convergence of sequences of positive operators. Proceedings of the National Academy of Sciences U.S.A., 60, 1196–1200.
31. Sonker, S., Jena, B.B., Jindal, R., Paikray, S.K.: A generalized theorem on double absolute factorable matrix summability. Applied Mathematics and Information Sciences **16**(2), 313–320 (2022)
32. Sonker, S., Jindal, R.: Approximation of signals by the triple product summability means of the Fourier series. Soft Computing: Theories and Applications **425**, 169–179 (2022)
33. Sonker, S., Sangwan, P.: (2021) Approximation of signals by harmonic-Euler triple product means. Journal of the Indian Mathematical Society **88**(1–2), 176–186 (2021)
34. Sonker, S., Jindal, R.: On approximation of signals belonging to some classes by (N, pm, qm)(E, θ)(E, θ) means of conjugate series of its Fourier series. The Journal of Analysis **2022**, 1–18 (2022)

35. Totik, V.: Approximation by Meyer-König and Zeller operators. Math. Z. **184**, 425–446 (1983)
36. Trif, T.: Meyer-König and Zeller operators based on the q-integers. Revista de Analiză Numerică **29**(2), 221–229 (2000)
37. Wu, C., Danghang, L.: A fuzzy variant Weierstrass approximation theorem. The Journal of Fuzzy Mathematics **7**(1), 101–104 (1999)
38. Wu, C., Ming, M.: On embedding problem of fuzzy number space Part I. Fuzzy Sets Syst. **44**, 33–38 (1991)
39. Wu, C., Ming, M.: On embedding problem of fuzzy number space Part II. Fuzzy Sets Syst. **45**, 189–202 (1992)
40. Yang, C.-J., Zhang, C.-G.: Fuzzy asymptotic expansions of Meyer-König and Zeller operators. Fuzzy Information and Engineering **7**(4), 499–506 (2015)
41. Zareh, L.A.: Fuzzy sets. Inf. Control. **8**, 338–353 (1965)

Chapter 3
On Approximation of Signals in the Generalized Zygmund Class Using $(E, s)(\overline{N}, q_n)$ Mean

B. P. Padhy, P. Baliarsingh, S. K. Paikray, and U. K. Misra

Abstract In the recent past, various researchers have worked on approximation of signals realizing its importance in different fields of science and engineering. Summability method is the one among the methods which approximates the signals in a better way. Looking at the growing popularity of this method, in this chapter we have approximated the signals in the Zygmund class using $(E, s)(\overline{N}, q_n)$ mean.

Keywords Degree of approximation · Fourier series · Conjugate fourier series · Derived conjugate fourier series · $(E, s)(\overline{N}, q_n)$-mean

3.1 Introduction

The development of approximation of signals (or functions) was initiated long back by P. L. Chebyshev and K. Weirstrass. The necessity to approximate signals arose in various situations when it is needed to replace a signal by a smoother signal or by one which is simpler and more convenient in computations. The approximation of signals in Lipschitz class have been studied by [8–12, 18, 19] and in Hölder space the degrees of approximation of signals have been studied by [1, 3, 7, 15, 16, 20, 22]. Similarly, in Zygmund class the degrees of approximation of signals have been studied by [2, 6, 13, 14, 16, 21]. Very recently, Das et al. [2], Nigam [17]

B. P. Padhy (✉)
Department of Mathematics, School of Applied Sciences, KIIT Deemed to be University, Bhubaneswar, Odisha, India
e-mail: birupakhya.padhyfma@kiit.ac.in

P. Baliarsingh
Institute of Mathematics and Applications, Bhubaneswar, Odisha, India

S. K. Paikray
Department of Mathematics, Veer Surendra Sai University of Technology, Burla, Odisha, India
e-mail: skpaikray_math@vssut.ac.in

U. K. Misra
Department of Mathematics, National Institute of Science and Technology, Pallur Hills, Berhampur, Odisha, India

B. C. Tripathy et al. (eds.), *Operators, Inequalities and Approximation*, Industrial and Applied Mathematics, https://doi.org/10.1007/978-981-97-3238-8_3

and Singh et al. [23] proved some results concerning approximation of signals in the generalized Zygmund class by using different summability means. This motivated us to investigate on the degree of approximation of a signal in the generalized Zygmund class by $(E, s)(\overline{N}, q_n)$-product mean of the Fourier series, conjugate Fourier series and derived conjugate Fourier series.

Suppose $\delta(\beta)$ be a signal, which is periodic in $[0, 2\pi]$ such that

$$\int_0^{2\pi} |\delta(\beta)|^p d\beta < \infty.$$

Let us denote

$$L_p[0, 2\pi] = \left\{\delta : [0, 2\pi] \to \mathbb{R} : \int_0^{2\pi} |\delta(\beta)|^p d\beta < \infty\right\}, p \geq 1.$$

Then the Fourier series of $\delta(\beta)$ is given by

$$\sum_{n=0}^{\infty} u_n(\beta) = \frac{a_0}{2} + \sum_{n=1}^{\infty}(a_n \cos n\beta + b_n \sin n\beta). \tag{3.1}$$

The conjugate series of (3.1) is

$$\sum_{n=1}^{\infty}(b_n \cos n\beta - a_n \sin n\beta) \tag{3.2}$$

and the conjugate derived Fourier series is

$$\sum_{n=1}^{\infty} n(a_n \cos n\beta + b_n \sin n\beta). \tag{3.3}$$

Assuming $S_k(\delta; \beta)$, $\overline{S_k}(\delta; \beta)$ and $\overline{S_k}'(\delta; \beta)$ as the k-th partial sum of (3.1), (3.2) and (3.3) respectively, we can write

$$S_k(\delta; \beta) - \delta(\beta) = \frac{1}{2\pi}\int_0^{\pi} \varphi(\beta, \lambda)\frac{\sin\left(k + \frac{1}{2}\right)\lambda}{\sin\frac{\lambda}{2}} d\lambda$$

$$\overline{S_k}(\delta; \beta) - \overline{\delta}(\beta) - = \frac{1}{2\pi}\int_0^{\pi} \Psi(\beta; \lambda)\frac{\cos\left(k + \frac{1}{2}\right)\lambda - \cos\frac{\lambda}{2}}{\sin\frac{\lambda}{2}} d\lambda$$

$$\overline{S_k'}(\delta;\beta) - \overline{\delta}'(\beta) - = -\frac{2}{\pi}\int_0^{\pi} \frac{\psi(\beta;\lambda)}{4\sin\frac{\lambda}{2}}\left(k+\frac{1}{2}\right)\sin\left(k+\frac{1}{2}\right)\lambda\, d\lambda$$
$$-\frac{1}{\pi}\int_0^{\pi}\frac{\psi(\beta;\lambda)}{4\sin\frac{\lambda}{2}}\frac{\sin\left(k+\frac{1}{2}\right)\lambda}{\tan\frac{\lambda}{2}}\, d\lambda$$

where

$$\overline{\delta}(\beta) = -\frac{1}{2\pi}\int_0^{\pi}\Psi(\beta;\lambda)\,\cot\frac{\lambda}{2}\, d\lambda$$

and

$$\overline{\delta}'(\beta) = -\frac{1}{\pi}\int_0^{\pi}\psi(\beta;\lambda)\, cosec^2\frac{\lambda}{2}\, d\lambda.$$

We define $\|.\|_p$ by

$$\|\delta\|_p = \left(\frac{1}{2\pi}\int_0^{2\pi}|\delta(\beta)|^p d\beta\right)^{\frac{1}{p}}, 1 \le p < \infty$$

and

$$\|\delta\|_p = ess \sup_{0\le\beta\le 2\pi}|\delta(\beta)|,\; p = \infty.$$

Let

$$Z_p^{(\sigma)} = \left\{\delta \in L_p : 1 \le p < \infty, \sup_{\lambda\ne 0}\frac{\|\delta(.+\lambda)+\delta(.-\lambda)-2\delta(.)\|_p}{\sigma(\lambda)} < \infty\right\}$$

and $\|\delta\|_p^{(\sigma)}$ on $Z_p^{(\sigma)}$ is defined by

$$\|\delta\|_p^{(\sigma)} = \|\delta\|_p + \sup_{\lambda\ne 0}\frac{\|\delta(.+\lambda)+\delta(.-\lambda)-2\delta(.)\|_p}{\sigma(\lambda)}, p \ge 1.$$

Suppose

$$\Delta_p^{(\sigma)} = \left\{\delta \in L_p : 1 \le p < \infty, \sup_{\lambda\ne 0}\frac{\|\delta(.+\lambda)-\delta(.-\lambda)\|_p}{\sigma(\lambda)} < \infty\right\}$$

and $\|\delta\|_p^{(\sigma)}$ on $\Delta_p^{(\sigma)}$ is defined by

$$\|\delta\|_p^{(\sigma)} = \|\delta\|_p + \sup_{\lambda\ne 0}\frac{\|\delta(.+\lambda)-\delta(.-\lambda)\|_p}{\sigma(\lambda)}, p \ge 1.$$

Again, suppose

$$\Omega_p^{(\sigma)} = \left\{\delta \in L_p : 1 \leq p < \infty, \sup_{\lambda \neq 0} \frac{\|\delta(.+\lambda) + \delta(.-\lambda)\|_p}{\sigma(\lambda)} < \infty\right\}$$

and $\|\delta\|_p^{(\sigma)}$ on $\Omega_p^{(\sigma)}$ is defined by

$$\|\delta\|_p^{(\sigma)} = \|\delta\|_p + \sup_{\lambda \neq 0} \frac{\|\delta(.+\lambda) + \delta(.-\lambda)\|_p}{\sigma(\lambda)}, \ p \geq 1.$$

Here, $\sigma : [0, 2\pi] \to \mathbb{R}$ is a function with $\sigma(\lambda) > 0$ for $0 \leq \lambda < 2\pi$ and

$$\lim_{\lambda \to 0^+} \sigma(\lambda) = \sigma(0) = 0.$$

The spaces $Z_p^{(\sigma)}$, $\Delta_p^{(\sigma)}$ and $\Omega_p^{(\sigma)}$ are complete since L_p, $(p \geq 1)$ is complete. So, $Z_p^{(\sigma)}$, $\Delta_p^{(\sigma)}$ and $\Omega_p^{(\sigma)}$ are Banach spaces under $\|.\|_p^{(\sigma)}$.

If $\sigma(\lambda)$ and $\omega(\lambda)$ represents the Zygmund moduli of continuity (see [24]) such that $\left(\frac{\sigma(\lambda)}{\omega(\lambda)}\right)$ is positive and non-decreasing then

$$\|\delta\|_p^{(\omega)} \leq \max.\left(1, \frac{\sigma(2\pi)}{\omega(2\pi)}\right)\|\delta\|_p^{(\sigma)} \leq \infty. \tag{3.4}$$

Clearly,

$$Z_p^{(\sigma)} \subseteq Z_p^{(\omega)} \subseteq L_p, (p \geq 1);$$
$$\Delta_p^{(\sigma)} \subseteq \Delta_p^{(\omega)} \subseteq L_p, (p \geq 1);$$
and
$$\Omega_p^{(\sigma)} \subseteq \Omega_p^{(\omega)} \subseteq L_p, (p \geq 1).$$

The Zygmund modulus of continuity σ satisfies

(a) $\sigma(0) = 0$
(b) $\sigma(\lambda_1 + \lambda_2) \leq \sigma(\lambda_1) + \sigma(\lambda_2)$.

Let $\sum u_n$ be an infinite series with sequence of partial sums $\{r_n\}$.

Let $\{q_k\}$ represents the sequence of non-negative integers such that

$$Q_n = \sum_{k=0}^{n} q_k \to \infty \quad \text{as} \quad n \to \infty. \tag{3.5}$$

By $(\overline{N}, q_n)$ method, the series $\sum u_n$ is said to be summable to $'r'$ if

$$\lim_{n\to\infty} t_n^{\overline{N}} \to r,$$

where

$$t_n^{\overline{N}} = \frac{1}{Q_n}\sum_{k=0}^{n} q_k r_k, \quad n = 0, 1, 2, \ldots \tag{3.6}$$

represents the $(\overline{N}, q_n)$ mean of $\{r_n\}$ generated by the sequence $\{q_n\}$.

We know, $(\overline{N}, q_n)$ method is regular [4].

The (E, s) transform of $\{r_n\}$ is given by

$$E_n^s = \frac{1}{(1+s)^n}\sum_{k=0}^{n}\binom{n}{k} s^{n-k} r_k. \tag{3.7}$$

If $E_n^s \to r$ as $n \to \infty$ then $\sum u_n$ is summable to 'r' by (E, s) summability. Also, (E, s) method is regular [4].

The $(E, s)(\overline{N}, q_n)$ transform of $\{r_n\}$ is given by

$$\tau_n^{E_s,\overline{N}} = \frac{1}{(1+s)^n}\sum_{k=0}^{n}\binom{n}{k}\left\{\frac{1}{Q_k}\sum_{\nu=0}^{k} q_\nu r_\nu\right\}. \tag{3.8}$$

The series $\sum u_n$ is summable to r by the $(E, s)(\overline{N}, q_n)$ transform if $\tau_n^{E_s,\overline{N}} \to r$ as $n \to \infty$.

The following notations are used in the rest part of our chapter:

$$\varphi(\beta, \lambda) = \delta(\beta+\lambda) + \delta(\beta-\lambda) - 2\delta(\beta)$$
$$\Psi(\beta, \lambda) = \delta(\beta+\lambda) - \delta(\beta-\lambda)$$
$$\psi(\beta, \lambda) = \delta(\beta+\lambda) + \delta(\beta-\lambda)$$
$$\kappa_n^{E_s,\overline{N}}(\lambda) = \frac{1}{2\pi(1+s)^n}\sum_{k=0}^{n}\binom{n}{k}\left\{\frac{1}{Q_k}\sum_{\nu=0}^{k} q_\nu \frac{\sin\left(\nu+\frac{1}{2}\right)\lambda}{\sin\left(\frac{\lambda}{2}\right)}\right\}$$
$$\overline{\kappa}_n^{E_s,\overline{N}}(\lambda) = \frac{1}{2\pi(1+s)^n}\sum_{k=0}^{n}\binom{n}{k}\left\{\frac{1}{Q_k}\sum_{\nu=0}^{k} q_\nu \frac{\cos\left(\nu+\frac{1}{2}\right)\lambda - \cos\left(\frac{\lambda}{2}\right)}{\sin\left(\frac{\lambda}{2}\right)}\right\}$$
$$\overline{Y_1}'(\lambda) = \frac{-2}{4\pi(1+s)^n}\sum_{k=0}^{n} k\binom{n}{k} s^{n-k}\left\{\frac{1}{Q_k}\sum_{\nu=0}^{k} q_\nu \frac{\sin(\nu+\frac{1}{2})\lambda}{\sin\frac{\lambda}{2}}\right\}$$
$$\overline{Y_2}'(\lambda) = \frac{-1}{4\pi(1+s)^n}\sum_{k=0}^{n}\binom{n}{k} s^{n-k}\left\{\frac{1}{Q_k}\sum_{\nu=0}^{k} q_\nu \frac{\cos \nu\lambda}{\sin^2\frac{\lambda}{2}}\right\}.$$

3.2 Known Results

Using Matrix-Euler summability means, Lal and Shireen [5] proved the following theorems:

Theorem 3.1 *Let the lower triangular matrix $A = (a_{n,k})$ satisfy the following conditions:*

$$\sum_{k=0}^{n} a_{n,k} = 1, a_{n,k} \geq 0 \ (n = 0, 1, 2, \ldots; \ k = 0, 1, 2, \ldots) \tag{3.9}$$

$$\sum_{k=0}^{n} |\Delta a_{n,k}| = O\left(\frac{1}{(n+1)}\right) \text{ and } (n+1)a_{n,n} = O(1). \tag{3.10}$$

The best approximation of the Fourier series (3.1) by Matrix-Euler mean is given by

$$E_n(\delta) = \inf_{t_n^{\Delta,E}} \|t_n{}^{\Delta,E} - \delta\|_p^{\omega} = O\left(\frac{1}{(n+1)} \int_{\frac{1}{(n+1)}}^{\pi} \frac{\sigma(\lambda)}{\lambda^2 \omega(\lambda)} d\lambda\right) \tag{3.11}$$

where

$$t_n{}^{\Delta,E} = \sum_{k=0}^{n} a_{n,k} \ \frac{1}{2^k} \sum_{\nu=0}^{k} \binom{n}{k} \ r_\nu$$

represents the Matrix-Euler mean of a 2π periodic and Lebesgue integrable function $\delta : [0, 2\pi] \rightarrow \mathbb{R}$, that belong to $Z_p^{(\sigma)}$, $(p \geq 1)$. Here, σ and ω are the Zygmund moduli of continuity and $\frac{\sigma(\lambda)}{\omega(\lambda)}$ is positive and non-decreasing.

Using Hausdorff mean, Nigam [17] proved the following theorems:

Theorem 3.2 *Error approximation of a conjugate function $\overline{\delta}$ of a 2π periodic function $h \in \Delta_p^{(\sigma)}$, $p \geq 1$, using $H = (\theta_{j,\alpha})$ of conjugate Fourier series is given by*

$$\|\overline{M}_j^H(\delta; .) - \overline{\delta}(.)\|_p^{(\sigma)} = O\left(\frac{1}{j+1} \int_{\frac{1}{j+1}}^{\pi} \frac{\sigma(\lambda)}{\lambda^2 \, \omega(\lambda)} d\lambda\right),$$

where $\sigma(\lambda)$ and $\omega(\lambda)$ are Zygmund moduli of continuity, provided

$$\int_0^{\pi} \frac{\sigma(\lambda)}{\lambda \omega(\lambda)} d\lambda = O\left(\frac{m(\xi)}{\omega(\xi)}\right), \ 0 < \xi < \pi.$$

Theorem 3.3 *Error approximation of a conjugate derived function $\overline{\delta'}$ of a 2π periodic function $\delta \in \Omega_p^{(\sigma)}$, $p \geq 1$, using $H = (\theta_{j,\alpha})$ of conjugate derived Fourier series is given by*

$$\|\overline{M'}_j^H(\delta;.) - \overline{\delta'}(.)\|_p^{(\sigma)} = O\left(\frac{1}{j+1}\int_{\frac{1}{j+1}}^{\pi} \frac{(\lambda+1)\,\sigma(\lambda)}{\lambda^3\,\omega(\lambda)}d\lambda\right),$$

where $\sigma(\lambda)$ and $\omega(\lambda)$ are Zygmund moduli of continuity, provided

$$\int_0^{\pi} \frac{\sigma(\lambda)}{\lambda^2\omega(\lambda)}d\lambda = O\left(\frac{\sigma(\xi)}{\xi\,\omega(\xi)}\right),\ 0 < \xi < \pi.$$

In what follows, we have proved the following theorems for approximation of signals (or functions) in Zygmund class associated with Fourier series, conjugate Fourier series and derived conjugate Fourier series using $(E, s)(\overline{N}, q_n)$-mean.

3.3 Main Theorems

Theorem 3.4 *Let $\delta : [0, 2\pi] \to \mathbb{R}$ be a periodic function (with period 2π) belonging to $Z_p^{(\sigma)}$, $(p \geq 1)$, which is integrable in the sense of Lebesgue. Then the degree of approximation of h by using $(E, s)(\overline{N}, q_n)$ mean of (3.1) is given by*

$$E_n(\delta) = \inf_{\tau_n^{E_s,\,\overline{N}}} \|\tau_n^{E_s,\,\overline{N}} - \delta\|_p^{\omega} = O\left(\frac{1}{(n+1)}\int_{\frac{1}{(n+1)}}^{\pi} \frac{\sigma(\lambda)}{\lambda\,\omega(\lambda)}d\lambda\right) \tag{3.12}$$

where $\sigma(\lambda)$ and $\omega(\lambda)$ are the Zygmund moduli of continuity and $\frac{\sigma(\lambda)}{\lambda\,\omega(\lambda)}$ is positive and non-decreasing.

Theorem 3.5 *Let $\delta : [0, 2\pi] \to \mathbb{R}$ be a periodic function (with period 2π) belonging to $\Delta_p^{(\sigma)}$, $(p \geq 1)$, which is integrable in the sense of Lebesgue. Then the degree of approximation of $\overline{\delta}$ by using $(E, s)(\overline{N}, q_n)$ mean of (3.2) is given by*

$$E_n(\overline{\delta}) = \inf_n \|F_n(.)\|_p^{\omega} = O\left(\int_{\frac{1}{(n+1)}}^{\pi} \frac{\sigma(\lambda)}{\lambda\,\omega(\lambda)}d\lambda\right) \tag{3.13}$$

where $\sigma(\lambda)$ and $\omega(\lambda)$ are the Zygmund moduli of continuity and $\frac{\sigma(\lambda)}{\lambda\,\omega(\lambda)}$ is positive and non-decreasing.

Theorem 3.6 *The degree of approximation of a conjugate derived function $\overline{\delta'}$ of a 2π periodic function $\delta \in \Omega_p^{(\sigma)}$, $p \geq 1$, using $(E, s)(\overline{N}, q_n)$- mean of conjugate derived Fourier series (3.3) is given by*

$$E_n(\overline{\delta'}) = \inf_n \|G_n{}'(.)\|_p^\omega = O\left(\int_{\frac{1}{n+1}}^{\pi} \frac{\sigma(\lambda)}{\lambda^2\, \omega(\lambda)} d\lambda\right)$$

where $\sigma(\lambda)$ *and* $\omega(\lambda)$ *are the Zygmund moduli of continuity and* $\frac{\sigma(\lambda)}{\lambda\, \omega(\lambda)}$ *is positive and non-decreasing, provided*

$$\int_0^{\xi} \frac{\sigma(\lambda)}{\lambda\, \omega(\lambda)} d\lambda = O\left(\frac{\sigma(\xi)}{\omega(\xi)}\right).$$

We need the following lemmas to prove our main theorems.

3.4 Lemmas

Lemma 3.1

$$(i)|\kappa_n^{E_s,\overline{N}}| = O(n) \ \ for \ \ 0 \le \lambda \le \frac{1}{(n+1)}.$$

$$(ii)|\kappa_n^{E_s,\overline{N}}| = O\left(\frac{1}{\lambda}\right) \ \ for \ \ \frac{1}{(n+1)} \le \lambda \le \pi.$$

Lemma 3.2 *Let* $\delta \in Z_p^{(\sigma)}$ *then for* $0 < \lambda \le \pi$,

(i) $\|\varphi(., \lambda)\|_l = O(\sigma(\lambda))$

(ii) $\|\varphi(. + \gamma, \lambda) + \varphi(. - \gamma, \lambda) - 2\varphi(., \lambda)\|_p = O(\sigma(\lambda))$ *or* $O(\sigma(\gamma))$

(iii) *If* $\sigma(\lambda)$ *and* $\omega(\lambda)$ *are as defined in Theorem-3.1, then*

$$\|\varphi(. + \gamma, \lambda) + \varphi(. - \gamma, \lambda) - 2\varphi(., \lambda)\|_p = O\left(\sigma(\gamma)\frac{\sigma(\lambda)}{\omega(\lambda)}\right)$$

where $\varphi(\beta, \lambda) = \delta(\beta + \lambda) + \delta(\beta - \lambda) - 2\delta(\beta)$.

Lemma 3.3

$$(i)|\overline{\kappa}_n^{E_s,\overline{N}}| = O(n) \ \ for \ \ 0 \le \lambda \le \frac{1}{(n+1)}.$$

$$(ii)|\overline{\kappa}_n^{E_r,\overline{N}}| = O\left(\frac{1}{\lambda}\right) \ \ for \ \ \frac{1}{(n+1)} \le \lambda \le \pi.$$

Lemma 3.4 *Let* $\delta \in \Delta_p^{(\sigma)}$ *then for* $0 < \lambda \leq \pi$,

$$(i)\ \|\Psi(.,\lambda)\|_p = O(\sigma(\lambda))$$
$$(ii)\ \|\Psi(.+\gamma,\lambda) - \Psi(.-\gamma,\lambda)\|_p = O(\sigma(\lambda))\ or\ O(\sigma(\gamma))$$
(iii) *If* $\sigma(\lambda)$ *and* $\omega(\lambda)$ *are as defined in Theorem-3.2, then*
$$\|\Psi(.+\gamma,\lambda) - \Psi(.-\gamma,\lambda)\|_p = O\left(\omega(\gamma)\frac{\sigma(\lambda)}{\omega(\lambda)}\right)$$

where $\Psi(\beta,\lambda) = \delta(\beta+\lambda) - \delta(\beta-\lambda)$.

Lemma 3.5

$$(i)|\overline{Y_1}^{'}(\lambda)| = O(n^2)\ for\ 0 < \lambda \leq \pi.$$
$$(ii)|\overline{Y_2}^{'}(\lambda)| = O\left(\frac{1}{\lambda^2}\right)\ for\ 0 < \lambda \leq \pi.$$

Lemma 3.6 *Let* $\delta \in \Omega_p^{(\sigma)}$ *then for* $0 < \lambda \leq \pi$,

$$(i)\ \|\psi(.,\lambda)\|_p = O(\sigma(\lambda))$$
$$(ii)\ \|\psi(.+\gamma,\lambda) + \psi(.-\gamma,\lambda)\|_p = O(\sigma(\lambda))\ or\ O(\sigma(\gamma))$$
(iii) *If* $\sigma(\lambda)$ *and* $\omega(\lambda)$ *are as defined in Theorem-3.3, then*
$$\|\psi(.+\gamma,\lambda) + \psi(.-\gamma,\lambda)\|_p = O\left(\omega(\gamma)\frac{\sigma(\lambda)}{\omega(\lambda)}\right)$$

where
$$\psi(\beta,\lambda) = \delta(\beta+\lambda) + \delta(\beta-\lambda).$$

3.5 Proof of the Lemmas

Proof of Lemma 3.1 For $0 \leq \lambda \leq \frac{1}{n+1}$ and $\sin n\lambda \leq n\ \sin\lambda$, we have

$$\begin{aligned}
|\kappa_n^{E_s,\overline{N}}(\lambda)| &= \frac{1}{2\pi(1+s)^n}\left|\sum_{k=0}^{n}\binom{n}{k}\ s^{n-k}\left\{\frac{1}{Q_k}\sum_{\nu=0}^{k} q_\nu\ \frac{\sin\left(\nu+\frac{1}{2}\right)\lambda}{\sin\frac{\lambda}{2}}\right\}\right| \\
&\leq \frac{1}{2\pi(1+s)^n}\left|\sum_{k=0}^{n}\binom{n}{k}\ s^{n-k}\left\{\frac{1}{Q_k}\sum_{\nu=0}^{k} q_\nu\ \frac{(2\nu+1)\sin\frac{\lambda}{2}}{\sin\frac{\lambda}{2}}\right\}\right| \\
&\leq \frac{1}{2\pi(1+s)^n}\left|\sum_{k=0}^{n}\binom{n}{k}\ s^{n-k}\ (2k+1)\left\{\frac{1}{Q_k}\sum_{\nu=0}^{k} q_\nu\right\}\right| \\
&\leq \frac{(2n+1)}{2\pi(1+s)^n}\left|\sum_{k=0}^{n}\binom{n}{k}\ s^{n-k}\right| \\
&= O(n)
\end{aligned} \tag{3.14}$$

which completes the proof of (i).

Again, by Jordan's lemma

$$\sin\left(\frac{\lambda}{2}\right) \geq \frac{\lambda}{\pi}, \quad \sin n\lambda \leq 1, \quad \left(\frac{1}{n+1} \leq \lambda \leq \pi\right).$$

Now,

$$\begin{aligned}
|\kappa_n^{E_s,\overline{N}}(\lambda)| &= \frac{1}{2\pi(1+s)^n}\left|\sum_{k=0}^{n}\binom{n}{k}\ s^{n-k}\left\{\frac{1}{Q_k}\sum_{\nu=0}^{k} q_\nu\ \frac{\sin\left(\nu+\frac{1}{2}\right)\lambda}{\sin\frac{\lambda}{2}}\right\}\right| \\
&\leq \frac{1}{2\pi(1+s)^n}\left|\sum_{k=0}^{n}\binom{n}{k}\ s^{n-k}\left\{\frac{1}{Q_k}\sum_{\nu=0}^{k}\frac{\pi}{\lambda}\ q_\nu\right\}\right| \\
&= \frac{1}{2\lambda(1+s)^n}\left|\sum_{k=0}^{n}\binom{n}{k}\ s^{n-k}\left\{\frac{1}{Q_k}\sum_{\nu=0}^{k} q_\nu\right\}\right| \\
&= \frac{1}{2\lambda(1+s)^n}\left|\sum_{k=0}^{n}\binom{n}{k}\ s^{n-k}\right| \\
&= O\left(\frac{1}{\lambda}\right)
\end{aligned} \tag{3.15}$$

which completes the proof of (ii).

Proof of Lemma 3.3 For $0 \leq \lambda \leq \frac{1}{n+1}$ and $\sin\ n\lambda \leq n\ \sin\ \lambda$, we have

$$\begin{aligned}
|\overline{\kappa}_n^{E_s,\overline{N}}(\lambda)| &= \frac{1}{2\pi(1+s)^n}\left|\sum_{k=0}^{n}\binom{n}{k}\ s^{n-k}\left\{\frac{1}{Q_k}\sum_{\nu=0}^{k} q_\nu\ \frac{\cos\left(\nu+\frac{1}{2}\right)\lambda-\cos\frac{\lambda}{2}}{\sin\frac{\lambda}{2}}\right\}\right| \\
&= \frac{1}{2\pi(1+s)^n}\left|\sum_{k=0}^{n}\binom{n}{k}\ s^{n-k}\left\{\frac{1}{Q_k}\sum_{\nu=0}^{k} q_\nu \frac{(\cos(\nu\lambda)\cos\frac{\lambda}{2}-\sin(\nu\lambda)\sin\frac{\lambda}{2}-\cos\frac{\lambda}{2}}{\sin\frac{\lambda}{2}}\right\}\right| \\
&= \frac{1}{2\pi(1+s)^n}\left|\sum_{k=0}^{n}\binom{n}{k}\ s^{n-k}\left\{\frac{1}{Q_k}\sum_{\nu=0}^{k} q_\nu\left(\frac{2\sin^2\frac{\nu\lambda}{2}.\cos\frac{\lambda}{2}}{\sin\frac{\lambda}{2}}+\sin(\nu\lambda)\right)\right\}\right| \\
&\leq \frac{1}{2\pi(1+s)^n}\left|\sum_{k=0}^{n}\binom{n}{k}\ s^{n-k}\left\{\frac{1}{Q_k}\sum_{\nu=0}^{k} q_\nu\left(O(\nu)+O(\nu)\right)\right\}\right| \\
&\leq \frac{1}{\pi(1+s)^n}\left|\sum_{k=0}^{n}\binom{n}{k}\ s^{n-k}\left\{\frac{O(k)}{Q_k}\sum_{\nu=0}^{k} q_\nu\right\}\right| \\
&= O(n).
\end{aligned} \tag{3.16}$$

This completes the proof of (i).

By Jordan's lemma

$$\sin\left(\frac{\lambda}{2}\right) \geq \frac{\lambda}{\pi}, \quad \sin n\lambda \leq 1, \quad \left(\frac{1}{n+1} \leq \lambda \leq \pi\right).$$

Now,

$$\begin{aligned}
|\overline{K}_n^{E_s,\overline{N}}(\lambda)| &= \frac{1}{2\pi(1+s)^n}\left|\sum_{k=0}^{n}\binom{n}{k}\; s^{n-k}\left\{\frac{1}{Q_k}\sum_{\nu=0}^{k} q_\nu \;\frac{\cos\left(\nu+\frac{1}{2}\right)\lambda - \cos\frac{\lambda}{2}}{\sin\frac{\lambda}{2}}\right\}\right| \\
&\leq \frac{1}{2\pi(1+s)^n}\left|\sum_{k=0}^{n}\binom{n}{k}\; s^{n-k}\left\{\frac{1}{Q_k}\sum_{\nu=0}^{k}\frac{\pi}{\lambda}\, q_\nu\right\}\right| \\
&= \frac{1}{2\lambda(1+s)^n}\left|\sum_{k=0}^{n}\binom{n}{k}\; s^{n-k}\left\{\frac{1}{Q_k}\sum_{\nu=0}^{k} q_\nu\right\}\right| \\
&\leq \frac{1}{2\lambda(1+s)^n}\left|\sum_{k=0}^{n}\binom{n}{k}\; s^{n-k}\right| \\
&= O\left(\frac{1}{\lambda}\right).
\end{aligned} \tag{3.17}$$

This completes the proof of (ii).

Proof of Lemma 3.5 For $\lambda \in \left(0, \frac{1}{n+1}\right]$ and $\sin n\lambda \leq n\ \sin\lambda$, we have

$$\begin{aligned}
|\overline{Y_1}'(\lambda)| &= \left|\frac{-2}{4\pi(1+s)^n}\sum_{k=0}^{n} k\binom{n}{k} s^{n-k}\left\{\frac{1}{Q_k}\sum_{\nu=0}^{k} q_\nu \frac{\sin\left(\nu+\frac{1}{2}\right)\lambda}{\sin\frac{\lambda}{2}}\right\}\right| \\
&\leq \frac{1}{2\pi(1+s)^n}\left|\sum_{k=0}^{n} k\binom{n}{k}\; s^{n-k}\left\{\frac{1}{Q_k}\sum_{\nu=0}^{k} q_\nu \;\frac{(2\nu+1))\sin\left(\nu+\frac{1}{2}\right)\lambda}{\sin\frac{\lambda}{2}}\right\}\right| \\
&\leq \frac{1}{2\pi(1+s)^n}\left|\sum_{k=0}^{n} k\binom{n}{k}\; s^{n-k}\,(2k+1)\left\{\frac{1}{Q_k}\sum_{\nu=0}^{k} q_\nu\right\}\right| \\
&= O(n^2).
\end{aligned}$$

For $\lambda \in \left[\frac{1}{n+1}, \pi\right]$, $\frac{1}{\sin\left(\frac{\lambda}{2}\right)} \leq \frac{\pi}{\lambda}$, $\sin\lambda \leq \lambda$

$$\begin{aligned}|\overline{Y_1}'(\lambda)| &= \left|\frac{-2}{4\pi(1+s)^n}\sum_{k=0}^{n} k\binom{n}{k}\, s^{n-k}\left\{\frac{1}{Q_k}\sum_{\nu=0}^{k} q_\nu \frac{\sin(\nu+\frac{1}{2})\lambda}{\sin\frac{\lambda}{2}}\right\}\right| \\ &\le \frac{1}{4\pi(1+s)^n}\left|\sum_{k=0}^{n} k\binom{n}{k}\, s^{n-k}\left\{\frac{1}{Q_k}\sum_{\nu=0}^{k}\frac{\pi}{\lambda}(2\nu+1)\lambda\, q_\nu\right\}\right| \\ &\le \frac{1}{2\pi(1+s)^n}\left|\sum_{k=0}^{n} k\binom{n}{k}\, s^{n-k}\,(2k+1)\left\{\frac{1}{Q_k}\sum_{\nu=0}^{k} q_\nu\right\}\right| \\ &= O(n^2).\end{aligned}$$

This completes the proof of (i).

Again, we know, $\frac{1}{\sin\left(\frac{\lambda}{2}\right)} \le \frac{\pi}{\lambda}, \ \left(0 < \lambda \le \pi\right);\ \ \sin\lambda \le \lambda, \lambda > 0;$

and $|\sin\lambda| \le 1,\ |\cos\lambda| \le 1$ for all λ.

Clearly, for $\lambda \in (0, \pi]$,

$$\begin{aligned}|\overline{Y_2}'(\lambda)| &= \left|\frac{-1}{4\pi(1+s)^n}\sum_{k=0}^{n}\binom{n}{k}\, s^{n-k}\left\{\frac{1}{Q_k}\sum_{\nu=0}^{k} q_\nu\, \frac{\cos(\nu\lambda)}{\sin^2\frac{\lambda}{2}}\right\}\right| \\ &= O\left(\frac{1}{\lambda^2}\right).\end{aligned}$$

This completes the proof of (ii).

For the proofs of **Lemmas** 3.2, 3.4 **and** 3.6 (see [5]).

3.6 Proof of Main Theorems

Proof of Theorem 3.4 Denoting the $(E, s)(\overline{N}, q_n)$ transform of $S_k(\delta; \beta)$ by $\tau_n^{E_s,\overline{N}}$, we have,

$$\begin{aligned}\tau_n^{E_s,\overline{N}} - \delta(\beta) &= \frac{1}{2\pi(1+s)^n}\int_0^{\pi}\varphi(\beta,\lambda)\sum_{k=0}^{n}\binom{n}{k}\, s^{n-k}\left\{\frac{1}{Q_n}\sum_{k=0}^{n} q_k \frac{\sin\left(k+\frac{1}{2}\right)\lambda}{\sin\frac{\lambda}{2}}\right\} d\lambda \\ &= \int_0^{\pi}\varphi(\beta;\lambda)\,\kappa_n^{E_s,\overline{N}}(\lambda)\, d\lambda \\ &= \chi_n(\beta),\ \ \text{(say)}\end{aligned}$$

Then,

$$\chi_n(\beta+\gamma)+\chi_n(\beta-\gamma)-2\chi_n(\beta) = \int_0^{\pi}\left\{\varphi(\beta+\gamma,\lambda)+\varphi(\beta-\gamma,\lambda)-2\varphi(\beta,\lambda)\right\}\kappa_n^{E_s,\overline{N}}(\lambda)\, d\lambda$$

Using Minkowski's inequality, we have

$$
\begin{aligned}
&\|\chi_n(.+\gamma)+\chi_n(.-\gamma)-2\chi_n(.)\|_p \\
&=\left\{\frac{1}{2\pi}\int_0^{2\pi}|\chi_n(\beta+\gamma)+\chi_n(\beta-\gamma)-2\chi_n(\beta)|^p\,d\beta\right\}^{\frac{1}{p}} \\
&=\frac{1}{2\pi}\left[\int_0^{2\pi}\left|\int_0^{\pi}\left\{\varphi(\beta+\gamma,\lambda)+\varphi(\beta-\gamma,\lambda)-2\varphi(\beta,\lambda)\right\}\kappa_n^{E_s,\overline{N}}(\lambda)\,d\lambda\right|^p d\beta\right]^{\frac{1}{p}} \\
&\le\int_0^{\pi}|\kappa_n^{E_s,\overline{N}}(\lambda)|\left\{\frac{1}{2\pi}\int_0^{2\pi}|\varphi(\beta+\gamma,\lambda)+\varphi(\beta-\gamma,\lambda)-2\varphi(\beta,\lambda)|^p d\beta\right\}^{\frac{1}{p}}d\lambda \\
&=\int_0^{\pi}\|\varphi(.+\gamma,\lambda)+\varphi(.-\gamma,\lambda)-2\varphi(.,\lambda)\|_p|\kappa_n^{E_s,\overline{N}}(\lambda)|\,d\lambda
\end{aligned}
$$

$$
\begin{aligned}
&=\int_0^{\frac{1}{n+1}}\|\varphi(.+\gamma,\lambda)+\varphi(.-\gamma,\lambda)-2\varphi(.,\lambda)\|_p|\kappa_n^{E_s,\overline{N}}(\lambda)|\,d\lambda \\
&\qquad+\int_{\frac{1}{n+1}}^{\pi}\|\varphi(.+\gamma,\lambda)+\varphi(.-\gamma,\lambda)-2\varphi(.,\lambda)\|_p|\kappa_n^{E_s,\overline{N}}(\lambda)|\,d\lambda \\
&=\Gamma_1+\Gamma_2,\quad \text{(say)}
\end{aligned}
\tag{3.18}
$$

Using **Lemma** 3.1(i), **Lemma** 3.2 and monotonicity of $\left(\frac{\sigma(\lambda)}{\omega(\lambda)}\right)$ with respect to 'λ', we have

$$
\begin{aligned}
\Gamma_1&=\int_0^{\frac{1}{n+1}}\|\varphi(.+\gamma,\lambda)+\varphi(.-\gamma,\lambda)-2\varphi(.,\lambda)\|_p|\kappa_n^{E_s,\overline{N}}(\lambda)|\,d\lambda \\
&=\int_0^{\frac{1}{n+1}}O\left(\omega(\gamma)\frac{\sigma(\lambda)}{\omega(\lambda)}\right)O(n)\,d\lambda.
\end{aligned}
$$

By using the second mean value theorem of integral, we have

$$
\begin{aligned}
\Gamma_1&\le O\left(n\,\omega(\gamma)\,\frac{\sigma\left(\frac{1}{n+1}\right)}{\omega\left(\frac{1}{n+1}\right)}\int_0^{\frac{1}{n+1}}d\lambda\right) \\
&=O\left(\frac{n}{n+1}\,\omega(\gamma)\,\frac{\sigma\left(\frac{1}{n+1}\right)}{\omega\left(\frac{1}{n+1}\right)}\right) \\
&=O\left(\omega(\gamma)\,\frac{\sigma\left(\frac{1}{n+1}\right)}{\omega\left(\frac{1}{n+1}\right)}\right).
\end{aligned}
\tag{3.19}
$$

Again, by using **Lemma** 3.1(ii) and **Lemma** 3.2, we get

$$\Gamma_2 = \int_{\frac{1}{n+1}}^{\pi} \|\varphi(.+\gamma,\lambda) + \varphi(.-\gamma,\lambda) - 2\varphi(.,\lambda)\|_p |\kappa_n^{E_s,\overline{N}}(\lambda)|\, d\lambda$$
$$\leq \int_{\frac{1}{n+1}}^{\pi} O\left(\omega(\gamma)\frac{\sigma(\lambda)}{\omega(\lambda)}\right)\frac{1}{\lambda}\, d\lambda$$
$$= O\left(\omega(\gamma)\int_{\frac{1}{n+1}}^{\pi} \frac{\sigma(\lambda)}{\lambda\,\omega(\lambda)}\, d\lambda\right). \tag{3.20}$$

By (3.6), (3.19) and (3.20), we have

$$\|\chi_n(.+\gamma) + \chi_n(.-\gamma) - 2\chi_n(.)\|_p$$
$$= O\left(\omega(\gamma)\,\frac{\sigma\left(\frac{1}{n+1}\right)}{\omega\left(\frac{1}{n+1}\right)}\right) + O\left(\omega(\gamma)\int_{\frac{1}{n+1}}^{\pi} \frac{\sigma(\lambda)}{\lambda\,\omega(\lambda)}\, d\lambda\right).$$

Therefore, we have

$$\sup_{\gamma\neq 0} \frac{\|\chi_n(.+\gamma) + \chi_n(.-\gamma) - 2\chi_n(.)\|_p}{\omega(\gamma)}$$
$$= O\left(\frac{\sigma\left(\frac{1}{n+1}\right)}{\omega\left(\frac{1}{n+1}\right)}\right) + O\left(\int_{\frac{1}{n+1}}^{\pi} \frac{\sigma(\lambda)}{\lambda\,\omega(\lambda)}\, d\lambda\right). \tag{3.21}$$

As,

$$\varphi(\beta,\lambda) = |\delta(\beta+\lambda) + \delta(\beta-\lambda) - 2\delta(\beta)|,$$

by applying Minkowski's inequality,

$$\|\varphi(\beta,\lambda)\|_p = \|\delta(\beta+\lambda) + \delta(\beta-\lambda) - 2\delta(\beta)\|_p$$
$$= O\left(\sigma(\lambda)\right). \tag{3.22}$$

Now, using **Lemma** 3.1((i), (ii)) and (3.22), we get

$$\|\chi_n(.)\|_p \leq \left(\int_0^{\frac{1}{n+1}} + \int_{\frac{1}{n+1}}^{\pi}\right) \|\varphi(.,\lambda)\|_p |\ \kappa_n^{E_s,\overline{N}}(\lambda)|\, d\lambda$$
$$= O\left(n\int_0^{\frac{1}{n+1}} \sigma(\lambda)\, d\lambda\right) + O\left(\int_{\frac{1}{n+1}}^{\pi} \frac{\sigma(\lambda)}{\lambda} d\lambda\right)$$
$$= O\left(\sigma\left(\frac{1}{n+1}\right)\right) + O\left(\int_{\frac{1}{n+1}}^{\pi} \frac{\sigma(\lambda)}{\lambda} d\lambda\right). \tag{3.23}$$

From (3.21) and (3.23), we have

$$\begin{aligned}\|\chi_n(.)\|_p^{\omega} &= \|\chi_n(.)\|_p + \sup_{\gamma\neq 0}\frac{\|\chi_n(.+\gamma)+\chi_n(.-\gamma)-2\chi_n(.)\|_p}{\omega(\gamma)}\\ &= O\left(\sigma\left(\frac{1}{n+1}\right)\right) + O\left(\int_{\frac{1}{n+1}}^{\pi}\frac{\sigma(\lambda)}{\lambda}d\lambda\right) + O\left(\frac{\sigma\left(\frac{1}{n+1}\right)}{\omega\left(\frac{1}{n+1}\right)}\right)\\ &+ O\left(\int_{\frac{1}{n+1}}^{\pi}\frac{\sigma(\lambda)}{\lambda\,\omega(\lambda)}\,d\lambda\right) = \sum_{j=1}^{4} G_j.\end{aligned}$$

In view of monotonicity of $\omega(\lambda)$ for $0 < \lambda \leq \pi$, we have

$$\sigma(\lambda) = \frac{\sigma(\lambda)}{\omega(\lambda)}.\omega(\lambda) \leq \omega(\pi).\frac{\sigma(\lambda)}{\omega(\lambda)} = O\left(\frac{\sigma(\lambda)}{\omega(\lambda)}\right).$$

Therefore,

$$G_1 = O(G_3).$$

Again, by using monotonicity of $\omega(\lambda)$,

$$G_2 = \int_{\frac{1}{n+1}}^{\pi}\frac{\sigma(\lambda)}{\lambda}d\lambda = \int_{\frac{1}{n+1}}^{\pi}\frac{\sigma(\lambda)}{\lambda\,\omega(\lambda)}\,\omega(\lambda)d\lambda \leq \omega(\pi)\int_{\frac{1}{n+1}}^{\pi}\frac{\sigma(\lambda)}{\lambda\,\omega(\lambda)}d\lambda = O(G_4).$$

Since $\frac{\sigma(\lambda)}{\omega(\lambda)}$ is positive and increasing

$$G_4 = \int_{\frac{1}{n+1}}^{\pi}\frac{\sigma(\lambda)}{\lambda\,\omega(\lambda)}d\lambda = \frac{\sigma\left(\frac{1}{n+1}\right)}{\omega\left(\frac{1}{n+1}\right)}\int_{\frac{1}{n+1}}^{\pi}\frac{d\lambda}{\lambda} \geq \frac{\sigma\left(\frac{1}{n+1}\right)}{\omega\left(\frac{1}{n+1}\right)}.$$

Therefore,

$$G_3 = O(G_4).$$

Thus,

$$\|\chi_n(.)\|_p^{\omega} = O(G_4) = O\left(\int_{\frac{1}{n+1}}^{\pi}\frac{\sigma(\lambda)}{\lambda\,\omega(\lambda)}d\lambda\right).$$

Hence,

$$E_n(\delta) = \inf_n \|\chi_n(.)\|_p^{\omega} = O\left(\int_{\frac{1}{n+1}}^{\pi}\frac{\sigma(\lambda)}{\lambda\,\omega(\lambda)}d\lambda\right).$$

This completes the proof of **Theorem** 3.4.

Proof of Theorem 3.5 Denoting the $(E, s)(\overline{N}, q_n)$ transform of $\overline{S_k}\Big(\delta; \beta\Big)$ by $\overline{\tau}_n^{E_s,\overline{N}}$, then

$$\begin{aligned}
&\overline{\tau}_n^{E_s,\overline{N}} - \overline{\delta}(\beta) \\
&= \frac{1}{2\pi(1+s)^n}\int_0^{\pi} \Psi(\beta;\lambda) \sum_{k-0}^{n} \binom{n}{k} s^{n-k} \left\{ \frac{1}{Q_k} \sum_{\nu=0}^{k} q_\nu \frac{\cos\left(\nu+\frac{1}{2}\right)\lambda - \cos\frac{\lambda}{2}}{\sin\frac{\lambda}{2}} \right\} d\lambda \\
&= \int_0^{\pi} \Psi(\beta;\lambda)\, \overline{k}_n^{E_s,\overline{N}}(\lambda)\, d\lambda \\
&= F_n(\beta) \ \text{(say)}
\end{aligned}$$

Then

$$F_n(\beta+\gamma) - F_n(\beta-\gamma) = \int_0^{\pi} \{\Psi(\beta+\gamma;\lambda) - \Psi(\beta-\gamma;\lambda)\}\, \overline{\kappa}_n^{E_s,\overline{N}}(\lambda)\, d\lambda.$$

Using Minkowski's inequality, we have

$$\begin{aligned}
\|F_n(.+\gamma) - F_n(.-\gamma)\|_p &= \left\{ \frac{1}{2\pi} \int_0^{2\pi} |F_n(\beta+\gamma) - F_n(\beta-\gamma)|^p\, d\beta \right\}^{\frac{1}{p}} \\
&= \frac{1}{2\pi}\left[\int_0^{2\pi} \left| \int_0^{\pi} \{\Psi(\beta+\gamma;\lambda) - \Psi(\beta-\gamma;\lambda)\}\, \overline{\kappa}_n^{E_s,\overline{N}}(\lambda)\, d\lambda \right|^p d\beta \right]^{\frac{1}{p}} \\
&\le \int_0^{\pi} \left\{ \frac{1}{2\pi} \int_0^{2\pi} |\Psi(\beta+\gamma;\lambda) - \Psi(\beta-\gamma;\lambda)|^p\, d\beta \right\}^{\frac{1}{p}} |\overline{\kappa}_n^{E_s,\overline{N}}|\, d\lambda \\
&= \int_0^{\pi} \|\Psi(.+\gamma;\lambda) - \Psi(.-\gamma;\lambda)\|_p\, |\overline{\kappa}_n^{E_s,\overline{N}}(\lambda)|\, d\lambda \\
&= \int_0^{\frac{1}{n+1}} \|\Psi(.+\gamma;\lambda) - \Psi(.-\gamma;\lambda)\|_p\, |\overline{\kappa}_n^{E_s,\overline{N}}(\lambda)|\, d\lambda \\
&\quad + \int_{\frac{1}{n+1}}^{\pi} \|\Psi(.+\gamma;\lambda) - \Psi(.-\gamma;\lambda)\|_p\, |\overline{\kappa}_n^{E_s,\overline{N}}(\lambda)|\, d\lambda \\
&= I_1 + I_2 \ \text{(say)}
\end{aligned} \tag{3.24}$$

Using **Lemma** 3.3(i), **Lemma** 3.4 and monotonicity of $\left(\frac{\sigma(\lambda)}{\omega(\lambda)}\right)$ with respect to 'λ', we have

$$I_1 = \int_0^{\frac{1}{n+1}} \|\Psi(.+\gamma,\lambda) - \Psi(.-\gamma,\lambda)\|_p \, |\overline{\kappa}_n^{E_s,\overline{N}}(\lambda)| \, d\lambda$$
$$= \int_0^{\frac{1}{n+1}} O\left(\omega(\gamma)\frac{\sigma(\lambda)}{\omega(\lambda)}\right) O(n) \, d\lambda.$$

By using the second mean value theorem of integral, we have

$$I_1 \leq O\left(n\,\omega(\gamma)\,\frac{\sigma\left(\frac{1}{n+1}\right)}{\omega\left(\frac{1}{n+1}\right)} \int_0^{\frac{1}{n+1}} d\lambda\right)$$
$$= O\left(\frac{n}{n+1}\,\omega(\gamma)\,\frac{\sigma\left(\frac{1}{n+1}\right)}{\omega\left(\frac{1}{n+1}\right)}\right)$$
$$= O\left(\omega(\gamma)\,\frac{\sigma\left(\frac{1}{n+1}\right)}{\omega\left(\frac{1}{n+1}\right)}\right). \tag{3.25}$$

Again, by using **Lemma** 3.3(ii) and **Lemma** 3.4, we get

$$I_2 = \int_{\frac{1}{n+1}}^{\pi} \|\Psi(.+\gamma,\lambda) - \Psi(.-\gamma,\lambda)\|_p |\overline{\kappa}_n^{E_s,\overline{N}}(\lambda)| \, d\lambda$$
$$\leq \int_{\frac{1}{n+1}}^{\pi} O\left(\omega(\gamma)\frac{\sigma(\lambda)}{\omega(\lambda)}\right)\frac{1}{\lambda} \, d\lambda$$
$$= O\left(\omega(\gamma) \int_{\frac{1}{n+1}}^{\pi} \frac{\sigma(\lambda)}{\lambda\,\omega(\lambda)} \, d\lambda\right). \tag{3.26}$$

By (3.24), (3.25) and (3.26), we have

$$\|F_n(.+\gamma) - F_n(.-\gamma)\|_p$$
$$= O\left(\omega(\gamma)\,\frac{\sigma\left(\frac{1}{n+1}\right)}{\omega\left(\frac{1}{n+1}\right)}\right) + O\left(\omega(\gamma) \int_{\frac{1}{n+1}}^{\pi} \frac{\sigma(\lambda)}{\lambda\,\omega(\lambda)} \, d\lambda\right).$$

Therefore, we have

$$\sup_{\gamma\neq 0} \frac{\|F_n(.+\gamma) - F_n(.-\gamma)\|_p}{\omega(\gamma)}$$
$$= O\left(\frac{\sigma\left(\frac{1}{n+1}\right)}{\omega\left(\frac{1}{n+1}\right)}\right) + O\left(\int_{\frac{1}{n+1}}^{\pi} \frac{\sigma(\lambda)}{\lambda\,\omega(\lambda)} \, d\lambda\right). \tag{3.27}$$

Since $\delta \in \Delta_p^{(\sigma)}$ and $\Psi(\beta;\lambda) = |\delta(\beta+\lambda) - \delta(\beta-\lambda)|$, by Minkowski's inequality, we have

$$\|\Psi(\beta,\lambda)\|_p = \|\delta(\beta+\lambda) - \delta(\beta-\lambda)\|_p. \\ = O\Big(\sigma(\lambda)\Big) \tag{3.28}$$

Now, using **Lemmas** 3.3 and (3.28), we get

$$\begin{aligned} \|F_n(.)\|_p &\le \left(\int_0^{\frac{1}{n+1}} + \int_{\frac{1}{n+1}}^{\pi}\right) \|\Psi(.,\lambda)\|_p | \overline{\mathcal{K}}_n^{E_s,\overline{N}}(\lambda)|\, d\lambda \\ &= O\left(n\int_0^{\frac{1}{n+1}} \sigma(\lambda)\, d\lambda\right) + O\left(\int_{\frac{1}{n+1}}^{\pi} \frac{\sigma(\lambda)}{\lambda} d\lambda\right) \\ &= O\left(\sigma\left(\frac{1}{n+1}\right)\right) + O\left(\int_{\frac{1}{n+1}}^{\pi} \frac{\sigma(\lambda)}{\lambda} d\lambda\right). \end{aligned} \tag{3.29}$$

From (3.27) and (3.29), we have

$$\|F_n(.)\|_p^{\omega} = \|F_n(.)\|_p + \sup_{\gamma\neq 0} \frac{\|F_n(.+\gamma) - F_n(.-\gamma)\|_p}{\omega(\gamma)}$$

$$= O\left(\sigma\left(\frac{1}{n+1}\right)\right) + O\left(\int_{\frac{1}{n+1}}^{\pi} \frac{\sigma(\lambda)}{\lambda} d\lambda\right) + O\left(\frac{\sigma\left(\frac{1}{n+1}\right)}{\omega\left(\frac{1}{n+1}\right)}\right) \\ + O\left(\int_{\frac{1}{n+1}}^{\pi} \frac{\sigma(\lambda)}{\lambda\,\omega(\lambda)}\, d\lambda\right) = \sum_{j=1}^{4} B_j.$$

In view of monotonicity of $\omega(\lambda)$ for $0 < \lambda \le \pi$, we have

$$\sigma(\lambda) = \frac{\sigma(\lambda)}{\sigma(\lambda)}.\omega(\lambda) \le \omega(\pi).\frac{\sigma(\lambda)}{\omega(\lambda)} = O\left(\frac{\sigma(\lambda)}{\omega(\lambda)}\right).$$

Therefore,

$$B_1 = O(B_3).$$

Again, by using monotonicity of $\omega(\lambda)$,

$$B_2 = \int_{\frac{1}{n+1}}^{\pi} \frac{\sigma(\lambda)}{\lambda} d\lambda = \int_{\frac{1}{n+1}}^{\pi} \frac{\sigma(\lambda)}{\lambda\,\omega(\lambda)}\, \omega(\lambda) d\lambda \le \omega(\pi) \int_{\frac{1}{n+1}}^{\pi} \frac{\sigma(\lambda)}{\lambda\,\omega(\lambda)} d\lambda = O(B_4).$$

Since $\frac{\sigma(\lambda)}{\omega(\lambda)}$ is positive and increasing

$$B_4 = \int_{\frac{1}{n+1}}^{\pi} \frac{\sigma(\lambda)}{\lambda\,\omega(\lambda)} d\lambda = \frac{\sigma\left(\frac{1}{n+1}\right)}{\omega\left(\frac{1}{n+1}\right)} \int_{\frac{1}{n+1}}^{\pi} \frac{d\lambda}{\lambda} \geq \frac{\sigma\left(\frac{1}{n+1}\right)}{\omega\left(\frac{1}{n+1}\right)}$$

Therefore,

$$B_3 = O(B_4).$$

Thus,

$$\| F_n(.)\|_l^{\omega} = O(G_4) = O\left(\int_{\frac{1}{n+1}}^{\pi} \frac{\sigma(\lambda)}{\lambda\,\omega(\lambda)} d\lambda\right)$$

Hence,

$$E_n(\overline{\delta}) = \inf_n \| F_n(.)\|_p^{\omega} = O\left(\int_{\frac{1}{n+1}}^{\pi} \frac{\sigma(\lambda)}{\lambda\,\omega(\lambda)} d\lambda\right)$$

This completes the proof of **Theorem** 3.5.

Proof of Theorem 3.6 Denoting the $(E, s)(\overline{N}, q_n)$ transform of $\overline{S_k}'(\delta; \beta)$ by $\overline{\tau_n}'^{E_s,\overline{N}}$. Then,

$$\begin{aligned}
\overline{\tau_n}'^{E_s,\overline{N}} - \overline{\delta}'(\beta) &= \frac{-2k}{\pi(1+s)^n} \int_0^{\pi} \rho(\beta,\lambda) \sum_{k=0}^{n} \binom{n}{k} s^{n-k} \left\{ \frac{1}{Q_k} \sum_{\nu=0}^{k} q_{\nu} \sin\left(\nu + \frac{1}{2}\right)\lambda \right\} d\lambda \\
&- \frac{1}{\pi(1+s)^n} \int_0^{\pi} \rho(\beta,\lambda) \sum_{k=0}^{n} \binom{n}{k} s^{n-k} \left\{ \frac{1}{Q_k} \sum_{\nu=0}^{k} q_{\nu} \frac{\cos(\nu\lambda)}{\sin\frac{\lambda}{2}} \right\} d\lambda \\
\Rightarrow \overline{\tau_n}'^{E_s,A} - \overline{\delta}'(\beta) &= \frac{-2k}{4\pi(1+s)^n} \int_0^{\pi} \psi(\beta,\lambda) \sum_{k=0}^{n} \binom{n}{k} s^{n-k} \left\{ \frac{1}{Q_k} \sum_{\nu=0}^{k} q_{\nu} \frac{\sin\left(\nu + \frac{1}{2}\right)\lambda}{\sin\frac{\lambda}{2}} \right\} d\lambda \\
&- \frac{1}{4\pi(1+s)^n} \int_0^{\pi} \psi(\beta,\lambda) \sum_{k=0}^{n} \binom{n}{k} s^{n-k} \left\{ \frac{1}{Q_k} \sum_{\nu=0}^{k} q_{\nu} \frac{\cos \nu\lambda}{\sin^2\frac{\lambda}{2}} \right\} d\lambda \\
&= \int_0^{\pi} \psi(\beta,\lambda)\{\overline{Y_1'}(\lambda) + \overline{Y_2'}(\lambda)\}\, d\lambda \\
&= G_n'(\beta), \quad \text{(say)}
\end{aligned}$$

Here $\rho(\beta, \lambda) = \frac{\psi(\beta;\lambda)}{4\sin\frac{\lambda}{2}}$.

Now,

$$G_n^{'}(\beta+\gamma)+G_n^{'}(\beta-\gamma)=\int_0^{\pi}\Big\{\psi(\beta+\gamma,\lambda)+\psi(\beta-\gamma,\lambda)\Big\}\,\{\overline{Y_1^{'}}(\lambda)+\overline{Y_2^{'}}(\lambda)\}\,d\lambda$$

Using Minkowski's inequality, we have

$$\begin{aligned}
\|G_n^{'}(.+\gamma)+G_n^{'}(.-\gamma)\|_p &= \left\{\frac{1}{2\pi}\int_0^{2\pi}|G_n^{'}(\beta+\gamma)+G_n^{'}(\beta-\gamma)|^p\,d\beta\right\}^{\frac{1}{p}}\\
&\le \int_0^{\pi}\left\{\frac{1}{2\pi}\int_0^{2\pi}|\psi(\beta+\gamma,\lambda)+\psi(\beta-\gamma,\lambda)|^p d\beta\right\}^{\frac{1}{p}}|\overline{Y_1^{'}}(\lambda)+\overline{Y_2^{'}}(\lambda)|\,d\lambda\\
&= \int_0^{\pi}\|\psi(.+\gamma,\lambda)+\psi(.-\gamma,\lambda)\|_p\,|\overline{Y_1^{'}}(\lambda)+\overline{Y_2^{'}}(\lambda)|\,d\lambda\\
&= \int_0^{\frac{1}{n+1}}\|\psi(.+\gamma,\lambda)+\psi(.-\gamma,\lambda)\|_p\,|\overline{Y_1^{'}}(\lambda)+\overline{Y_2^{'}}(\lambda)|\,d\lambda\\
&\quad+\int_{\frac{1}{n+1}}^{\pi}\|\psi(.+\gamma,\lambda)+\psi(.-\gamma,\lambda)\|_p\,|\overline{Y_1^{'}}(\lambda)+\overline{Y_2^{'}}(\lambda)|\,d\lambda\\
&= I_1^{'}+I_2^{'},\quad \text{(say)}
\end{aligned}\tag{3.30}$$

Further,

$$|\psi(\beta+\gamma;\lambda)+\psi(\beta-\gamma;\lambda)|\le\|\delta(\beta+\gamma+\lambda)+\delta(\beta+\gamma-\lambda)\|+\|\delta(\beta-\gamma+\lambda)+\delta(\beta-\gamma-\lambda)\|$$

By Minkowski's inequality, we have

$$\begin{aligned}
|\psi(.+\gamma;\lambda)+\psi(.-\gamma;\lambda)|_p &\le \|\delta(.+\gamma+\lambda)+\delta(.+\gamma-\lambda)\|_p+\|\delta(.-\gamma+\lambda)+\delta(.-\gamma-\lambda)\|_p\\
&= O(\sigma(\lambda))\text{ or }O(\sigma(\gamma)).
\end{aligned}$$

Again, by using **Lemma** 3.5(i), **Lemma** 3.6 and monotonicity of $\frac{\sigma(\lambda)}{\omega(\lambda)}$, we get

$$\begin{aligned}
I_1^{'} &= \int_0^{\frac{1}{n+1}}\|\psi(.+\gamma,\lambda)+\psi(.-\gamma,\lambda)\|_p|\overline{Y_1^{'}}(\lambda)+\overline{Y_2^{'}}(\lambda)|\,d\lambda\\
&\le O\left(\int_0^{\frac{1}{n+1}}\omega(\gamma)\frac{\sigma(\lambda)}{\omega(\lambda)}\,n^2\,d\lambda\right)+O\left(\int_0^{\frac{1}{n+1}}\omega(\gamma)\frac{\sigma(\lambda)}{\omega(\lambda)}\,\frac{1}{\lambda^2}\,d\lambda\right)\\
&= O\left(n^2\omega(\gamma)\,\frac{\sigma\left(\frac{1}{n+1}\right)}{\omega\left(\frac{1}{n+1}\right)}\right)+O\left(\omega(\gamma)\int_0^{\frac{1}{n+1}}\frac{\sigma(\lambda)}{\omega(\lambda)}\,\frac{1}{\lambda^2}\,d\lambda\right).
\end{aligned}\tag{3.31}$$

Similarly, by using **Lemma** 3.5(ii) and **Lemma** 3.6, we have

$$I_2' = \int_{\frac{1}{n+1}}^{\pi} \|\psi(.+\gamma,\lambda)+\psi(.-\gamma,\lambda)\|_p |\overline{Y_1'}(\lambda)+\overline{Y_2'}(\lambda)|\,d\lambda$$

$$= O\left(n^2\omega(\gamma)\int_0^{\frac{1}{n+1}}\frac{\sigma(\lambda)\,d\lambda}{\omega(\lambda)}\right)+O\left((n+1)\omega(\gamma)\frac{\sigma\left(\frac{1}{n+1}\right)}{\omega\left(\frac{1}{n+1}\right)}\right). \quad (3.32)$$

By (3.30), (3.31 and (3.32), we get

$$\|G_n'(.+\gamma)+G_n'(.-\gamma)\|_p$$

$$= O\left(n^2\omega(\gamma)\frac{\sigma\left(\frac{1}{n+1}\right)}{\omega\left(\frac{1}{n+1}\right)}\right)+O\left(\omega(\gamma)\int_0^{\frac{1}{n+1}}\frac{\sigma(\lambda)}{\omega(\lambda)}\frac{1}{\lambda^2}\,d\lambda\right)$$

$$+O\left(n^2\omega(\gamma)\int_0^{\frac{1}{n+1}}\frac{\sigma(\lambda)\,d\lambda}{\omega(\lambda)}\right)+O\left((n+1)\omega(\gamma)\frac{\sigma\left(\frac{1}{n+1}\right)}{\omega\left(\frac{1}{n+1}\right)}\right).$$

Therefore, we have

$$\sup_{\gamma\neq 0}\frac{\|G_n'(.+\gamma)+G_n'(.-\gamma)\|_p}{\omega(\gamma)}$$

$$= O\left(n^2\frac{\sigma\left(\frac{1}{n+1}\right)}{\omega\left(\frac{1}{n+1}\right)}\right)+O\left(\int_0^{\frac{1}{n+1}}\frac{\sigma(\lambda)}{\omega(\lambda)}\frac{1}{\lambda^2}\,d\lambda\right)$$

$$+O\left(n^2\int_0^{\frac{1}{n+1}}\frac{\sigma(\lambda)\,d\lambda}{\omega(\lambda)}\right)+O\left((n+1)\frac{\sigma\left(\frac{1}{n+1}\right)}{\omega\left(\frac{1}{n+1}\right)}\right). \quad (3.33)$$

Since $\delta\in\Omega_p^{(\sigma)}$ and $\psi(\beta;\lambda)=|\delta(\beta+\lambda)+\delta(\beta-\lambda)|$, by Minkowski's inequality, we have

$$\|\psi(\beta,\lambda)\|_p=\|\delta(\beta+\lambda)+\delta(\beta-\lambda)\|_p = O\left(\sigma(\lambda)\right).$$

Therefore,

$$\|G_n^{'}(.)\|_p \leq \left(\int_0^{\frac{1}{n+1}} + \int_{\frac{1}{n+1}}^{\pi}\right) \|\psi(.,\lambda)\|_p |\ \overline{Y_1}^{'}(\lambda) + \overline{Y_2}^{'}(\lambda)|\ d\lambda$$

$$= O\left(n^2 \int_0^{\frac{1}{n+1}} \sigma(\lambda)\, d\lambda\right) + O\left(\int_0^{\frac{1}{n+1}} \frac{\sigma(\lambda)}{\lambda^2}\, d\lambda\right)$$

$$+ O\left(n^2 \int_{\frac{1}{n+1}}^{\pi} \sigma(\lambda)\, d\lambda\right) + O\left(\int_{\frac{1}{n+1}}^{\pi} \frac{\sigma(\lambda)}{\lambda^2} d\lambda\right). \tag{3.34}$$

From (3.33), (3.34) and by the monotonicity of $\omega(\lambda)$ we have

$$\|G_n^{'}(.)\|_p^{\omega} = \|G_n^{'}(.)\|_p + \sup_{\gamma \neq 0} \frac{\|Gi_n'(.+\gamma) + G_n^{'}(.-\gamma)\|_p}{\omega(\gamma)}$$

$$= O\left(\int_{\frac{1}{n+1}}^{\pi} \frac{\sigma(\lambda)}{\lambda^2\, \omega(\lambda)} d\lambda\right)$$

provided

$$\int_0^{\eta} \frac{\sigma(\lambda)}{\lambda\, \omega(\lambda)} d\lambda \;=\; O\left(\frac{\sigma(\xi)}{\omega(\xi)}\right).$$

Hence,

$$E_n(\overline{\delta'}) = \inf_n \|G_n^{'}(.)\|_p^{\omega} = O\left(\int_{\frac{1}{n+1}}^{\pi} \frac{\sigma(\lambda)}{\lambda^2\, \omega(\lambda)} d\lambda\right).$$

This completes the proof of **Theorem** 3.6.

3.7 Conclusion

In this chapter, we have obtained the degree of approximation of signals associated with Fourier series, conjugate Fourier series and derived conjugate Fourier series which not only generalizes the known theorems but also [2, 6, 8–14, 16, 18, 19, 21].

References

1. Chandra, P.: Degree of approximation of functions in the Hölder metric. J. Ind. Math. Soc. 99–114(1988)
2. Das, A.A., Paikray, S.K., Pradhan, T., Dutta, H.: Approximation of signals in the weighted Zygmund class via Euler-Hausdorff product summability mean of Fourier series. J. Indian Math. Soc. (New Ser.) **86**, 296–314 (2019)
3. Das, G., Ghosh, T., Ray, B.K.: Degree of approximation of function by their Fourier series in the generalized Hölder metric. Proc. Indian Acad. Sci. (Math. Sci.) **106**(2), 139–153 (1996)
4. Hardy, G.H.: Divergent Series, 1st edn. Oxford University Press (1949)
5. Lal, S., Shireen, F.: Best approximation of functions of generalized Zygmund class by Matrix-Euler summability mean of Fourier series. Bull. Math. Anal. Appl. **5**(4), 1–13 (2013)
6. Leindler, L.: Strong approximation and generalized Zygmund class. Acta. Sci. Math. **43**, 301–309 (1981)
7. Mahapatra, R.N., Chandra, P.: Degree of approximation of functions in the Hölder metric. Acta Math. Hung. **41**(1–2), 67–76 (1983)
8. Mishra, V.N., Khatri, K., Mishra, L.N.: Approximation of functions belonging to $Lip\Big(\xi(t), r\Big)$ class by $(N, p_n)(E, q)$-summability of conjugate series of Fourier series. J. Ineqal. Appl. **296** (2012)
9. Mishra, L.N., Mishra, V.N., Khatri, K., Deepmala, D.: On the Trigonometric approximation of signals belonging to generalized weighted Lipschitz class $W\Big(L^r, \xi(t)\Big)$, $(r \geq 1)$-class by matrix (C^1, N_p) operator of conjugate series of its Fourier series. Appl. Math. Comput. **237**, 252–263 (2014)
10. Mishra, V.N., Khatri, K., Mishra, L.N.: Product $(N, p_n)(C, 1)$-summability of a sequence of a sequence of Fourier coefficients. Math. Sci. (2012). https://doi.org/10.1186/2251-7456-6-38
11. Mishra, V.N., Khatri, K., Mishra, L.N., Deepmala, D.: Trigonometric approximation of periodic signals belonging to generalized weighted Lipschitz $W\Big(L_r, \xi(t)\Big)$, $(r \geq 1)$-class by Nörlund Euler $(N, p_n)(E, q)$ operator of conjugate series of its Fourier series. J. class. Anal. **5**, 91–105 (2014)
12. Mishra, V.N., Mishra, L.N.: Trigonometric approximation of signals (functions) in L^p, $(p \geq 1)$-norm. Int. J. Contemp. Math. Sci. **7**, 909–918 (2012)
13. Moricz, F.: Enlarged Lipschitz and Zygmund classes of functions and Fourier transforms. East. J. Approx. **16**(3), 259–271 (2010)
14. Moricz, F., Nemeth, J.: Generalized Zygmund classes of functions and strong approximation of Fourier series. Acta. Sci. Math. **73**, 637–647 (2007)
15. Nayak, L., Das, G., Ray, B.K.: An estimate of the rate of convergence of Fourier series in the generalized Hölder metric by deferred Cesàro mean, j. Math. Anal. App. **420**, 563–575 (2014)
16. Nigam, H.K., Hadish, M.: Best approximation of functions in generalized Hölder class. J. Ineq. App. **2018**(276), 1–15 (2018)
17. Nigam, H.K.: On approximation in generalized Zygmund class. Demonstr. Math. (De Gruyter) **52**, 370–387 (2019)
18. Parida, P., Paikray, S.K., Das, M., Misra, U.K.: Degree of approximation by product (N; pn; qn)(E; q) summability Fourier series of a signal belonging to Lip(Alpha; r)-class. TWMS J. App. Eng. Math. **9**(4), 901–908 (2019)
19. Pradhan, T., Paikray, S.K., Misra, U.K.: Approximation of signals belonging to generalized Lipschitz class using $(\overline{N}; p_n; q_n)(E; s)$-summability mean of Fourier series. Cogent Math. **1250343**, 1–9 (2016)
20. Pradhan, T., Jena, B.B., Paikray, S.K., Dutta, H., Misra, U.K.: On approximation of the rate of convergence of Fourier series in the generalized Hölder metric by deferred N orlund mean. Afrika Mat. **30**, 1119–1131 (2019)
21. Pradhan, T., Paikray, S.K., Das, A.A., Dutta, H.: On approximation of signals in the generalized Zygmund class via (E; 1)(N; pn) summability means of conjugate Fourier series. Proyecciones J. Math. **38**, 1015–1033 (2019)

22. Prosdorff, S.: Zur Konvergenz der Fourier richen Hölder stetiger Funktionen. Math. Nachar **69**, 7–14 (1975)
23. Singh, M.V., Mittal, M.L., Rhoades, B.E.: Approximation of functions in the generalized Zygmund class using Hausdroff means. J. Inequal. Appl. **101** (2017). https://doi.org/10.1186/s13600-017-1361-8
24. Zygmund, A.: Trigonometric Series, vol. 1, 2nd revised edn. Cambridge university Press (1939)
25. Vinti, G., Zampogni, L.: Approximation by means of nonlinear Kantorovich sampling type operators in Orlicz spaces. J. Approx. Theory **161**, 511–528 (2009)
26. Bardaro, C., Butzer, P.L., Stens, R.L., Vinti, G.: Kantorovich-type generalized sampling series in the setting of Orlicz spaces. Sampl. Theory Signal Image Process. **6**(1), 29–52 (2007)
27. Costarelli, D., Seracini, M., Vinti, G.: A comparison between the sampling Kantorovich algorithm for digital image processing with some interpolation and quasi-interpolation methods. Appl. Math. Comput. **374**, 125046 (2020)
28. Costarelli, D., Vinti, G.: Rate of approximation for multivariate sampling Kantorovich operators on some functions spaces. J. Int. Eq. Appl. **26**(4), 455–481 (2014)
29. Kadak, U.: Fractional type multivariate sampling operators. Revista de la Real Academia de Ciencias Exactas, Fısicas y Naturales. Serie A. Matemáticas, **115**(153) (2021)
30. Kadak, U.: Fractional type multivariate neural network operators. Math. Methods Applies Sci. (2021). https://doi.org/10.1002/mma.7460
31. Coroianu, L., Costarelli, D., Gal, S.G., Vinti, G.: The max-product generalized sampling operators: convergence and quantitative estimates. Appl. Math. Comput. **355**, 173–183 (2019)

Chapter 4
Trigonometric Approximation of Signals Belonging to $Lip(\xi(t), r)$ Class by $(C, 1)(N, p_m, q_m)(E, \theta)$ Means of Conjugate Fourier Series

Smita Sonker and Rozy Jindal

Abstract Various authors have studied the trigonometric or Fourier approximation in various function spaces using different means $(C, 2)(E, 1)$, (E, q) and $(N, p_n, q_n)(E, q)$, etc. In our study, we approximate the functions belonging to different classes by a more generalized $(C, 1)(N, p_m, q_m)(E, \theta)$ means of conjugate Fourier series. The discipline of Fourier approximation holds significant practical importance. Because the product summability means can be used to approximate functions and are stronger than the individual summability means. Additionally, approximation theory is used to identify the degree of approximation and to solve engineering problems by approximating the function and minimizing error. The established theorem builds, expands, and generalizes a number of previous approximation theory conclusions. Additionally, the outcome encourages researchers who are interested in working with product summability measures.

Keywords Summability means · Lipschitz $lip(\xi(t), r)$ class · Signal approximation · Fourier series · Hölder's inequality

Subject Classification MSC 2010: 41A25 · 42B05 · 42B08 · 40G05

4.1 Introduction

The $(C, 1)$ transformed mean of $\{s_n\}$, i.e., the partial sum of $\sum u_n$ is

$$C_r^1 = \frac{1}{r+1} \sum_{l=0}^{r} s_l, \quad r = 0, 1, 2, \ldots.$$

It is s.t.b. $(C, 1)$ summable to s if $lim_{r \to \infty} C_r^1 = s$.

Let $p = \{p_m\}$ be a positive monotonic and non-increasing $\in R$ as

S. Sonker (✉) · R. Jindal
Department of Mathematics, National Institute of Technology, Kurukshetra 136119, Haryana, India
e-mail: smitafma@nitkkr.ac.in

B. C. Tripathy et al. (eds.), *Operators, Inequalities and Approximation*, Industrial and Applied Mathematics, https://doi.org/10.1007/978-981-97-3238-8_4

$$P_m = \sum_{w=0}^{m} p_w \to \infty, \quad as \quad m \to \infty,$$

$$P_{-1} = p_{-1} = 0, \quad for \quad all \quad i \geq 1.$$

Let an increasing sequence $\{\mathcal{R}_m\}$ for a positive real sequence $q = \{q_m\}$, such that

$$\mathcal{R}_m = (p * q)_m = \sum_{w=0}^{m} p_{m-w} q_w \to \infty \quad as \quad m \to \infty$$

denotes the convolution product where

$$Q_m = \sum_{w=0}^{m} q_w \to \infty, \quad Q_{-1} = q_{-1} = 0, \quad for \quad all \quad i \geq 1.$$

Then

$$t_m^N = \frac{1}{\mathcal{R}_m} \sum_{w=0}^{m} p_{m-w} q_w s_w$$

describes the transformed (N, p_m, q_m) mean of $\{s_m\}$[5].

If $t_m^N \to s$, as $m \to \infty$, then $\sum u_m$ is s.t.b. (N, p_m, q_m) summable to s.

The necessary and sufficient conditions for the regularity of the (N, p_m, q_m) means are

$$\sum_{h=0}^{m} |p_{m-h} q_h| = \mathcal{O}(\mathcal{R}_m)$$

and $p_{m-h} = o(\mathcal{R}_m)$, as $m \to \infty$, for every $h > 0$, for which $q_h \neq 0$.

The (E, θ) transformed mean of $\{s_m\}$ is

$$E_m^\theta = \frac{1}{(1+\theta)^m} \sum_{k=0}^{m} \binom{m}{k} \theta^{m-k} s_k.$$

If $E_m^\theta \to s \ as \ m \to \infty$, then $\sum u_m$ is Euler summable to s, and this method is regular [7].

The $(C, 1)(N, p_m, q_m)$ transform of (E, θ) defines the $(C, 1)(N, p_m, q_m)(E, \theta)$ transform t_n^{CNE} of the nth partial sums s_n of $\sum_{n=0}^{\infty} u_n$ is defined as

$$t_n^{CNE} = \frac{1}{n+1} \sum_{n=0}^{m} \frac{1}{\mathcal{R}_m} \sum_{k=0}^{m} \frac{p_k q_{m-k}}{(1+\theta)^{m-k}} \sum_{v=0}^{m-k} \binom{m-k}{v} \theta^k s_v.$$

If $t_n^{CNE} \to s$ as $n \to \infty$, then $\sum u_n$ is $(C, 1)(N, p_m, q_m)(E, \theta)$ summable to the sum s if the limit $lim_{n\to\infty} t_n^{CNE}$ exist.

Particular cases:

1. $(C, 1)(N, p_m, q_m)(E, \theta)$ changes to $(C, 1)(N, p_m)(E, \theta)$ means if $q_m = 1\ \forall m$.
2. $(C, 1)(N, p_m, q_m)(E, \theta)$ changes to $(C, 1)(N, p_m, q_m)(E, 1)$ means if $\theta = 1$.
3. $(C, 1)(N, p_m, q_m)(E, \theta)$ changes to $(C, 1)(\overline{N}, q_m)(E, \theta)$ means if $p_m = 1\ \forall m$.
4. $(C, 1)(N, p_m, q_m)(E, \theta)$ changes to $(C, 1)(N, \frac{1}{m+1})(E, \theta)$ means if $p_m = \frac{1}{m+1}$ and $q_m = 1\ \forall m$.

Let $'h'$ be a signal having 2π periodic time and is integrable in the same way as of Lebesgue for the limit $(-\pi, \pi)$. Let

$$s_n(h; x) = \frac{a_0}{2} + \sum_{m=1}^{n} a_m \cos mx + \sum_{m=1}^{n} b_m \sin mx \tag{4.1}$$

denote the nth partial sum.

The conjugate Fourier series is written as

$$\sum_{m=1}^{\infty} (b_m \cos mx - a_m \sin mx), \tag{4.2}$$

and its nth partial sum is given by

$$\tilde{s_n}(h; x) = \sum_{l=1}^{n} (b_l \cos lx - a_l \sin lx). \tag{4.3}$$

Definition 4.1 For h, the L_∞-norm usually denoted as $\|h\|_\infty$ is

$$\|h\|_\infty = sup\{|h(x)| : x \in \mathbb{R}\},$$

and L_p-norm, usually denoted as $\|h\|_p$, described over $[0, 2\pi]$ as

$$\|h\|_p = \left\{ \int_0^{2\pi} |h(x)|^p dx \right\}^{1/p}, \quad p \geq 1.$$

Definition 4.2 The approximation of signal h by $\tau_n(x)$ under $\|.\|_\infty$ is described by Zygmund [2] as

$$\|\tau_n - h\|_\infty = sup\{|\tau_n(x) - h(x)| : x \in \mathbb{R}\}$$

and the error function $E_n(h)$ of $h \in L_p$ is

$$E_n(h) = \min_n \|\tau_n(h;x) - h(x)\|_p .$$

This is called Trigonometric Fourier Approximation [13].

Definition 4.3 A real valued signal h is of Lipschitz class if

$$|h(x+t) - h(x)| = \mathcal{O}(|t|^\beta), \ 0 < \beta \le 1, \ t > 0,$$

and is denoted by $h \in Lip\beta$.

Also, $h \in Lip(\beta, p)$ if

$$\left\{\int_0^{2\pi} |h(x+t) - h(x)|^p dx\right\}^{1/p} = \mathcal{O}(|t|^\beta) \ for \ 0 < \beta \le 1, \ p \ge 1, \ and \ t > 0.$$

For positive increasing function $\xi(t)$, $h \in Lip(\xi(t), p)$ if

$$\left\{\int_0^{2\pi} |h(x+t) - h(x)|^p dx\right\}^{1/p} = \mathcal{O}(\xi(t)) \ for \ p \ge 1 \ and \ t > 0.$$

It should be noted here that

$Lip(\xi(t), p) \xrightarrow{\xi(t)=t^\beta} Lip(\beta, p) \xrightarrow{p\to\infty} Lip\beta \ for \ 0 < \beta \le 1, \ p \ge 1, \ and$ $t > 0$.

Particular cases: **Notations**:

$$\Psi_x(t) := \Psi(x,t) = h(x+t) - h(x-t),$$

$$\Delta c_n = c_n - c_{n+1}, \ \ n \ge 0,$$

and

$$\widetilde{K_n^{CNE}}(t) = \frac{(n+1)^{-1}}{2\pi} \sum_{m=0}^{n} \frac{1}{\mathcal{R}_m} \sum_{k=0}^{m} p_k q_{m-k} \left(\frac{\mathcal{R}(t)}{1+\theta}\right)^k \frac{\cos\left[\frac{(k+1)t}{2} - k\tan^{-1}\left(\frac{\theta-1}{\theta+1}\tan\frac{t}{2}\right)\right]}{\sin\frac{t}{2}}$$

where

$$\mathcal{R}(t) = \sqrt{(\theta+1)^2 - 4\theta\sin^2\frac{t}{2}}.$$

A broad field of mathematics that is connected to the study of Fourier analysis and Operator Theory is approximation theory. Trigonometric polynomials were used to approximate later functions after beginning by estimating earlier functions with the aid of polynomials. However, to reduce approximation error today, trigonometric Fourier polynomials are utilized to approximate the functions. Several summability methods and functions from different classes are also employed to improve the assessments to a certain extent. Several academics worked on signal approximation under various constraints and standards (functions). Jadia [4], Pandey [8], McFadden [10], Singh [3], and Khare [18] studied the summability of the Fourier series and its respective conjugate series. Mittal & Kumar [14] employed the approach of matrix summability to investigate the Fourier series and its conjugate series, while almost (N, p, q) summability method was used by Singh & Singh [29] to examine the Fourier series and its conjugate series. The (N, p_n) and (N, p, q) techniques of summability are specific cases of the matrix method. Krasniqi [34] worked on product summability of $(C, 1)(E, q)$. Sonker [19] derived results on Fourier-Laguerre series. Mittal and Prasad [15] gave results using Fourier coefficients. Mursallen and Alotaibi [16] studied the generalized matrix summability of conjugate series. Saxena and Verma [26] worked on $(H, 1)(E, q)$ product summability. Krasniqi and Deepamala [35] worked on approximating signals by $(N, p_n, q_n)(E, \theta)$ means. Sonker and Singh [20], Sonker and Sangwan [21, 22] also gave different results on product summability. Sonker et al. [23] derive useful result using absolute matrix summability. Also, Sonker and Jindal [24, 25] worked on approximation of signals using summability means. You may view numerous further examples of product summability with different functions from different classes in [6, 11, 12, 17, 27, 28, 30, 32, 33].

Therefore, in this paper, $(C, 1)(N, p_m, q_m)(E, \theta)$ product summability is introduced and different results are established.

4.2 Known Results

Krasniqi and Deepmala [35] give the following results:

4.2.1 Theorem

Let (N, p_n, q_n) be a generalized regular Nörlund method of summability defined by two positive generating sequences (p_n) and (q_n). Let $h : [0, 2\pi] \to \mathbb{R}$ be a 2π periodic function, Lebesgue integrable, and $\in Lip(\alpha, r)$, $(r \geq 1)$ class. If either

$$(i)\ \ (k+1)^2 p_k q_{n-k} = \mathcal{O}(\mathcal{R}_k), \qquad (ii)\ \ \sum_{k=0}^{n-1} |\Delta p_k q_{n-k}| = \mathcal{O}\left(\frac{\mathcal{R}_n}{(n+1)^2}\right) \tag{4.4}$$

or

$$(i)'\ \ (k+1)^2 p_k q_{n-k} = \mathcal{O}(\mathcal{R}_k), \qquad (ii)'\ \ \sum_{k=0}^{n-1} \left|\Delta\left(\frac{\mathcal{R}_k}{k+1}\right)\right| = \mathcal{O}\left(\frac{\mathcal{R}_n}{(n+1)^2}\right) \tag{4.5}$$

holds true, then the function $\tilde{h}$ is approximated by $(N, p_n, q_n)(E, \theta)$ transform

$$\widetilde{t_n^{p,q,\theta}}(x) = \frac{1}{R_n}\sum_{k=0}^{n} \frac{p_k q_{n-k}}{(1+\theta)^k} \sum_{v=0}^{k} \binom{k}{v} \theta^{k-v} \tilde{s}_v(x)$$

of $\tilde{s}_n(h, x)$ of series (4.2) is described by the following estimation:

$$\left\| \widetilde{t_n^{p,q,\theta}}(x) - \tilde{h} \right\|_{L^r} = \mathcal{O}_\theta((n+1)^{-\alpha}), \;\; 0 < \alpha \leq 1,$$

for all $n \in \{0, 1, 2, \dots\}$ and $\theta \neq 3$.

4.2.2 Theorem

Let (N, p_n, q_n) be as defined in Theorem 4.2.1 satisfying conditions (4.4) or (4.5). Assume that $\xi(t)$ is a modulus of continuity such that

$$\int_0^v \frac{\xi(t)}{t} dt = \mathcal{O}(\xi(v)), \;\; 0 < v < \pi. \tag{4.6}$$

Let $h : [0, 2\pi] \to \mathbb{R}$ be a 2π periodic function, Lebesgue integrable, and belonging to the $Lip(\xi(t), r)$, $(r \geq 1)$ class. Then the function $\tilde{h}$ is approximated by $\widetilde{t_n^{p,q,\theta}}$ means of $\tilde{s}_n(h, x)$ of series (4.2) is described by

$$\left\| \widetilde{t_n^{p,q,\theta}} - \tilde{h} \right\|_{L^r} = \mathcal{O}_\theta\left(\frac{1}{(n+1)^2} \int_{\frac{\pi}{n+1}}^{\pi} \frac{\xi(t)}{t^3} dt \right)$$

for all $n \in \{0, 1, 2, \dots\}$.

4.2.3 Theorem

Let (N, p_n, q_n) be as defined in Theorem 4.2.1 satisfying conditions (4.4) or (4.5) and (4.6). Let $h : [0, 2\pi] \to \mathbb{R}$ be a 2π periodic function, Lebesgue integrable, and $\in Lip(\xi(t), r)$, $(r \geq 1)$ class, and let $\frac{\xi(t)}{t}$ be monotone decreasing in $(\pi/(n+1), \pi)$. Then the function $\tilde{h}$ is approximated by $\widetilde{t_n^{p,q,\theta}}$ means of $\tilde{s}_n(h, x)$ of series (4.2) is described by

$$\left\| \widetilde{t_n^{p,q,\theta}} - \tilde{h} \right\|_{L^r} = \mathcal{O}_\theta\left(\xi\left(\frac{\pi}{n+1} \right) \right)$$

for all $n \in \{0, 1, 2, \dots\}$ and $\theta \neq 3$.

Our aim is to approximate the function $\tilde{h}$ by t_n^{CNE} means, in the different cases to which h belongs. The results described above will be greatly expanded by our findings.

4.3 Useful Lemmas

The proof of our main results requires the following auxiliary assertions.

4.3.1 Lemma

If $0 \leq t \leq \frac{\pi}{n+1}$, then $\left|\widetilde{K_n^{CNE}(t)}\right| = \mathcal{O}\left(\frac{1}{(n+1)t}\right)$, $n \in \{0, 1, 2, \dots\}$.

Proof If $0 \leq t \leq \frac{\pi}{n+1}$ and $n \in \{0, 1, 2, \dots\}$, then according to Young's inequality $\sin\frac{t}{2} \geq \frac{t}{\pi}$, we get

$$
\begin{aligned}
\left|\widetilde{K_n^{CNE}(t)}\right| &\leq \frac{1}{2\pi(n+1)} \sum_{m=0}^{n} \frac{1}{\mathcal{R}_m} \sum_{k=0}^{m} p_k q_{m-k} \left|\frac{\mathcal{R}(t)}{1+\theta}\right|^k \\
&\times \frac{\left|\cos\left[\frac{(k+1)t}{2} - k\tan^{-1}\left(\frac{\theta-1}{\theta+1}\tan\frac{t}{2}\right)\right]\right|}{\left|\sin\frac{t}{2}\right|} \\
&\leq \frac{1}{2\pi(n+1)} \sum_{m=0}^{n} \frac{1}{\mathcal{R}_m} \sum_{k=0}^{m} p_k q_{m-k} \frac{|(\theta+1)^2 - 4\theta\sin^2\frac{t}{2}|^{\frac{k}{2}}}{(1+\theta)^k} \frac{\pi}{t} \\
&\leq \frac{1}{2(n+1)} \sum_{m=0}^{n} \frac{1}{\mathcal{R}_m} \sum_{k=0}^{m} p_k q_{m-k} \frac{[(\theta+1)^2]^{\frac{k}{2}}}{(1+\theta)^k} \frac{1}{t} \\
&\leq \frac{1}{2(n+1)} \sum_{m=0}^{n} \frac{1}{\mathcal{R}_m} \sum_{k=0}^{m} p_k q_{m-k} \frac{1}{t} \\
&= \mathcal{O}((n+1)t)^{-1}.
\end{aligned}
$$

4.3.2 Lemma

If $\frac{\pi}{n+1} < t \leq \pi$, $0 \leq k \leq n$, $\theta \neq 3$, and either (4.4) and (4.5) then

$$
|\widetilde{t_n^{CNE}(t)}| = \mathcal{O}_\theta\left(\frac{1}{(n+1)^3 t^3}\right), \quad n \in \{0, 1, 2, \dots\}.
$$

Proof Let the condition (4.4) be satisfied. We denote

$$\varphi(t;\theta) := \varphi := \tan^{-1}\left(\frac{\theta-1}{\theta+1}\tan\frac{t}{2}\right)$$

and

$$\begin{aligned} P_l(t) = {} & \cos\frac{t}{2} - \cos\varphi - 2\cos\frac{t}{2}\cos(l+1)\left(\frac{t}{2}-\varphi\right)\cos^2\left(\frac{t}{4}-\frac{\varphi}{2}\right) \\ & + \sin\left(\frac{t}{2}-\varphi\right)\sin\left[\frac{t}{2}+(l+1)\left(\frac{t}{2}-\varphi\right)\right] \\ & + 2\sin\frac{t}{2}\sin(l+1)\left(\frac{t}{2}-\varphi\right)\sin^2\left(\frac{t}{4}-\frac{\varphi}{2}\right). \end{aligned}$$

Then, using summation by parts, we get

$$\begin{aligned} \left|\widetilde{t_n^{CNE}(t)}\right| = {} & \frac{1}{2\pi(n+1)}\left|\sum_{m=0}^{n}\frac{1}{\mathcal{R}_m}\sum_{k=0}^{m-1}\Delta(p_k q_{m-k})\right. \\ & \times \sum_{j=0}^{k}\frac{\left[(\theta+1)^2-4\theta\sin^2\frac{t}{2}\right]^{\frac{j}{2}}\cos\left[\frac{(j+1)t}{2}-j\varphi\right]}{(1+\theta)^j\sin\frac{t}{2}} \\ & \left. + p_m q_0\sum_{j=0}^{m}\frac{\left[(\theta+1)^2-4\theta\sin^2\frac{t}{2}\right]^{\frac{j}{2}}\cos\left[\frac{(j+1)t}{2}-j\varphi\right]}{(1+\theta)^j\sin\frac{t}{2}}\right| \\ \leq {} & \frac{1}{2\pi(n+1)}\left\{\sum_{m=0}^{n}\frac{1}{\mathcal{R}_m}\sum_{k=0}^{m-1}|\Delta(p_k q_{m-k})|\max_{0\leq l\leq k}\left|\sum_{j=0}^{l}\frac{\cos\left[\frac{(j+1)t}{2}-j\varphi\right]}{\sin\frac{t}{2}}\right|\right. \\ & \left. + p_m q_0\max_{0\leq l\leq m}\left|\sum_{j=0}^{l}\frac{\cos\left[\frac{(j+1)t}{2}-j\varphi\right]}{\sin\frac{t}{2}}\right|\right\} \\ \leq {} & \frac{1}{2t(n+1)}\left\{\sum_{m=0}^{n}\frac{1}{\mathcal{R}_m}\sum_{k=0}^{m-1}|\Delta(p_k q_{m-k})|\max_{0\leq l\leq k}\left|\mathbb{R}\sum_{j=0}^{l}e^{i\left[\frac{(j+1)t}{2}-j\varphi\right]}\right|\right. \\ & \left. + p_m q_0\max_{0\leq l\leq m}\left|\mathbb{R}\sum_{j=0}^{l}e^{i\left[\frac{(j+1)t}{2}-j\varphi\right]}\right|\right\} \\ \leq {} & \frac{1}{2t(n+1)}\left\{\sum_{m=0}^{n}\frac{1}{\mathcal{R}_m}\sum_{k=0}^{m-1}|\Delta(p_k q_{m-k})|\max_{0\leq l\leq k}\left|\mathbb{R}e^{\frac{it}{2}}\frac{1-e^{i(l+1)(\frac{t}{2}-\varphi)}}{1-e^{i(\frac{t}{2}-\varphi)}}\right|\right. \\ & \left. + p_m q_0\max_{0\leq l\leq m}\left|\mathbb{R}e^{\frac{it}{2}}\frac{1-e^{i(l+1)(\frac{t}{2}-\varphi)}}{1-e^{i(\frac{t}{2}-\varphi)}}\right|\right\} \end{aligned}$$

$$\leq \frac{1}{2t(n+1)}\left\{\sum_{m=0}^{n}\frac{1}{\mathcal{R}_m}\sum_{k=0}^{m-1}|\Delta(p_k q_{m-k})| \max_{0\leq l\leq k}\left|\frac{P_l(t)}{4\sin^2\left(\frac{t}{4}-\frac{\varphi}{2}\right)}\right|\right.$$

$$\left. + p_m q_0 \max_{0\leq l\leq m}\left|\frac{P_l(t)}{4\sin^2\left(\frac{t}{4}-\frac{\varphi}{2}\right)}\right|\right\}$$

$$= \frac{7}{8t(n+1)}\sum_{m=0}^{n}\frac{1}{\mathcal{R}_m}\left\{\sum_{k=0}^{m-1}\frac{|\Delta(p_k q_{m-k})|}{\sin^2\left(\frac{t}{4}-\frac{\varphi}{2}\right)}+\frac{p_m q_0}{\sin^2\left(\frac{t}{4}-\frac{\varphi}{2}\right)}\right\}$$

$$= \frac{7\pi^2}{32t(n+1)}\sum_{m=0}^{n}\frac{1}{\mathcal{R}_m}\left\{\sum_{k=0}^{m-1}\frac{|\Delta(p_k q_{m-k})|}{\left|\frac{t}{4}-\frac{\varphi}{2}\right|^2}+\frac{p_m q_0}{\left|\frac{t}{4}-\frac{\varphi}{2}\right|^2}\right\}.$$

Using the inequality of [9] on pp. 288–289,

$$\tan^{-1} u < u \quad for \quad u > 0,$$

we get

$$\left|\widetilde{t_n^{CNE}(t)}\right| \leq \frac{7\pi^2}{32t(n+1)}\sum_{m=0}^{n}\frac{1}{\mathcal{R}_m\left[\frac{t}{4}-\frac{\theta-1}{2(\theta+1)}\tan\frac{t}{2}\right]^2}\left\{\sum_{k=0}^{m-1}|\Delta(p_k q_{m-k})|+p_m q_0\right\}$$

$$\leq \frac{7\pi^2}{32t(n+1)}\sum_{m=0}^{n}\frac{1}{\mathcal{R}_m\left[\frac{t}{4}-\frac{\theta-1}{2(\theta+1)}\tan\frac{t}{2}\right]^2}\mathcal{O}\left(\frac{R_m}{(m+1)^2}+\frac{R_m}{(m+1)^2}\right)$$

$$= \mathcal{O}\left(\frac{(n+1)^{-3}}{t\left(t-\frac{2(\theta-1)}{\theta+1}\tan\frac{t}{2}\right)^2}\right)$$

$$= \mathcal{O}\left(\frac{(n+1)^{-3}}{t^3\left(\frac{(3-\theta)}{\theta+1}\right)^2}\right)$$

$$= \mathcal{O}_\theta\left(\frac{(n+1)^{-3}}{t^3}\right).$$

Similar methods can be used to prove this lemma under the condition (4.5). We shall skip it, and the lemma's proof is now complete.

4.3.3 Lemma [31]

Let $h \in Lip(\alpha, r)$, $0 < \alpha \leq 1$, and $r \geq 1$. Then

$$\left[\int_o^{2\pi} |\Psi(x,t)|^r dx\right]^{\frac{1}{r}} = \mathcal{O}(|t^\alpha|).$$

4.3.4 Lemma [31]

Let $h \in Lip(\xi, r)$, and $r \geq 1$. Then

$$\left[\int_o^{2\pi} |\Psi(x,t)|^r dx\right]^{\frac{1}{r}} = \mathcal{O}(\xi(t)).$$

4.4 Main Results

We prove firstly the following main result.

4.4.1 Theorem

Let (N, p_m, q_m) be a generalized regular Nörlund method of summability defined by two positive generating sequences (p_m) and (q_m). Let h be as defined in Theorem 4.2.1. If either (4.4) or (4.5) holds true, then the function $\tilde{h}$ is approximated by $(C, 1)(N, p_m, q_m)(E, \theta)$ transform

$$\widetilde{t_n^{CNE}(x)} = (n+1)^{-1} \sum_{m=0}^{n} \frac{1}{R_m} \sum_{k=0}^{m} \frac{p_k q_{m-k}}{(1+\theta)^k} \sum_{v=0}^{k} \binom{k}{v} \theta^{k-v} \tilde{s}_v(x)$$

of the series (4.2) is given by the following estimation:

$$\left\|\widetilde{t_n^{CNE}(x)} - \tilde{h}\right\|_{L^r} = \mathcal{O}_\theta((n+1)^{-\alpha-1}), \;\; 0 < \alpha \leq 1,$$

for all $n \in \{0, 1, 2, \ldots\}$ and $\theta \neq 3$.

Proof For $\tilde{s}_n(h, x)$ defined in (4.3),

$$\tilde{s}_n(h,x) - \tilde{h}(x) = \frac{1}{2\pi} \int_0^{\pi} \Psi(x,t) \frac{\cos(v+\frac{1}{2})t}{\sin\frac{t}{2}} dt.$$

Therefore, the (E, θ) means of $\tilde{s}_n(h, x) - \tilde{h}(x)$ have this form

$$\widetilde{E_k^{\theta}}(x) - \tilde{h}(x) = \frac{1}{2\pi(1+\theta)^k} \int_0^{\pi} \frac{\Psi(x,t)}{\sin\frac{t}{2}} \sum_{v=0}^{k} \binom{k}{v} \theta^{k-v} \cos\left(v+\frac{1}{2}\right) t dt,$$

which can be rewritten as follows:

$$\begin{aligned}
\widetilde{E_k^{\theta}}(x) - \tilde{h}(x) &= \frac{1}{2\pi(1+\theta)^k}\int_0^{\pi}\frac{\Psi(x,t)}{\sin\frac{t}{2}}\mathbb{R}\left\{e^{\frac{it}{2}}\sum_{v=0}^{k}\binom{k}{v}\theta^{k-v}e^{ivt}\right\}dt\\
&= \frac{1}{2\pi(1+\theta)^k}\int_0^{\pi}\frac{\Psi(x,t)}{\sin\frac{t}{2}}\mathbb{R}\left\{e^{\frac{it}{2}}(\theta+e^{it})^k\right\}dt\\
&= \frac{1}{2\pi(1+\theta)^k}\int_0^{\pi}\frac{\Psi(x,t)}{\sin\frac{t}{2}}\mathbb{R}\left\{e^{\frac{i(k+1)t}{2}}\left(\theta e^{-\frac{it}{2}}+e^{\frac{it}{2}}\right)^k\right\}dt\\
&= \frac{1}{2\pi(1+\theta)^k}\int_0^{\pi}\frac{\Psi(x,t)}{\sin\frac{t}{2}}\mathbb{R}\left\{e^{\frac{i(k+1)t}{2}}\left[(\theta+1)\cos\frac{t}{2}-i(\theta-1)\sin\frac{t}{2}\right]^k\right\}dt.
\end{aligned}$$

Denoting

$$\mathcal{R}(t)\cos\varphi(t;\theta) = (\theta+1)\cos\frac{t}{2} \quad \textit{and} \quad \mathcal{R}(t)\sin\varphi(t;\theta) = (\theta-1)\sin\frac{t}{2},$$

we obtain that

$$\mathcal{R}(t) = \sqrt{(\theta+1)^2 - 4\theta\sin^2\frac{t}{2}},$$

and

$$\varphi(t;\theta) = \tan^{-1}\left(\frac{\theta-1}{\theta+1}\tan\frac{t}{2}\right).$$

So, we get

$$\begin{aligned}
\widetilde{E_k^{\theta}}(x) - \tilde{h}(x) &= \frac{1}{2\pi(1+\theta)^k}\int_0^{\pi}\frac{\Psi(x,t)}{\sin\frac{t}{2}}\mathbb{R}\left\{e^{\frac{i(k+1)t}{2}}R^k(t)e^{-i\varphi(t;\theta)k}\right\}dt\\
&= \frac{1}{2\pi(1+\theta)^k}\int_0^{\pi}\frac{\Psi(x,t)}{\sin\frac{t}{2}}\mathcal{R}^k(t)\cos\left[\frac{(k+1)t}{2} - k\tan^{-1}\left(\frac{\theta-1}{\theta+1}\tan\frac{t}{2}\right)\right]dt.
\end{aligned}$$

Also, the (N, p_m, q_m) means of $\widetilde{E_k^{\theta}}(x) - \tilde{f}(x)$ have this form

$$\widetilde{NE_k^{\theta}}(x) - \tilde{h}(x) = \frac{1}{R_m}\sum_{k=0}^{m}p_k q_{m-k}\left\{\widetilde{E_k^{\theta}}(x) - \tilde{h}(x)\right\},$$

i.e.,

$$\begin{aligned}
\widetilde{NE_k^{\theta}}(x) - \tilde{h}(x) = &\frac{1}{2\pi R_m}\sum_{k=0}^{m}\frac{p_k q_{m-k}}{(1+\theta)^k}\int_0^{\pi}\frac{\Psi(x,t)}{\sin\frac{t}{2}}R^k(t)\\
&\cos\left[\frac{(k+1)t}{2} - k\tan^{-1}\left(\frac{\theta-1}{\theta+1}\tan\frac{t}{2}\right)\right]dt.
\end{aligned}$$

Consequently, the $(C, 1)$ means of $\widetilde{NE_k^\theta}(x) - \tilde{h}(x)$ are

$$\begin{aligned}
&\frac{1}{n+1}\sum_{m=0}^{n}(\widetilde{NE_k^\theta}(x) - \tilde{h}(x)) \\
&= \frac{1}{2\pi(n+1)}\sum_{m=0}^{n}\frac{1}{R_m}\sum_{k=0}^{m}\frac{p_k q_{m-k}}{(1+\theta)^k}\int_0^\pi \frac{\Psi(x,t)}{\sin\frac{t}{2}}R^k(t) \\
&\cos\left[\frac{(k+1)t}{2} - k\tan^{-1}\left(\frac{\theta-1}{\theta+1}\tan\frac{t}{2}\right)\right]dt
\end{aligned}$$

or

$$\widetilde{t_n^{CNE}(x)} - \tilde{h}(x) = \int_0^\pi \Psi(x,t)\widetilde{K_n^{CNE}(t)}dt.$$

Now, using Lemma 4.3.3 and Minkowski's inequality ([1], p. 18),

$$\begin{aligned}
\left\|\widetilde{t_n^{CNE}(x)} - \tilde{h}(x)\right\|_{L^r} &= \int_0^\pi \left[\int_0^{2\pi}|\Psi(x,t)|^r dx\right]^{1/r}\left|\widetilde{K_n^{CNE}(t)}\right|dt \\
&= \int_0^{\frac{\pi}{n+1}} \mathcal{O}(t^\alpha)\left|\widetilde{K_n^{CNE}(t)}\right|dt + \int_{\frac{\pi}{n+1}}^{\pi} \mathcal{O}(t^\alpha)\left|\widetilde{K_n^{CNE}(t)}\right|dt \\
&= J_1 + J_2,
\end{aligned}$$

since by Lemma 4.3.3 the implication $h \in Lip(\alpha) \implies \Psi \in Lip(\alpha)$ holds true.

Based on Lemma 4.3.1,

$$\begin{aligned}
J_1 &= \int_0^{\frac{\pi}{n+1}} \mathcal{O}(t^\alpha)\mathcal{O}\left(\frac{1}{(n+1)t}\right)dt \\
&= \mathcal{O}((n+1)^{-\alpha-1}).
\end{aligned}$$

Using Lemma 4.3.2, we have

$$\begin{aligned}
J_2 &= \int_{\frac{\pi}{n+1}}^{\pi} \mathcal{O}(t^\alpha)\mathcal{O}_\theta\left(\frac{1}{(n+1)^3 t^3}\right)dt \\
&= \mathcal{O}_\theta\left(\frac{1}{(n+1)^3}\right)\left.\frac{t^{\alpha-2}}{\alpha-2}\right|_{\frac{\pi}{n+1}}^{\pi} \\
&= \mathcal{O}_\theta\left(\frac{1}{(n+1)^{\alpha+1}}\right).
\end{aligned}$$

Therefore, we have

$$\left\|\widetilde{t_n^{CNE}(x)} - \tilde{h}(x)\right\|_{L^r} = \mathcal{O}_\theta\left(\frac{1}{(n+1)^{\alpha+1}}\right), \quad \theta \neq 3, \ 0 < \alpha \leq 1.$$

4.4.2 Theorem

Let (N, p_m, q_m) be as defined in Theorem 4.4.1 satisfying conditions (4.4) or (4.5). Assume that $\xi(t)$ is a modulus of continuity such that

$$\int_0^{v} \frac{\xi(t)}{t} dt = \mathcal{O}((\xi)v), \quad 0 < v < \pi. \tag{4.7}$$

Let h be as defined in Theorem 4.2.2. Then the function $\tilde{h}$ is approximated by $\widetilde{t_n^{CNE}}$ means of $\tilde{s}_n(h, x)$ which is given by

$$\left\| \widetilde{t_n^{CNE}} - \tilde{h} \right\|_{L^r} = \mathcal{O}_\theta \left(\frac{1}{(n+1)^3} \int_{\frac{\pi}{n+1}}^{\pi} \frac{\xi(t)}{t^3} dt \right)$$

for all $n \in \{0, 1, 2, \dots\}$ and $\theta \neq 3$.

Proof Using Lemma 4.3.4, we have

$$\begin{aligned} \left\| \widetilde{t_n^{CNE}}(x) - \tilde{h}(x) \right\|_{L^r} &= \int_0^{\frac{\pi}{n+1}} \mathcal{O}(\xi(t)) \left| \widetilde{K_n^{CNE}}(t) \right| dt + \int_{\frac{\pi}{n+1}}^{\pi} \mathcal{O}(\xi(t)) \left| \widetilde{K_n^{CNE}}(t) \right| dt \\ &= J_1^{(1)} + J_2^{(1)}. \end{aligned}$$

Now applying Lemma 4.3.1 and condition (4.7),

$$\begin{aligned} J_1^{(1)} &= \mathcal{O}\left(\int_0^{\frac{\pi}{n+1}} \frac{\xi(t)}{(n+1)t} dt \right) \\ &= \mathcal{O}\left(\xi \left(\frac{\pi}{(n+1)^2} \right) \right) \\ &= \mathcal{O}\left(\frac{1}{(n+1)^3} \int_{\frac{\pi}{n+1}}^{\pi} \frac{\xi(t)}{t^3} dt \right), \end{aligned}$$

since

$$\begin{aligned} \frac{1}{(n+1)^3} \int_{\frac{\pi}{n+1}}^{\pi} \frac{\xi(t)}{t^3} dt &\geq \frac{\xi\left(\frac{\pi}{n+1}\right)}{(n+1)^3} \int_{\frac{\pi}{n+1}}^{\pi} \frac{dt}{t^3} \\ &\geq \frac{\xi\left(\frac{\pi}{n+1}\right)}{(n+1)^3} \left(\frac{1}{-2t^2} \right) \Bigg|_{\frac{\pi}{n+1}}^{\pi} \\ &\geq \frac{1}{4\pi^2} \xi \left(\frac{\pi}{(n+1)^2} \right). \end{aligned}$$

On the other hand, Lemma 4.3.2 gives

$$J_2^{(1)} = \mathcal{O}_\theta \left(\frac{1}{(n+1)^3} \int_{\frac{\pi}{n+1}}^{\pi} \frac{\xi(t)}{t^3} dt \right).$$

Consequently, we clearly have

$$\left\| \widetilde{t_n^{CNE}(x)} - \tilde{h}(x) \right\|_{L^r} = \mathcal{O}_\theta \left(\frac{1}{(n+1)^3} \int_{\frac{\pi}{n+1}}^{\pi} \frac{\xi(t)}{t^3} dt \right),$$

which proves the statement of the theorem.

4.4.3 *Theorem*

Let (N, p_m, q_m) be as defined in Theorem 4.4.1 satisfying conditions (4.4) or (4.5) and (4.6). Let h be as defined in Theorem 4.2.3, and let $\frac{\xi(t)}{t}$ be monotone decreasing in $(\pi/(n+1), \pi)$. Then the function $\tilde{h}$ is approximated by $\widetilde{t_n^{CNE}}$ means of $\tilde{s}_n(h, x)$ which is described by

$$\left\| \widetilde{t_n^{CNE}} - \tilde{h} \right\|_{L^r} = \mathcal{O}_\theta \left(\xi \left(\frac{\pi}{(n+1)^2} \right) \right)$$

for all $n \in \{0, 1, 2, \dots\}$ and $\theta \neq 3$.

Proof Using the conditions of the theorem and the prior theorem's proof, we have

$$\begin{aligned} \left\| \widetilde{t_n^{CNE}} - \tilde{h} \right\|_{L^r} &= \mathcal{O}_\theta \left(\frac{1}{(n+1)^3} \int_{\frac{\pi}{n+1}}^{\pi} \frac{\xi(t)}{t^3} dt \right) \\ &= \mathcal{O}_\theta \left(\frac{1}{(n+1)^3} \frac{\xi\left(\frac{\pi}{n+1}\right)}{\frac{\pi}{n+1}} \int_{\frac{\pi}{n+1}}^{\pi} \frac{dt}{t^2} \right) \\ &= \mathcal{O}_\theta \left(\xi \left(\frac{\pi}{(n+1)^2} \right) \right). \end{aligned}$$

4.5 Corollaries

The main results can be used to deduce a number of previously known results. Let's look at a few of them now.

4.5.1 Corollary

In Theorem 4.4.1, if $\theta = 1$, $p_m = 1 \; \forall m$, and $n = 1$, then the function $\tilde{h}$ is approximated by the $(N, q_m)(E, 1)$ transform

$$\widetilde{t_m^{NE}}(x) = \frac{1}{Q_m} \sum_{k=0}^{m} \frac{q_{m-k}}{2^k} \sum_{v=0}^{k} \binom{k}{v} \tilde{s_v}(x)$$

which is given by

$$\left\| \widetilde{t_m^{NE}} - \tilde{h} \right\|_{L^r} = \mathcal{O}\left((m+1)^{-\alpha}\right), \quad 0 < \alpha \leq 1.$$

4.5.2 Corollary

In Theorem 4.4.2, if $\theta = 1$, $p_m = 1 \; \forall m$, and $n = 1$, then the approximation of the function $\tilde{h}$ by the $(N, q_m)(E, 1)$ transform is given by

$$\left\| \widetilde{t_m^{NE}} - \tilde{h} \right\|_{L^r} = \mathcal{O}\left(\frac{1}{(m+1)^2} \int_{\frac{\pi}{m+1}}^{\pi} \frac{\xi(t)}{t^3} dt \right).$$

4.5.3 Corollary

In Theorem 4.4.3, if $\theta = 1$, $p_m = 1 \; \forall m$, and $n = 1$, then the approximation of the function $\tilde{h}$ by the $(N, q_m)(E, 1)$ transform is given by

$$\left\| \widetilde{t_m^{NE}} - \tilde{h} \right\|_{L^r} = \mathcal{O}\left(\xi\left(\frac{\pi}{m+1} \right) \right).$$

4.6 Conclusion

This paper focuses on the approximation of signal (function) belonging to various classes by $(C, 1)(N, p_m, q_m)(E, \theta)$ means of conjugate Fourier series to determine the degree of approximation (under some special conditions on the sequences $\{p_m\}$ and $\{q_m\}$) of periodic functions belonging to generalized Lipschitz classes. We deduced several new corollaries from these results, which are more general than those established earlier by others.

4.7 Applications and Uses

The approximation theory is a field of great practical significance. Analysis of periodic functions is important because of its applications in the engineering fields like digital communication, digital signal processing, mechanical engineering, etc. These functions are derived using a polynomial approximation function and a Fourier truncated series. The closest estimate can be carried out using either a Fourier approximation or a polynomial approximation. The polynomial approximation of the function is done using Taylor series expansion, and the quality of the approximation is dependent on the number of terms utilized. Naturally, a function must be infinite times differentiable in some interval in order to have a Taylor series, which is a fairly strict requirement. Nevertheless, sine and cosine functions are used in Fourier approximation and act as far more adaptable components than powers of any variable. Sines and cosines are useful in approximating non-analytical functions as well as wildly discontinuous ones. Due to their numerous uses, Fourier approximation has taken on significant new dimensions in signal analysis.

Error of Approximation The term "error estimation of signals using summability techniques" refers to the computation of the error function $E_n(f)$, defined. If $E_n(f) > 0$, the approximation is greater than the actual value. The approximation is too large if $E_n(f) < 0$. Most of the time, the size of the error |E| is all that we are interested in (or can detect). We can lessen the inaccuracy by using an approximation value that is close to the exact value.

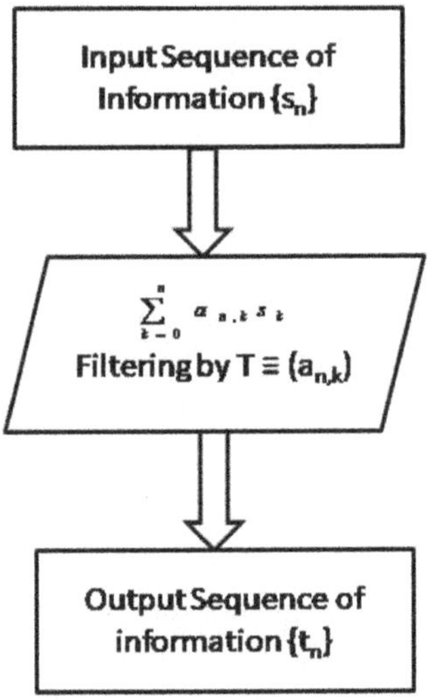

Digital filters like FIR filters and IIR filters, which have various uses including voice and picture processing, radar, sonar, and medical signal processing, frequently use signal approximation. Several engineering problems benefit greatly from approximation techniques based on Fourier coefficients and various summability techniques.

So, it will help us to find the least conditions to solve these problems with minimized error and more accuracy by using different summability product means. So that the output data is filtered according to our requirements of the modeled problem.

In summary, the aim of all these measures is to minimize approximation errors and improve accuracy. The more the authors reduce the error, the stronger the results will be. The concept of product summarizability is very useful.

Acknowledgements The authors offer their true thanks to the Science and Engineering Research Board for giving financial support through Project No.: EEQ/2018/000393.

References

1. Zygmund, A.: Trigonometric Series, 2nd edn. Cambridge University Press, London-New York (1968)
2. Zygmund, A., Stegun, I.A. (eds.): Trigonometric Series, 2nd edn. Cambridge University Press (1959)
3. Singh, A.N.: Nörlund summability of Fourier series and its conjugate series. Bull. Call. Math. Soc. **82**, 99–105 (1990)
4. Jadiya, B.L.: On Nörlund summability of conjugate Fourier series. Indian J. Pure Appl. Math. **13**(11), 1354–1359 (1982)
5. Borwein, D.: On product of sequences. J. Lond. Math. Soc. **33**, 352–357 (1958)
6. Sonker, S., Jena, B.B., Jindal, R., Paikray, S.K.: A generalized theorem on double absolute factorable matrix summability. Appl. Math. Inf. Sci. **16**(2), 315–322 (2022)
7. Hardy, G.H.: Divergent Series. Oxford University Press (1949)
8. Pandey, G.S.: On Nörlund summability of Fourier series. Indian J. Pure Appl. Math. **8**, 412–417 (1977)
9. Kuang, JCh.: Applied Inequalities, Shandong Science and Technology Press, 3rd edn. Ji'nan City, Shandong Province, China (2004)
10. McFadden, L.: Absolute Nörlund summability. Duke Math., J. **9**, 168–207 (1942)
11. Mishra, L.N., Mishra, V.N., Khatri, K., Deepmala, D.: On the trigonometric approximation of signals belonging to generalized weighted Lipschitz $W(L^r, \xi(t))(r \geq 1)$- class by matrix $(C^1.N_p)$ operator of conjugate series of its Fourier series. Appl. Math. Comput. **237**, 252–263 (2014)
12. Sonker, S., Jindal, R.: Approximation of signals by the triple product summability means of the Fourier series. Soft Comput.: Theor. Appl. **425**, 169–179 (2022)
13. Mittal, M.L., Rhoades, B.E., Mishra, V.N.: Approximations of signals (functions) belonging to the weighted $W(L_p, \xi(t)), (p \geq 1)$ class by linear operators. Int. J. Math. Math. Sci., ID **5353**, 1–10 (2006)
14. Mittal, M.L., Kumar, R.: Matrix summability of Fourier series and its conjugate series. Bull. Call. Math. Soc. **82**, 362–368 (1990)
15. Mittal, M.L., Prasad, G.: On a sequence of Fourier coefficients. Indian J. Pure Appl. Math. **23**(3), 235–241 (1992)
16. Mursaleen, M., Alotaibi, A.: Generalized matrix summability of a conjugate derived Fourier series. J. Ineq. Appl. **2017**, 273 (2017)
17. Mittal, M.L., Rhoades, B.E., Sonker, S., Singh, U.: Approximation of signals of class Lip (α, p) by linear operators. Appl. Math. Comput. **217**(9), 4483–4489 (2011)
18. Khare, S.P.: Generalized Nörlund summability of Fourier series and its conjugate series. Indian J. Pure Appl. Math. **21**(5), 457–467 (1990)

19. Sonker, S.: Approximation of Functions by means of its Fourier-Laguerre series. Proc. ICMS-2014 **1**, 125–128 (2014)
20. Sonker, S., Singh, U.: Degree of approximation of the conjugate of signals (functions) belonging to Lip(α, r)-class by (C, 1)(E, q) means of conjugate trigonometric Fourier series. J. Inequalities Appl. **2012**(1), 1–7 (2012)
21. Sonker, S., Sangwan, P.: Approximation of Fourier and its conjugate series by triple Euler product summability. J. Phys.: Conf. Ser. (2021)
22. Sonker, S., Sangwan, P.: Approximation of signals by Harmonic-Euler triple product means. J. Indian Math. Soc. **88**(1–2), 176–186 (2021)
23. Sonker, S., Jena, B.B., Jindal, R., Paikray, S.K.: A generalized theorem on double absolute factorable matrix summability. Appl. Math. Inf. Sci. **16**(2), 315–322 (2022)
24. Sonker, S., Jindal, R.: Approximation of signals by the triple product summability means of the Fourier series. Soft Comput.: Theor. Appl. **425**, 169–179 (2022)
25. Sonker, S., Jindal, R.: On approximation of signals belonging to some classes by $(N, p_m, q_m)(E, \theta)(E, \theta)$ means of conjugate series of its Fourier series. J. Anal. 1–18 (2022)
26. Verma, S., Saxena, K.: A study on $(H, 1)(E, q)$ product summability of Fourier series and its conjugate series. Math. Theory Model. **7**(5) (2017)
27. Mishra, V.N., Khan, H.H., Khatri, K., Mishra, L.N.: Degree of approximation of conjugate of signals (functions) belonging to the generalized weighted Lipschitz $W(L^r, \xi(t))(r \geq 1)$-class by $(C, 1)(E, q)$ means of conjugate trigonometric Fourier series. Bull. Math. Anal. Appl. **5**(4), 40–53 (2013)
28. Mishra, V.N., Khatri, K., Mishra, L.N.: Approximation of functions belonging to $Lip(\xi, (t), r)$ class by $(N, p_n)(E, q)$ summability of conjugate series of Fourier series. J. Inequal. Appl. **2012**(296), 10 (2012)
29. Singh, U.N., Singh, V.S.: Almost Nörlund summability of Fourier series and its conjugate series. Bull. Call. Math. Soc. **87**, 57–62 (1995)
30. Mishra, V.N., Mishra, L.N.: Trigonometric approximation of signals (functions) in L_p-norm. Int. J. Contemp. Math. Sci. **7**(19), 909–918 (2012)
31. Mishra, V.N., Sonavane, V.: Approximation of functions of Lipschitz class by $(N, p_n)(E, 1)$ summability means of conjugate series of Fourier series. J. Class. Anal. **6**(2), 137–151 (2015)
32. Mishra, V.N., Khatri, K., Mishra, L.N.: Using linear operators to approximate signals of $Lip(\alpha, p)$, $(p \geq 1)$-class. Filomat **27**(2), 353–363 (2013). https://doi.org/10.2298/FIL1302353M
33. Krasniqi, X.Z.: Applications of some classes of sequences on approximation of functions (signals) by almost generalized Nörlund means of their Fourier series. Int. J. Anal. Appl. **9**(1), 45–53 (2015)
34. Krasniqi, X.Z.: On the degree of approximation of a function by $(C, 1)(E, q)$ means of its Fourier-Laguerre series. Int. J. Anal. Appl. **1**(1), 33–39 (2013)
35. Krasniqi, X.Z., Deepmala, D.: On approximation of functions belonging to some classes of functions by $(N, p_n, q_n)(E, \theta)$ means of conjugate series of its Fourier series. Khayyam J. Math. (KJM) **6**(1), 73–86 (2020)

Chapter 5
Turán Type Inequalities For the (p, k)−Generalization of the Mittag-Leffler Function

Asena Çetinkaya, Şahsene Altınkaya, and Oya Mert

Abstract The Turán type inequalities attract attention in a variety of research fields such as orthogonal polynomials and special functions. Using the newly established (p, k)-generalization of the Mittag-Leffler function, we derive new Turán type inequalities. These outcomes are the generalization of several findings in the literature.

Keywords Turán type inequalities · (p, k)-Gamma function · (p, k)-Mittag-Leffler function · (p, k)-Pochhammer symbol · Legendre polynomial

2010 AMS Mathematics Subject Classification 33E12 · 26D07

5.1 Introduction

The Mittag-Leffler function acts a crucial role in many branches of engineering sciences and applied mathematics. The Mittag-Leffler function appears as the solution of fractional order differential or fractional order integral equations. In 1930, Hille and Tamarkin proved a prominent formula for the solution of Riemann-Liouville fractional integral which is the most known result in this field. More properties and applications of the Mittag-Leffler function to fractional calculus can be found in [19, 38]. Due to its applications in applied mathematics, the Mittag-Leffler function has recently taken the interest of the scientific community. Numerous generalizations,

A. Çetinkaya
Department of Mathematics and Computer Science, Istanbul Kültür University, Istanbul, Türkiye
e-mail: asnfigen@hotmail.com

Ş. Altınkaya (✉)
Department of Mathematics, Faculty of Arts and Science, Istanbul Beykent University, 34500 Istanbul, Türkiye
e-mail: sahsenealtinkaya@beykent.edu.tr

O. Mert
Department of Mathematics, Tekirdağ Namık Kemal University, Tekirdağ, Türkiye
e-mail: oyamert@nku.edu.tr

B. C. Tripathy et al. (eds.), *Operators, Inequalities and Approximation*, Industrial and Applied Mathematics, https://doi.org/10.1007/978-981-97-3238-8_5

properties, and applications of the Mittag-Leffler function were examined by Srivatava and Tomovski [63], Gorenflo and Mainardi [30] (see also [28, 29, 37] and references therein).

The Pochhammer $k-$symbol and the $k-$gamma function were introduced by Diaz and Pariguan [18]. They proved new identities generalizing those fulfilled by the classical Pochhammer symbol and the gamma function. Later, Mansour [46] determined the $k-$gamma function by a combination of some functional equations. In 2012, Dorrego and Cerutti [20] defined the $k-$Mittag-Leffler function, proved its properties and evaluated its Laplace transform. In the same year, Gehlot [25] defined the generalized $k-$Mittag-Leffler function. For more studies on these functions, one may refer to [15, 44].

In 2017, Gehlot [26] defined the (p, k)-Pochhammer symbol and the (p, k)-gamma function, respectively. He derived the limit formula, recurrence formula, integral representation, and infinite products for two-parameter gamma function. These generalizations led Gehlot [24] to define the $(p, k)-$Mittag-Leffler function which consists of the previous generalizations of the Mittag-Leffler as special cases.

In 1941, while investigating the zeros of Legendre polynomials, Turán type inequality was expressed by the Hungarian mathematician Paul Turán [68]. Although this inequality was solved by Turán in a letter to Szegö in 1946, he published it in 1950. In 1948, Szegö [66] expressed four different proofs of the Turán type inequalities and expanded this outcome to Gegenbauer, Laguerre, and Hermite polynomials. One of the studies on Legendre polynomials for Turán type inequalities was done by Alzer et al. [2]. They were interested in a notable result published by Constantinescu [16], which proposed a new progress and a reverse of Turán type inequality. There is an extensive literature available today related to the Turán type inequalities and their applications in several special functions and classical orthogonal polynomials [12, 67].

The Turan type inequalities were applied to many functions but it is fairly new for implementing the Mittag-Leffler function. In [49, 50], Mehrez and Sitnik proved new Turán type inequalities for series of the Mittag-Leffler function by using monotonicity for the appropriate ratio of sections. In 2018, Mehrez [53] derived these inequalities for the functions named as $q-$Mittag-Leffler and $q-$Wright functions. He attempted to get some of the most important conclusions by employing methods developed for the classical Mittag-Leffler and the Wright functions. Moreover, these inequalities were considered for the generalized Mittag-Leffler functions in [3, 21, 22].

This chapter is organized as follows. In Sect. 5.2, we give properties of the (p, k)-Pochhammer symbol and the (p, k)-gamma function. Section 5.3 is dedicated to the Mittag-Leffler function and its generalizations. In Sect. 5.4, we give some literature associated with the Turán type inequalities and emphasize some studies of such inequalities for several Mittag-Leffler functions. In Sect. 5.5, we collect some lemmas and deduce new Turán type inequalities for the $(p, k)-$Mittag-Leffler function. At the end of the paper, a conclusion section is included.

5.2 Pochhammer Symbol and Gamma Function with Two-Parameter

The section presents the (p, k)-Pochhammer symbol and the (p, k)-gamma function, respectively.

5.2.1 Two-Parameter Pochhammer Symbol

Gehlot in [26, Definition 2.1] defined the (p, k)-Pochhammer symbol (or two-parameter Pochhammer symbol) as below.

Definition 5.1 Let $p, k \in \mathbb{R}^{+}\backslash\{0\}$, $\varkappa \in \mathbb{C}$, $\mathrm{Re}(\varkappa) > 0$, $\ell \in \mathbb{N}$. The (p, k)-Pochhammer symbol ${}_{p}(\varkappa)_{\ell,k}$ is expressed by

$$ {}_{p}(\varkappa)_{\ell,k} = \left(\frac{\varkappa p}{k}\right)\left(\frac{\varkappa p}{k} + p\right)\left(\frac{\varkappa p}{k} + 2p\right)\ldots\left(\frac{\varkappa p}{k} + (\ell - 1)\,p\right). \qquad (5.1) $$

Gehlot also derived the (p, k)-Pochhammer symbol in terms of the symmetric function, and evaluated its derivative properties.

In [35], the generalized (p, k)-Pochhammer symbol is given by

$$ {}_{p}(\varkappa)_{\ell q,k} = \begin{cases} \frac{{}_{p}\Gamma_{k}(\varkappa+\ell q k)}{{}_{p}\Gamma_{k}(\varkappa)}, & (p, k \in \mathbb{R}^{+};\ \varkappa \in \mathbb{C}) \\ \left(\frac{\varkappa p}{k}\right)\left(\frac{\varkappa p}{k} + p\right)\ldots\left(\frac{\varkappa p}{k} + (\ell q - 1)\,p\right), & (\ell, q \in \mathbb{N};\ \varkappa \in \mathbb{C}) \end{cases}. \qquad (5.2) $$

The generalized (p, k)-Pochhammer symbol is the deformation of the usual Pochhammer symbol. From (5.2), we observe the following:

1. For $p = k, q = 1$, we observe that ${}_{k}(\varkappa)_{\ell q,k} = (\varkappa)_{\ell,k}$ is the Pochhammer $k-$symbol expressed by Diaz and Pariguan [18] as

$$ (\varkappa)_{\ell,k} = \varkappa(\varkappa + k)(\varkappa + 2k)\ldots(\varkappa + (\ell - 1)k), $$

where $k \in \mathbb{R}$, $\varkappa \in \mathbb{C}$, $\ell \in \mathbb{N}^{+}$.

2. For $p = k = 1$, we observe that ${}_{1}(\varkappa)_{\ell q,1} = (\varkappa)_{\ell q}$ is the generalized Pochhammer symbol expressed by Shukla and Prajabati [59] as

$$ (\varkappa)_{\ell q} = \frac{\Gamma(\varkappa + \ell q)}{\Gamma(\varkappa)} = q^{\ell q}\prod_{r=1}^{q}\left(\frac{\varkappa + r - 1}{q}\right)_{\ell}, \quad (q \in \mathbb{N}). \qquad (5.3) $$

3. For $p = k = 1, q = 1$, we observe that ${}_{1}(\varkappa)_{\ell,1} = (\varkappa)_{\ell}$ is the classical Pochhammer symbol [58] expressed by

$$(\varkappa)_\ell = \frac{\Gamma(\varkappa+\ell)}{\Gamma(\varkappa)} = \begin{cases} 1 & (\ell = 0;\ \varkappa \in \mathbb{C}\backslash\{0\}) \\ \varkappa(\varkappa+1)\ldots(\varkappa+\ell-1) & (\ell \in \mathbb{N};\ \varkappa \in \mathbb{C}) \end{cases}.$$

The following theorem gives a relation between the (p,k)-Pochhammer symbol and it's special cases.

Theorem 5.1 [26, Theorem 2.10] *Let* $p, k \in \mathbb{R}^+\backslash\{0\},\ \varkappa \in \mathbb{C}\backslash k\mathbb{Z}^-,\ \mathrm{Re}(\varkappa) > 0,\ \ell, q \in \mathbb{N}$. *Then, we have*

$$_p(\varkappa)_{\ell,k} = (\frac{p}{k})^\ell (\varkappa)_{\ell,k} = (p)^\ell (\frac{\varkappa}{k})_\ell.$$

For the generalized (p,k)-Pochhammer symbol, we arrive

$$_p(\varkappa)_{\ell q,k} = (\frac{p}{k})^{\ell q} (\varkappa)_{\ell q,k} = (p)^{\ell q} (\frac{\varkappa}{k})_{\ell q} = (pq)^{\ell q} \prod_{r=1}^{q} (\frac{\frac{\varkappa}{k}+r-1}{q})_\ell.$$

More details on the generalized two-parameter Pochhammer symbol can be found in [15] and references therein.

5.2.2 Two-Parameter Gamma Function

The (p,k)-gamma function (or two-parameter gamma function) is closely related to the (p,k)-Pochhammer symbol. Based on the works in [18], Gehlot [26, Definition 2.2] defined the (p,k)-gamma function as below.

Definition 5.2 Let $p, k \in \mathbb{R}^+\backslash\{0\},\ \varkappa \in \mathbb{C}\backslash k\mathbb{Z}^-,\ \mathrm{Re}(\varkappa) > 0,\ \ell \in \mathbb{N}$. The (p,k)-gamma function $_p\Gamma_k$ is defined as

$$_p\Gamma_k(\varkappa) = \frac{1}{k} \lim_{\ell\to\infty} \frac{\ell!\, p^{\ell+1} (\ell p)^{\frac{\varkappa}{k}}}{_p(\varkappa)_{\ell+1,k}}, \tag{5.4}$$

or

$$_p\Gamma_k(\varkappa) = \frac{1}{k} \lim_{\ell\to\infty} \frac{\ell!\, p^{\ell+1} (\ell p)^{\frac{\varkappa}{k}-1}}{_p(\varkappa)_{\ell,k}}. \tag{5.5}$$

In similar fashion, the integral representation of the (p,k)-gamma function is defined as

$$_p\Gamma_k(\varkappa) = \int_0^\infty e^{-\frac{t^k}{p}} t^{\varkappa-1} dt$$

for $p, k \in \mathbb{R}^+\backslash\{0\},\ \varkappa \in \mathbb{C}\backslash k\mathbb{Z}^-,\ \mathrm{Re}(\varkappa) > 0$.

The (p, k)-gamma function is the deformation of the usual gamma function. From Eq. (5.5), we conclude the following results.

1. For $p = k$, we observe that ${}_k\Gamma_k(\varkappa) = \Gamma_k(\varkappa)$ is the $k-$gamma function expressed by Diaz and Pariguan [18] as

$$\Gamma_k(\varkappa) = \lim_{\ell\to\infty} \frac{\ell! k^\ell (\ell k)^{\frac{\varkappa}{k}-1}}{(\varkappa)_{\ell,k}},$$

where $k > 0$, $\varkappa \in \mathbb{C}\backslash k\mathbb{Z}^-$.
2. For $p = k = 1$, we observe that ${}_1\Gamma_1(\varkappa) = \Gamma(\varkappa)$ is the Euler's gamma function.

The connection between the (p, k)-gamma function and its special cases is presented by the next theorem.

Theorem 5.2 [26, Theorem 2.9] *Let* $p, k \in \mathbb{R}^+ \backslash \{0\}$, $\varkappa \in \mathbb{C}\backslash k\mathbb{Z}^-$, $\operatorname{Re}(\varkappa) > 0$. *Then, we have*

$$ {}_p\Gamma_k(\varkappa) = (\frac{p}{k})^{\frac{\varkappa}{k}} \Gamma_k(\varkappa) = \frac{p^{\frac{\varkappa}{k}}}{k} \Gamma(\frac{\varkappa}{k}).$$

For comprehensive details, one may also refer to [27, 42] and references given therein.

5.3 Generalizations of the Mittag-Leffler Function

The section deals with the generalizations of the Mittag-Leffler function. We here present the most known Mittag-Leffler functions in chronological order. Next, we give the (p, k)–Mittag-Leffler function and its special cases.

5.3.1 *The Mittag-Leffler Function*

In 1903, the Mittag-Leffler function $E_\alpha(z)$ defined by

$$E_\alpha(z) = \sum_{\ell=0}^{\infty} \frac{z^\ell}{\Gamma(\alpha\ell + 1)}, \quad (\alpha, z \in \mathbb{C}; \operatorname{Re}(\alpha) > 0) \tag{5.6}$$

was introduced by a Swedish mathematician Gosta Mittag-Leffler [55]. The Mittag-Leffler function is an entire function of z with order $[\operatorname{Re}(\alpha)]^{-1}$ and is a generalization of the exponential function to which it reduces for $\alpha = 1$. It interpolates between pure exponential and a hypergeometric function $\frac{z}{1-z}$ when $0 < \alpha < 1$.

In 1905, the Mittag-Leffler function with two parameters defined by

$$E_{\alpha,\beta}(z) = \sum_{\ell=0}^{\infty} \frac{z^\ell}{\Gamma(\alpha\ell+\beta)}, \quad (\alpha, \beta, z \in \mathbb{C};\ \mathrm{Re}(\alpha) > 0; \mathrm{Re}(\beta) > 0) \tag{5.7}$$

was expressed by Wiman [69].

In 1971, the Mittag-Leffler function with three parameters defined by

$$E^{\gamma}_{\alpha,\beta}(z) = \sum_{\ell=0}^{\infty} \frac{(\gamma)_\ell\, z^\ell}{\Gamma(\alpha\ell+\beta)\,\ell!}, \quad (\alpha, \beta, \gamma, z \in \mathbb{C};\ \mathrm{Re}(\alpha) > 0;\ \mathrm{Re}(\beta) > 0;\ \mathrm{Re}(\gamma) > 0) \tag{5.8}$$

was introduced by Prabhakar [57]. It is observed that $E^1_{\alpha,1}(z) =: E_\alpha(z)$ and $E^1_{\alpha,\beta}(z) =: E_{\alpha,\beta}(z)$.

Later on, Shukla and Prajabati [59] investigated the function $E^{\gamma,q}_{\alpha,\beta}(z)$ which is defined as

$$E^{\gamma,q}_{\alpha,\beta}(z) = \sum_{\ell=0}^{\infty} \frac{(\gamma)_{q\ell}\, z^\ell}{\Gamma(\alpha\ell+\beta)\,\ell!}, \tag{5.9}$$

where $\alpha, \beta, \gamma \in \mathbb{C},\ \mathrm{Re}(\alpha) > 0,\ \mathrm{Re}(\beta) > 0,\ \mathrm{Re}(\gamma) > 0,\ \ q \in (0, 1) \cup \mathbb{N}$ and $(\gamma)_{q\ell}$ is the generalized Pochhammer symbol defined by (5.3). This function is an entire function of order $[\mathrm{Re}\,\alpha]^{-1}$. The function $E^{\gamma,q}_{\alpha,\beta}(z)$ converges absolutely for all z when $q < \mathrm{Re}\,\alpha + 1$, and $|z| < 1$ when $q = \mathrm{Re}\,\alpha + 1$.

The function $E^{\gamma,q}_{\alpha,\beta}(z)$ is a generalization of the Mittag-Leffler function $E_\alpha(z)$, the Wiman's function $E_{\alpha,\beta}(z)$ and the function $E^{\gamma}_{\alpha,\beta}(z)$.

5.3.2 The (p, k)−Mittag-Leffler Function

Let $p, k \in \mathbb{R}^+\backslash\{0\}$, $\alpha, \beta, \gamma \in \mathbb{C}\backslash k\mathbb{Z}^-$ with $\mathrm{Re}(\alpha) > 0$, $\mathrm{Re}(\beta) > 0$, $\mathrm{Re}(\gamma) > 0$, $q \in (0, 1) \cup \mathbb{N}$, $\ell \in \mathbb{N}$. The (p, k)−Mittag-Leffler function defined as

$${}_pE^{\gamma,q}_{k,\alpha,\beta}(z) = \sum_{\ell=0}^{\infty} \frac{{}_p(\gamma)_{\ell q,k}\, z^\ell}{{}_p\Gamma_k(\alpha\ell+\beta)\ell!}, \tag{5.10}$$

where ${}_p\Gamma_k(\varkappa)$ is the two-parameter Gamma function expressed by (5.5) and ${}_p(\gamma)_{\ell q,k}$ is the generalized two-parameter Pochhammer symbol expressed by (5.2). This generalized function was defined and studied by Gehlot [24] in 2018.

The function ${}_pE^{\gamma,q}_{k,\alpha,\beta}(z)$ has the following special cases.

1. For $p = k$, the Eq. (5.10) yields the generalized k−Mittag-Leffler function defined as (see [25]):

$$ {}_kE^{\gamma,q}_{k,\alpha,\beta}(z) := E^{\gamma,q}_{k,\alpha,\beta}(z) = \sum_{\ell=0}^{\infty} \frac{(\gamma)_{\ell q,k}\, z^\ell}{\Gamma_k(\alpha\ell+\beta)\ell!}. $$

2. For $p = k, q = 1$, the Eq. (5.10) yields the $k-$Mittag-Leffler function defined as (see [20]):

$$ {}_kE^{\gamma,1}_{k,\alpha,\beta}(z) := E^{\gamma}_{k,\alpha,\beta}(z) = \sum_{\ell=0}^{\infty} \frac{(\gamma)_{\ell,k}\, z^\ell}{\Gamma_k(\alpha\ell+\beta)\ell!}. $$

3. For $p = k = 1$, the Eq. (5.10) yields the generalized Mittag-Leffler function defined in (5.9) (Shukla and Prajabati [59]):

$$ {}_1E^{\gamma,q}_{1,\alpha,\beta}(z) := E^{\gamma,q}_{\alpha,\beta}(z). $$

For comprehensive studies on the generalizations, we refer to [28, 29] and references given therein.

5.4 Turán Type Inequalities

In 1941, while examining the zeros of Legendre polynomials

$$ P_\ell(\varkappa) = \frac{1}{\ell!2^\ell}\frac{d^\ell}{d\varkappa^\ell}(\varkappa^2-1)^\ell = \frac{1}{2^\ell}\sum_{\mu=0}^{[\ell/2]}(-1)^\mu \frac{(2\ell-2\mu)!}{\mu!(\ell-\mu)!(\ell-2\mu)!}x^{\ell-2\mu}, $$

the Hungarian mathematician Paul Turán [68] established that the Legendre polynomials $P_\ell(x)$ fulfill the determinantal inequality

$$ \begin{vmatrix} P_\ell(\varkappa) & P_{\ell+1}(\varkappa) \\ P_{\ell+1}(\varkappa) & P_{\ell+2}(\varkappa) \end{vmatrix} \le 0, \quad (-1 \le \varkappa \le 1) $$

where $\ell \in \mathbb{N}$, and equality occurs only if $\varkappa = \mp 1$. Although this inequality was proved in 1946 in a letter to Szegö, it was published by Turán [68] in 1950. In 1948, Szegö [66] presented four different proofs of this inequality and extended the result to Laguerre, Hermite and ultraspherical polynomials. The proof of the inequalities

$$ [P_\ell(\varkappa)]^2 - P_{\ell+1}(\varkappa)P_{\ell-1}(\varkappa) \ge 0, $$

where $\ell \in \mathbb{N}$, $-1 \le \varkappa \le 1$ introduced by Turán is based upon the recurrence relation [65, p. 81]

$$\begin{cases} (\ell+1)P_{\ell+1}(\varkappa) = (2\ell+1)\varkappa P_\ell(\varkappa) - \ell P_{\ell-1}(\varkappa), \ \ell \geq 1 \\ P_{-1}(\varkappa) = 0, \quad P_0(\varkappa) = 1, \end{cases}$$

and the differential relation [65, p. 83]

$$(1-\varkappa^2)P'_\ell(\varkappa) = \ell P_{\ell-1}(\varkappa) - \ell\varkappa P_\ell(\varkappa).$$

These inequalities named by Karlin and Szegö [36] as Turán type inequalities because these type of inequalities was first introduced by Turán [68].

After the paper of Szegö [66], many authors discussed outcomes of the Turán type inequalities for special functions and orthogonal polynomials. Most of the proofs are based on recurrence relations. However, of special interest is a proof given by Szegö [66], which does not use recurrence relations. It is based on an inequality deduced from a theorem of Pólya and Schur for a certain class of entire functions.

Turán type inequalities are associated with the log-convex and log-concave functions. A function $f : [a, b] \subseteq \mathbb{R} \to \mathbb{R}$ is referred to as log-convex if its natural logarithm $\log f$ is convex. Thus, we have

$$f(t\varkappa + (1-t)y) \leq [f(\varkappa)]^t [f(y)]^{1-t}$$

for every $\varkappa, y \in [a, b]$ and $t \in [0, 1]$. If the above inequality is reversed, then the function f is said to be log-concave. The differentiable function f is log-convex (log-concave) if and only if f'/f is increasing (decreasing). If f is twice differentiable, then it is log-convex if and only if $f''(\varkappa)f(\varkappa) \geq (f'(\varkappa))^2$ for all $\varkappa$. There is a vast literature on log-convexity (log-concavity) related to the Turán type inequalities, and generally, is used in proofs of such inequalities [32]. Also, several researchers studied on log-convexity (log-concavity) for series in gamma functions and related topics (see [33, 34]).

The studies on Turán type inequalities have a common literature because these type inequalities have applications in number theory, complex analysis, and combinatorics. For more details, one may refer to the recent papers [5, 6, 8, 17, 43, 62] and references therein.

Today Turán type inequalities are proved for many classes of special functions:

- Kummer and q-Kummer hypergeometric functions [11, 51, 60]
- Bessel functions [4, 7, 9]
- Hypergeometric functions [10, 51].

and some others. Further, this approach is extended to various classical polynomials:

- Laguerre and Hermite polynomials [66]
- Jacobi polynomials [23]
- Lommel polynomials [14]
- Askey-Wilson polynomials [1]

and the list is far from being complete.

5.4.1 Turán Type Inequalities for the Mittag-Leffler Function

A series of recent papers in respect of Turán type inequalities for the Mittag-Leffler function was discussed by Mehrez and Sitnik [49, 50]. These authors found certain inequalities for the Mittag-Leffler function by examining monotonicity for the specific ratio of sections for series of the Mittag-Leffler function. They employed a new method, which is known as the Mehrez-Sitnik method, during their investigation [54, 61].

The Wright and Fox's H functions are two generalizations of the Mittag-Leffler function, and these both functions include "multi-indexed Mittag-Leffler functions" as special case [39–41, 47, 48, 64]. The Fox-Wright function is a generalization of the generalized hypergeometric function derived by presenting arbitrary positive scaling factors into the arguments of the gamma functions in the summand. The Fox-Wright function has an important role in fractional calculus [31, 45]. In [52], Mehrez and Sitnik used the same strategy to the Fox-Wright function and obtained many new Turán type inequalities. Furthermore, Mehrez [53] derived various Turán type inequalities for the $q-$Mittag-Leffler and $q-$Wright functions and established the monotonicity of ratios for sections of series of these functions with the results being linked to the Turán type inequalities. He used the approaches described in the case of the classical Mittag-Leffler and Wright functions to reach some of the main results in [53]. For further studies, we may also refer to [3, 21, 22, 70].

5.5 Main Results

In this section, analogues to the previous works, we establish several new Turán type inequalities for the function ${}_pE^{\gamma,q}_{k,\alpha,\beta}(z)$ defined by (5.10). Before moving forward, we recall several pivotal lemmas.

Lemma 5.1 ([13]) *Assume that* $\{a_\ell\}$ *and* $\{b_\ell\}$ $(\ell \in \mathbb{N} \cup \{0\})$ *are real numbers such that* $b_\ell > 0$, *and* $\left\{\frac{a_\ell}{b_\ell}\right\}_{\ell\geq 0}$ *is increasing (or decreasing). Then,* $\left\{\frac{a_0+a_1+\cdots+a_\ell}{b_0+b_1+\cdots+b_\ell}\right\}$ *is increasing (or decreasing).*

Lemma 5.2 ([56]) *Assume that* $\{a_\ell\}$ *and* $\{b_\ell\}$ $(n \in \mathbb{N} \cup \{0\})$ *are real numbers, and the power series* $A(\varkappa) = \sum\limits_{\ell=0}^{\infty} a_\ell \varkappa^\ell$ *and* $B(\varkappa) = \sum\limits_{\ell=0}^{\infty} b_\ell \varkappa^\ell$ *are convergent for* $|\varkappa| < r$. *If* $b_n > 0$, *and* $\left\{\frac{a_\ell}{b_\ell}\right\}_{\ell\geq 0}$ *is (strictly) increasing (or decreasing), then* $\frac{A(\varkappa)}{B(\varkappa)}$ *is also (strictly) increasing (or decreasing) on* $[0, r)$.

Theorem 5.3 *Let* $p, k \in \mathbb{R}^+\backslash\{0\}$, $\alpha, \beta, \gamma > 0$, $q \in (0, 1) \cup \mathbb{N}$. *Then, the following items are valid:*

(a) The function $\beta \mapsto {}_p\mathbb{E}^{\gamma,q}_{k,\alpha,\beta}(z) = {}_p\Gamma_k(\beta)\,{}_pE^{\gamma,q}_{k,\alpha,\beta}(z)$ *is log-convex on* $(0,\infty)$.
(b) The Turán type inequality

$$({}_p\mathbb{E}^{\gamma,q}_{k,\alpha,\beta+1}(z))^2 \leq {}_p\mathbb{E}^{\gamma,q}_{k,\alpha,\beta}(z)\,{}_p\mathbb{E}^{\gamma,q}_{k,\alpha,\beta+2}(z) \tag{5.11}$$

holds for every $z \in (0,\infty)$.
(c) Define the function

$$\begin{aligned} {}_pE^{\gamma,q,\ell}_{k,\alpha,\beta}(z) &= {}_pE^{\gamma,q}_{k,\alpha,\beta}(z) - \sum_{m=0}^{\ell} \frac{{}_p(\gamma)_{mq,k}}{{}_p\Gamma_k(\alpha m+\beta)}\frac{z^m}{m!} \\ &= \sum_{m=\ell+1}^{\infty} \frac{{}_p(\gamma)_{mq,k}}{{}_p\Gamma_k(\alpha m+\beta)}\frac{z^m}{m!} \end{aligned}$$

for $\ell \in \mathbb{N}$. *Then, the Turán type inequality*

$${}_pE^{\gamma,q,\ell}_{k,\alpha,\beta}(z)\,{}_pE^{\gamma,q,\ell+2}_{k,\alpha,\beta}(z) \leq ({}_pE^{\gamma,q,\ell+1}_{k,\alpha,\beta}(z))^2 \tag{5.12}$$

is valid for $p, k \in \mathbb{R}^+\backslash\{0\}$, $\alpha, \beta, \gamma > 0$, $q \in (0,1) \cup \mathbb{N}$ *and* $z > 0$.

Proof (a) To prove the log-convexity of $\beta \mapsto {}_p\mathbb{E}^{\gamma,q}_{k,\alpha,\beta}(z)$, we use the fact that the summation of log-convex functions is also log-convex. Hence, for all $\ell \geq 0$ we need to show that

$$\frac{\partial}{\partial\beta}\left(\log\frac{{}_p\Gamma_k(\beta)}{{}_p\Gamma_k(\alpha\ell+\beta)}\right) = {}_p\psi_k(\beta) - {}_p\psi_k(\alpha\ell+\beta),$$

and

$$\frac{\partial^2}{\partial\beta^2}\left(\log\frac{{}_p\Gamma_k(\beta)}{{}_p\Gamma_k(\alpha\ell+\beta)}\right) = {}_p\psi_k'(\beta) - {}_p\psi_k'(\alpha\ell+\beta) \geq 0,$$

where ${}_p\psi_k(z) = {}_p\Gamma_k'(z)/{}_p\Gamma_k(z)$. Since ${}_p\psi_k(z)$ is a concave function on $(0,\infty)$, then we conclude that $\beta \to \frac{{}_p\Gamma_k(\beta)}{{}_p\Gamma_k(\alpha\ell+\beta)}$ is log-convex on the interval $(0,\infty)$.

(b) Since $\beta \mapsto {}_p\mathbb{E}^{\gamma,q}_{k,\alpha,\beta}(z)$ is a log-convex function, we have

$${}_p\mathbb{E}^{\gamma,q}_{k,\alpha,t\beta_1+(1-t)\beta_2}(z) \leq ({}_p\mathbb{E}^{\gamma,q}_{k,\alpha,\beta_1}(z))^t({}_p\mathbb{E}^{\gamma,q}_{k,\alpha,\beta_2}(z))^{1-t}$$

for $\beta_1, \beta_2 > 0$, $z > 0$, $t \in [0,1]$. Setting $\beta_1 = \beta$, $\beta_2 = \beta+2$ and $t = 1/2$, we get the inequality given in (5.11).

(c) Using the formulas

$$ {}_pE^{\gamma,q,\ell}_{k,\alpha,\beta}(z) = {}_pE^{\gamma,q,\ell+1}_{k,\alpha,\beta}(z) + \frac{{}_p(\gamma)_{(\ell+1)q,k}}{{}_p\Gamma_k(\alpha(\ell+1)+\beta)}\frac{z^{\ell+1}}{(\ell+1)!}, $$

and

$$ {}_pE^{\gamma,q,\ell+2}_{k,\alpha,\beta}(z) = {}_pE^{\gamma,q,\ell+1}_{k,\alpha,\beta}(z) - \frac{{}_p(\gamma)_{(\ell+2)q,k}}{{}_p\Gamma_k(\alpha(\ell+2)+\beta)}\frac{z^{\ell+2}}{(\ell+2)!}, $$

we get

$$ \begin{aligned} &{}_pE^{\gamma,q,\ell}_{k,\alpha,\beta}(z)\ {}_pE^{\gamma,q,\ell+2}_{k,\alpha,\beta}(z) - ({}_pE^{\gamma,q,\ell+1}_{k,\alpha,\beta}(z))^2 \\ &= {}_pE^{\gamma,q,\ell+1}_{k,\alpha,\beta}(z)\left[\frac{{}_p(\gamma)_{(\ell+1)q,k}}{{}_p\Gamma_k(\alpha(\ell+1)+\beta)}\frac{z^{\ell+1}}{(\ell+1)!} - \frac{{}_p(\gamma)_{(\ell+2)q,k}}{{}_p\Gamma_k(\alpha(\ell+2)+\beta)}\frac{z^{\ell+2}}{(\ell+2)!}\right] \\ &\quad - \frac{{}_p(\gamma)_{(\ell+1)q,k}\ {}_p(\gamma)_{(\ell+2)q,k}\ z^{2\ell+3}}{{}_p\Gamma_k(\alpha(\ell+1)+\beta)\ {}_p\Gamma_k(\alpha(\ell+2)+\beta)(\ell+1)!(\ell+2)!} \\ &= \frac{{}_p(\gamma)_{(\ell+1)q,k}z^{\ell+1}\ {}_pE^{\gamma,q,\ell+2}_{k,\alpha,\beta}(z)}{{}_p\Gamma_k(\alpha(\ell+1)+\beta)(\ell+1)!} - \frac{{}_p(\gamma)_{(\ell+2)q,k}z^{\ell+2}\ {}_pE^{\gamma,q,\ell+1}_{k,\alpha,\beta}(z)}{{}_p\Gamma_k(\alpha(\ell+2)+\beta)(\ell+2)!} \\ &= \frac{{}_p(\gamma)_{(\ell+1)q,k}}{{}_p\Gamma_k(\alpha(\ell+1)+\beta)(\ell+1)!}\sum_{m=\ell+3}^{\infty}\frac{{}_p(\gamma)_{mq,k}z^{m+\ell+1}}{{}_p\Gamma_k(\alpha m+\beta)m!} \\ &\quad - \frac{{}_p(\gamma)_{(\ell+2)q,k}}{{}_p\Gamma_k(\alpha(\ell+2)+\beta)(\ell+2)!}\sum_{m=\ell+2}^{\infty}\frac{{}_p(\gamma)_{mq,k}z^{m+\ell+2}}{{}_p\Gamma_k(\alpha m+\beta)m!} \\ &= \sum_{m=\ell+3}^{\infty}\left[\frac{{}_p(\gamma)_{(\ell+1)q,k}\ {}_p(\gamma)_{mq,k}}{{}_p\Gamma_k(\alpha(\ell+1)+\beta)\ {}_p\Gamma_k(\alpha m+\beta)(\ell+1)!m!}\right. \\ &\quad \left. - \frac{{}_p(\gamma)_{(\ell+2)q,k}\ {}_p(\gamma)_{(m-1)q,k}}{{}_p\Gamma_k(\alpha(\ell+2)+\beta)\ {}_p\Gamma_k(\alpha(m-1)+\beta)(\ell+2)!(m-1)!}\right]z^{m+\ell+1} \\ &= \sum_{m=\ell+3}^{\infty}\frac{{}_p\Delta^{\gamma}_k(\alpha,\beta,\gamma,q)\ z^{m+\ell+1}}{{}_p\Gamma_k(\alpha(\ell+2)+\beta)\ {}_p\Gamma_k(\alpha(\ell+1)+\beta)\ {}_p\Gamma_k(\alpha m+\beta)\ {}_p\Gamma_k(\alpha(m-1)+\beta)(\ell+2)!m!}, \end{aligned} $$

where

$$ \begin{aligned} &{}_p\Delta^{\gamma}_k(\alpha,\beta,\gamma,q) \\ &= (\ell+2)\ {}_p(\gamma)_{(\ell+1)q,k}\ {}_p(\gamma)_{mq,k}\ {}_p\Gamma_k(\alpha(\ell+2)+\beta)\ {}_p\Gamma_k(\alpha(m-1)+\beta) \\ &\quad -m\ {}_p(\gamma)_{(\ell+2)q,k}\ {}_p(\gamma)_{(m-1)q,k}\ {}_p\Gamma_k(\alpha(\ell+1)+\beta)\ {}_p\Gamma_k(\alpha m+\beta). \end{aligned} $$

Thus, we find that

$$
\begin{aligned}
&{}_p\Delta_k^{\gamma}(\alpha,\beta,\gamma,q)\\
&= (\ell+2)\ {}_p(\gamma)_{(\ell+1)q,k}\ {}_p(\gamma)_{mq,k}\Bigg(\,{}_p\Gamma_k(\alpha(\ell+2)+\beta)\ {}_p\Gamma_k(\alpha(m-1)+\beta)\\
&\quad -\tfrac{m\ {}_p(\gamma)_{(\ell+2)q,k}\ {}_p(\gamma)_{(m-1)q,k}}{(\ell+2)\ {}_p(\gamma)_{(\ell+1)q,k}\ {}_p(\gamma)_{mq,k}}\ {}_p\Gamma_k(\alpha(\ell+1)+\beta)\ {}_p\Gamma_k(\alpha m+\beta)\Bigg)\\
&= (\ell+2)\ {}_p(\gamma)_{(\ell+1)q,k}\ {}_p(\gamma)_{mq,k}\Bigg(\,{}_p\Gamma_k(\alpha(\ell+2)+\beta)\ {}_p\Gamma_k(\alpha(m-1)+\beta)\\
&\quad -\tfrac{m\ {}_p\Gamma_k(\gamma+(\ell+2)qk)\ {}_p\Gamma_k(\gamma+(m-1)qk)}{(\ell+2)\ {}_p\Gamma_k(\gamma+(\ell+1)qk)\ {}_p\Gamma_k(\gamma+mqk)}\ {}_p\Gamma_k(\alpha(\ell+1)+\beta)\ {}_p\Gamma_k(\alpha m+\beta)\Bigg)\\
&\le (\ell+2)\ {}_p(\gamma)_{(\ell+1)q,k}\ {}_p(\gamma)_{mq,k}\ \big(\,{}_p\Gamma_k(\alpha(\ell+2)+\beta)\ {}_p\Gamma_k(\alpha(m-1)+\beta)\\
&\quad -\ {}_p\Gamma_k(\alpha(\ell+1)+\beta)\ {}_p\Gamma_k(\alpha m+\beta)\big).
\end{aligned}
$$

The ratio $\varkappa \mapsto \frac{{}_p\Gamma_k(\varkappa+a)}{{}_p\Gamma_k(\varkappa)}$ is increasing on $(0,\infty)$ when $a>0$ because the function ${}_p\Gamma_k(\varkappa)$ is log-convex on $(0,\infty)$. Hence, for $\ell \ge 0$ and $m \ge \ell+3$, setting $a=\alpha$, $\varkappa = \alpha(\ell+1)+\beta < \alpha(\ell+1)+\beta+\alpha(m-(\ell+2))$, we arrive at

$$
\frac{{}_p\Gamma_k(\alpha(\ell+1)+\beta+\alpha)}{{}_p\Gamma_k(\alpha(\ell+1)+\beta)} \le \frac{{}_p\Gamma_k(\alpha(\ell+1)+\beta+\alpha(m-(\ell+2))+\alpha)}{{}_p\Gamma_k(\alpha(\ell+1)+\beta+\alpha(m-(\ell+2))}
$$

that is,

$$
\frac{{}_p\Gamma_k(\alpha(\ell+2)+\beta)}{{}_p\Gamma_k(\alpha(\ell+1)+\beta)} \le \frac{{}_p\Gamma_k(\alpha m+\beta)}{{}_p\Gamma_k(\alpha(m-1)+\beta)}.
$$

It means that ${}_p\Delta_k^{\gamma}(\alpha,\beta,\gamma,q) \le 0$, thus the Turán type inequality given by (5.12) holds. □

Corollary 5.1 *Let* $\alpha,\beta>0$ *and* $\ell\in\mathbb{N}$. *Then*

$$
E_{\alpha,\beta}^{1,1,\ell}(z)E_{\alpha,\beta}^{1,1,\ell+2}(z) \le (E_{\alpha,\beta}^{1,1,\ell+1}(z))^2.
$$

Proof Letting $p=q=k=\gamma=1$ in item (c) of Theorem 5.3, we get the above Turán type inequality, which was obtained by Mehrez and Sitnik [49]. □

Theorem 5.4 *Let* $p,k\in\mathbb{R}^+\backslash\{0\}$, $\alpha,\beta_1,\beta_2,\gamma>0$ *and* $q\in(0,1)\cup\mathbb{N}$.

(a) If $\beta_2>\beta_1$ $(\beta_2<\beta_1)$, *then the function* $z\mapsto {}_pE_{k,\alpha,\beta_1}^{\gamma,q}(z)/{}_pE_{k,\alpha,\beta_2}^{\gamma,q}(z)$ *is increasing (or decreasing) on* $(0,\infty)$.

(b) If $\beta_2 > \beta_1$, then the Turán type inequality

$$ {}_pE^{\gamma,q}_{k,\alpha,\beta_1-1}(z)\,{}_pE^{\gamma,q}_{k,\alpha,\beta_2}(z) - {}_pE^{\gamma,q}_{k,\alpha,\beta_2-1}(z)\,{}_pE^{\gamma,q}_{k,\alpha,\beta_1}(z) + (\beta_2-\beta_1)\,{}_pE^{\gamma,q}_{k,\alpha,\beta_1}(z)\,{}_pE^{\gamma,q}_{k,\alpha,\beta_2}(z) \geq 0 $$

holds.

Proof (a) By using (5.10), we get

$$ {}_pE^{\gamma,q}_{k,\alpha,\beta_1}(z)/{}_pE^{\gamma,q}_{k,\alpha,\beta_2}(z) = \sum_{\ell=0}^{\infty} \frac{{}_p(\gamma)_{\ell q,k}\, z^{\ell}}{{}_p\Gamma_k(\alpha\ell+\beta_1)\ell!} \Bigg/ \sum_{\ell=0}^{\infty} \frac{{}_p(\gamma)_{\ell q,k}\, z^{\ell}}{{}_p\Gamma_k(\alpha\ell+\beta_2)\ell!}. $$

According to the Lemma 5.2, we need to show the monotonicity of the sequence $(u_\ell)_{\ell\geq 0}$ expressed by

$$ u_\ell = \frac{{}_p\Gamma_k(\alpha\ell+\beta_2)}{{}_p\Gamma_k(\alpha\ell+\beta_1)}. $$

Hence

$$ \frac{u_{\ell+1}}{u_\ell} = \frac{{}_p\Gamma_k(\alpha\ell+\alpha+\beta_2)\ {}_p\Gamma_k(\alpha\ell+\beta_1)}{{}_p\Gamma_k(\alpha\ell+\alpha+\beta_1)\ {}_p\Gamma_k(\alpha\ell+\beta_2)}. $$

If $\beta_2 > \beta_1$, we find that the sequence $(u_\ell)_{\ell\geq 0}$ is increasing on $(0,\infty)$ because $u_{\ell+1} \geq u_\ell$ for all $\ell \geq 0$. Thus, by Lemma 5.2 we conclude that the function $z \mapsto {}_pE^{\gamma,q}_{k,\alpha,\beta_1}(z)/{}_pE^{\gamma,q}_{k,\alpha,\beta_2}(z)$ is increasing on $(0,\infty)$.

If $\beta_1 > \beta_2$, we find that the sequence $(u_\ell)_{\ell\geq 0}$ is decreasing on $(0,\infty)$ because $u_{\ell+1} \leq u_\ell$ for all $\ell \geq 0$. Thus, by Lemma 5.2 we conclude that $z \mapsto {}_pE^{\gamma,q}_{k,\alpha,\beta_1}(z)/{}_pE^{\gamma,q}_{k,\alpha,\beta_2}(z)$ is decreasing on $(0,\infty)$.

(b) In [59], Shukla and Prajabati proved that

$$ \frac{d}{dz}E_{\alpha,\beta}(z) = \frac{E_{\alpha,\beta-1}(z) - (\beta-1)E_{\alpha,\beta}(z)}{\alpha z}. $$

Thus, for $\beta_2 > \beta_1$ we observe

$$ \left(\frac{{}_pE^{\gamma,q}_{k,\alpha,\beta_1}(z)}{{}_pE^{\gamma,q}_{k,\alpha,\beta_2}(z)}\right)' = $$

$$ \frac{{}_pE^{\gamma,q}_{k,\alpha,\beta_1-1}(z)\,{}_pE^{\gamma,q}_{k,\alpha,\beta_2}(z) - {}_pE^{\gamma,q}_{k,\alpha,\beta_2-1}(z)\,{}_pE^{\gamma,q}_{k,\alpha,\beta_1}(z) + (\beta_2-\beta_1)\,{}_pE^{\gamma,q}_{k,\alpha,\beta_1}(z)\,{}_pE^{\gamma,q}_{k,\alpha,\beta_2}(z)}{\alpha z({}_pE^{\gamma,q}_{k,\alpha,\beta_2}(z))^2} \geq 0. $$

This completes the proof. □

Corollary 5.2 *Let $\alpha, \beta_1, \beta_2, \gamma > 0$. If $\beta_2 > \beta_1$, then the Turán type inequality*

$$E^{\gamma,q}_{\alpha,\beta_1-1}(z)E^{\gamma,q}_{\alpha,\beta_2}(z) - E^{\gamma,q}_{\alpha,\beta_2-1}(z)E^{\gamma,q}_{\alpha,\beta_1}(z) + (\beta_2-\beta_1)E^{\gamma,q}_{\alpha,\beta_1}(z)E^{\gamma,q}_{\alpha,\beta_2}(z) \geq 0$$

holds.

Proof By taking $p = k = 1$ in item (b) of Theorem 5.4, we get the proof, which was obtained by Mehrez and Sitnik [50]. □

Theorem 5.5 *Let $p, k \in \mathbb{R}^+\backslash\{0\}$, $\alpha, \beta, \gamma > 0$, $q \in (0,1) \cup \mathbb{N}$, $z > 0$ and $\ell \in \mathbb{N}$, then the function*

$${}_p\Omega^{\gamma,q,\ell}_{k,\alpha,\beta}(z) = \frac{{}_pE^{\gamma,q,\ell}_{k,\alpha,\beta}(z)\,{}_pE^{\gamma,q,\ell+2}_{k,\alpha,\beta}(z)}{({}_pE^{\gamma,q,\ell+1}_{k,\alpha,\beta}(z))^2}$$

is increasing on $(0,\infty)$. Thus the Turán type inequality

$$\begin{aligned}
&\frac{(\ell+1)\ {}_p\Gamma_k(\gamma+(\ell+1)qk)\ {}_p\Gamma_k(\gamma+(\ell+3)qk)}{(\ell+2)\ {}_p\Gamma_k^2(\gamma+(\ell+2)qk)}\\
&\times \frac{{}_p\Gamma_k^2(\alpha(\ell+2)+\beta)}{{}_p\Gamma_k(\alpha(\ell+1)+\beta)\ {}_p\Gamma_k(\alpha(\ell+3)+\beta)}\left({}_pE^{\gamma,q,\ell+1}_{k,\alpha,\beta}(z)\right)^2 \qquad (5.13)\\
&\leq {}_pE^{\gamma,q,\ell}_{k,\alpha,\beta}(z)\,{}_pE^{\gamma,q,\ell+2}_{k,\alpha,\beta}(z)
\end{aligned}$$

is valid. The constant on the LHS of (5.13) is sharp.

Proof According to the Cauchy product formula, we get

$$\begin{aligned}
&{}_p\Omega^{\gamma,q,\ell}_{k,\alpha,\beta}(z)\\
&= \frac{\sum_{m=0}^{\infty}\left(\sum_{j=0}^{m}\frac{{}_p(\gamma)_{(\ell+1+j)q,k}\ {}_p(\gamma)_{(\ell+3+m-j)q,k}}{{}_p\Gamma_k(\alpha(\ell+1+j)+\beta)\ {}_p\Gamma_k(\alpha(\ell+3+m-j)+\beta)(\ell+1+j)!(\ell+3+m-j)!}\right)z^{2\ell+m+4}}{\sum_{m=0}^{\infty}\left(\sum_{j=0}^{m}\frac{{}_p(\gamma)_{(\ell+2+j)q,k}\ {}_p(\gamma)_{(\ell+2+m-j)q,k}}{{}_p\Gamma_k(\alpha(\ell+2+j)+\beta)\ {}_p\Gamma_k(\alpha(\ell+2+m-j)+\beta)(\ell+2+j)!(\ell+2+m-j)!}\right)z^{2\ell+m+4}}. \qquad (5.14)
\end{aligned}$$

Let us consider the sequence ${}_p\lambda_{k,j}(\alpha,\beta,\gamma,q)$ expressed by

$$\begin{aligned}
&{}_p\lambda_{k,j}(\alpha,\beta,\gamma,q)\\
&= \frac{{}_p(\gamma)_{(\ell+1+j)q,k}\ {}_p(\gamma)_{(\ell+3+m-j)q,k}\ (\ell+2+j)!(\ell+2+m-j)!}{{}_p(\gamma)_{(\ell+2+j)q,k}\ {}_p(\gamma)_{(\ell+2+m-j)q,k}\ (\ell+1+j)!(\ell+3+m-j)!}\left({}_pU_{k,j}(\alpha,\beta)\right),
\end{aligned}$$

where

$${}_pU_{k,j}(\alpha,\beta) = \frac{{}_p\Gamma_k(\alpha(\ell+2+j)+\beta)\ {}_p\Gamma_k(\alpha(\ell+2+m-j)+\beta)}{{}_p\Gamma_k(\alpha(\ell+1+j)+\beta)\ {}_p\Gamma_k(\alpha(\ell+3+m-j)+\beta)}.$$

Because the function ${}_p\Gamma_k(\varkappa)$ is log-convex on $(0, \infty)$, the ratio $\frac{{}_p\Gamma_k(\varkappa+a)}{{}_p\Gamma_k(\varkappa)}$ $(a > 0)$ is increasing on $(0, \infty)$, thus we find $\frac{{}_pU_{k,j+1}(\alpha,\beta)}{{}_pU_{k,j}(\alpha,\beta)} \geq 1$ and hence, the sequence ${}_pU_{k,j}(\alpha, \beta)$ is also increasing. Therefore

$$\frac{{}_p\lambda_{k,j+1}(\alpha, \beta, \gamma, q)}{{}_p\lambda_{k,j}(\alpha, \beta, \gamma, q)} = {}_pK_{k,j}(\gamma, q)\frac{{}_pU_{k,j+1}(\alpha, \beta)}{{}_pU_{k,j}(\alpha, \beta)} \geq {}_pK_{k,j}(\gamma, q)$$

with

$${}_pK_{k,j}(\gamma, q) = \frac{{}_p(\gamma)^2_{(\ell+2+j)q,k}\ {}_p(\gamma)^2_{(\ell+2+m-j)q,k}}{{}_p(\gamma)_{(\ell+1+j)q,k}\ {}_p(\gamma)_{(\ell+3+j)q,k}\ {}_p(\gamma)_{(\ell+1+m-j)q,k}\ {}_p(\gamma)_{(\ell+3+m-j)q,k}}$$
$$\times \frac{(\ell+3+j)(\ell+3+m-j)}{(\ell+2+j)(\ell+2+m-j)}.$$

By using the inequality

$$\frac{{}_p(\gamma)^2_{(\ell+2+j)q,k}}{{}_p(\gamma)_{(\ell+1+j)q,k}\ {}_p(\gamma)_{(\ell+3+j)q,k}}\frac{(\ell+3+j)}{(\ell+2+j)} \geq 1,$$

we conclude that

$${}_pK_{k,j}(\gamma, q) = \frac{{}_p\Gamma_k^2(\gamma+(\ell+2+j)qk)(\ell+3+j)}{{}_p\Gamma_k(\gamma+(\ell+1+j)qk)\ {}_p\Gamma_k(\gamma+(\ell+3+j)qk)(\ell+2+j)}$$
$$\times \frac{{}_p\Gamma_k^2(\gamma+(\ell+2+m-j)qk)(\ell+3+m-j)}{{}_p\Gamma_k(\gamma+(\ell+1+m-j)qk)\ {}_p\Gamma_k(\gamma+(\ell+3+m-j)qk)(\ell+2+m-j)} \geq 1,$$

where we have used

$${}_p(\gamma)_{\ell q,k} = \frac{{}_p\Gamma_k(\gamma + \ell qk)}{{}_p\Gamma_k(\gamma)}, \quad (\gamma > 0).$$

Since $\frac{{}_p\Gamma_k(\gamma+\ell qk)}{{}_p\Gamma_k(\gamma)}$ is increasing on $(0, \infty)$, it follows that ${}_pK_{k,j}(\gamma, q)$ is increasing. Thus, the sequence ${}_p\lambda_{k,j}(\alpha, \beta, \gamma, q)$ is increasing. Taking into account of Lemmas 5.1 and 5.2, we observe that the ratio ${}_p\Omega^{\gamma,q,\ell}_{k,\alpha,\beta}(z)$ given by (5.14) is also increasing and

$$\lim_{z\to 0}\ {}_p\Omega^{\gamma,q,\ell}_{k,\alpha,\beta}(z) = \frac{(\ell+1)\ {}_p\Gamma_k(\gamma+(\ell+1)qk)\ {}_p\Gamma_k(\gamma+(\ell+3)qk)}{(\ell+2)\ {}_p\Gamma_k^2(\gamma+(\ell+2)qk)}$$
$$\times \frac{{}_p\Gamma_k^2(\alpha(\ell+2)+\beta)}{{}_p\Gamma_k(\alpha(\ell+1)+\beta)\ {}_p\Gamma_k(\alpha(\ell+3)+\beta)}.$$

The constant given above is the best possible for which the inequality in (5.13) holds. □

5.6 Concluding Remarks

In this research work, utilizing the $(p, k)-$Mittag-Leffler function, we have found some new Turán type inequalities. We concluded that the outcomes are an improvement of several findings proved by Mehrez and Sitnik. We also note that many other outcomes like Lazarević type and Wilker type inequalities can be derived as a future work for the $(p, k)-$Mittag-Leffler function.

References

1. Abreu, L.D., Bustoz, J.: Turán inequalities for symmetric Askey-Wilson polynomials. Rocky Mt. J. Math. **30**(2), 401–409 (2000)
2. Alzer, H., Gerhold, S., Kauers, M., Lupaş, A.: On Turán's inequality for Legendre polynomials. Expo. Math. **25**, 181–186 (2007)
3. Bansal, D., Mehrez, K.: On a new class of functions related with Mittag-Leffler and Wright functions and their properties. Commun. Korean Math. Soc. **35**(4), 1123–1132 (2020)
4. Baricz, Á.: Functional inequalities involving Bessel and modified Bessel functions of the first kind. Expo. Math. **26**, 279–293 (2008)
5. Baricz, Á., Ismail, M.E.H.: Turán type inequalities for Tricomi confluent hypergeometric functions. Constr. Approx. **37**(2), 195–221 (2013)
6. Baricz, Á., Pogány, T.K.: Turán determinants of Bessel functions. Forum Math. **26**, 295–322 (2014)
7. Baricz, Á., Ponnusamy, S.: On Turán type inequalities for modified Bessel functions. Proc. Am. Math. Soc. **141**(2), 523–532 (2013)
8. Baricz, Á., Ponnusamy, S., Singh, S.: Turán type inequalities for Struve functions. J. Math. Anal. Appl. **445**(1), 971–984 (2017)
9. Baricz, Á., Ponnusamy, S., Singh, S.: Turán type inequalities for general Bessel functions. Math. Inequal. Appl. **19**(2), 709–719 (2016)
10. Baricz, Á., Ponnusamy, S., Singh, S.: Turán type inequalities for confluent hypergeometric functions of the second kind. Stud. Sci. Math. Hung. **53**(1), 74–92 (2016)
11. Barnard, R.W., Gordy, M.B., Richards, K.C.: A note on Turán type and mean inequalities for the Kummer function. J. Math. Anal. Appl. **349**, 259–263 (2009)
12. Berg, C., Szwarc, R.: Bounds on Turán determinants. J. Approx. Theory **161**(1), 127–141 (2009)
13. Biernacki, M., Krzyz, J.: On the monotonicity of certain functionals in the theory of analytic functions. Ann. Univ. Mariae Curie-Skłodowska Sect. A **9**, 135–147 (1955)
14. Bustoz, J., Ismail, M.E.H.: Turán inequalities for symmetric orthogonal polynomials. Int. J. Math. Math. Sci. **20**, 1–8 (1997)
15. Chand, M., Prajabati, J.C., Bonyah, E.: Fractional integrals and solution of fractional kinetic equations involving $k-$Mittag-Leffler function. Trans. A. Razmadze Math. Inst. **171**, 144–166 (2017)
16. Constantinescu, E.: On the inequality of P. Turán for Legendre polynomials. JIPAM **6**(28) (2005)
17. Craven, T., Csordas, G.: Jensen polynomials and the Turán and Laguerre inequalities. Pacific J. Math. **136**(2), 241–260 (1989)
18. Díaz, R., Pariguan, E.: On hypergeometric functions and Pochhammer $k-$symbol. Divulgaciones Matemáticas **15**(2), 179–192 (2007)
19. Diethelm, K.: The analysis of fractional differential equations: an application-oriented exposition using differential operators of Caputo type. Lecture Notes in Mathematics. Springer, Berlin, Heidelberg (2010)

20. Dorrego, G.A., Cerutti, R.A.: The $k-$Mittag-Leffler function. Int. J. Contemp. Math. Sciences **7**, 705–716 (2012)
21. Dou, X.K., Yin, L.: Turán type inequalities for generalized Mittag-Leffler function. J. Inequal. Appl. **2017**, 200 (2017)
22. Dou, X.K., Yin, L., Lin, X.L.: Several Turán type inequalities for the generalized Mittag-Leffler function. J. Funct. Spaces **2020**(9509728), 6 (2020)
23. Gasper, G.: On the extension of Turán's inequality to Jacobi polynomials. Duke Math. J. **38**, 415–428 (1971)
24. Gehlot, K.S.: The $p-k$ Mittag-Leffler function. Palest. J. Math. **7**(2), 628–632 (2018)
25. Gehlot, K.S.: The generalized $k-$Mittag-Leffler function. Int. J. Contemp. Math. Sci. **7**, 2213–2219 (2012)
26. Gehlot, K.S.: Two parameter gamma function and its properties (2017). arXiv:1701.01052 [math.CA]
27. Gehlot, K.S.: Differential equation of $p-k$ Mittag-Leffler function. Palest. J. Math. **9**(2), 940–944 (2020)
28. Gorenflo, R., Kilbas, A.A., Mainardi, F., Rogosin, S.V.: Mittag-Leffler Functions, Related Topics and Applications. Springer, Berlin, Heidelberg (2014)
29. Gorenflo, R., Kilbas, A.A., Rogosin, S.V.: On the generalized Mittag-Leffler type functions. Integr. Transform. Spec. Funct. **7**, 215–224 (1998)
30. Gorenflo, R., Mainardi, F.: Fractional calculus: integral and differential equations of fractional order. In: Carpinteri, A., Mainardi, F. (eds.) Fractals and Fractional Calculus in Continuum Mechanics. Springer, Wien (1997)
31. Gorenflo, R., Luchko, Y., Mainardi, F.: Analytical properties and applications of the Wright function. Fract. Calc. Appl. Anal. **2**, 383–414 (1999)
32. Kalmykov, S.I., Karp, D.B.: Log-concavity and Turán type inequalities for the generalized hypergeometric function. Anal. Math. **43**(4), 567–580 (2017)
33. Kalmykov, S.I., Karp, D.B.: Log-concavity for series in reciprocal gamma functions and applications. Integr. Transform. Spec. Funct. **24**(11), 859–872 (2013)
34. Kalmykov, S.I., Karp, D.B.: Log-convexity and log-concavity for series in gamma ratios and applications. J. Math. Anal. Appl. **406**(2), 400–418 (2013)
35. Kamarujjama, M., Khan, N.U., Khan, O.: The generalized $p-k$ Mittag-Leffler function and solution of fractional kinetic equations. J. Anal. **27**(4), 1029–1046 (2019)
36. Karlin, S., Szegö, G.: On certain determinants whose elements are orthogonal polynomials. J. Anal. Math. **8**, 1–157 (1960/1961)
37. Kilbas, A.A., Saigo, M.: On Mittag-Leffler type function, fractional calculus operators and solutions of integral equations. Integr. Transform. Spec. Funct. **4**, 355–370 (1996)
38. Kilbas, A.A., Srivastava, H.M., Trujillo, J.J.: Theory and applications of fractional differential equations. North-Holland Mathematical Studies, vol. 204. Elsevier Science, Publishers, Amsterdam (2006)
39. Kilbas, A.A.: H-Transforms, Theory and Applications. CRC Press, Boca Raton (2004)
40. Kiryakova, V.: The multi-index Mittag-Leffler functions as an important class of special functions of fractional calculus. Comput. Math. Appl. **59**, 1885–1895 (2010)
41. Kiryakova, V.: Multiple (multiindex) Mittag-Leffler functions and relations to generalized fractional calculus. J. Comput. Appl. Math. **118**, 241–259 (2020)
42. Kokologiannaki, C.G., Krasniqi, V.: Some properties of the $k-$gamma function. Matematiche (Catania) **68**, 13–22 (2013)
43. Koumandos, S., Pedersen, H.L.: Turán type inequalities for the partial sums of the generating functions of Bernoulli and Euler numbers. Math. Nachr. **285**(17–18), 2129–2156 (2012)
44. Krasniqi, V.: Erratum to "A limit for the $k-$gamma and $k-$beta function". Int. Math. Forum **5**, 1613–1617 (2010)
45. Mainardi, F., Pagnini, G.: The role of the Fox-Wright functions in fractional sub-diffusion of distributed order. J. Comput. Appl. Math. **207**, 245–257 (2007)
46. Mansour, Z.: Determining the $k-$generalized gamma function $\Gamma_k(x)$ by functional equations. Int. J. Contemp. Math. Sci. **4**, 1037–1042 (2009)

47. Mathai, A.M., Saxena, R.: The H-Function with Applications in Statistics and Other Disciplines. Halsted Press (John Wiley and Sons), New York (1978)
48. Mathai, A.M., Saxena, R.K., Haubold, H.J.: The H-Function: Theory and Applications. Springer, New York (2010)
49. Mehrez, K., Sitnik, S.M.: Functional inequalities for the Mittag-Leffler functions. Results Math. **72**, 703–714 (2017)
50. Mehrez, K., Sitnik, S.M.: Turán type inequalities for classical and generalized Mittag-Leffler functions. Anal. Math. **44**, 521–541 (2018)
51. Mehrez, K., Sitnik, S.M.: Monotonicity of ratios of $q-$Kummer confluent hypergeometric and $q-$hypergeometric functions and associated Turán types inequalities (2014). arXiv:1412.1634 [math.CA]
52. Mehrez, K., Sitnik, S.M.: Functional inequalities for the Fox-Wright functions. Ramanujan J. **50**, 263–287 (2019)
53. Mehrez, K.: Turán type inequalities for $q-$Mittag-Leffler and $q-$Wright functions. Math. Inequal. Appl. **21**, 1135–1151 (2018)
54. Mehrez, K., Sitnik, S.M.: On monotonicity of ratios of some $q-$hypergeometric functions. Mat. Vesn. **68**, 225–231 (2016)
55. Mittag-Leffler, M.G.: Sur la nouvelle fonction $E_\alpha(x)$. C. R. Acad. Sci. Paris **137**, 554–558 (1903)
56. Ponnusamy, S., Vuorinen, M.: Asymptotic expansions and inequalities for hypergeometric function. Mathematika **44**(2), 278–301 (1997)
57. Prabhakar, T.R.: A singular integral equation with a generalized Mittag-Leffler function in the kernel. Yokohama Math. J. **19**(1), 7–15 (1971)
58. Rainville, E.D.: Special Functions. Macmillan, New York (1960)
59. Shukla, A.K., Prajabati, J.C.: On a generalization of Mittag-Leffler function and its properties. J. Math. Anal. Appl. **336**, 797–811 (2007)
60. Sitnik, S.M., Mehrez, K.: Proofs of some conjectures on monotonicity of ratios of Kummer, Gauss and generalized hypergeometric functions. Analysis **36**, 263–268 (2014)
61. Sitnik, S.M., Mehrez, K.: On monotonicity of ratios of some hypergeometric functions. Sib. Electron. Math. Rep. **13**, 260–268 (2016)
62. Skovgaard, H.: On inequalities of the Turán type. Math. Scand. **2**, 65–73 (1954)
63. Srivastava, H.M., Tomovski, Z.: Fractional calculus with an integral operator containing a generalized Mittag-Leffler function in the kernel. Appl. Math. Comput. **211**, 198–210 (2009)
64. Srivastava, H.M., Manocha, H.L.: A Treatise on Generating Functions. Halsted Press, Chichester, West Sussex, England, E. Horwood, New York (1984)
65. Szegö, G.: Orthogonal Polynomials. Colloquium Pub, New York (1939)
66. Szegö, G.: On an inequality of P. Turán concerning Legendre polynomials. Bull. Am. Math. Soc. **54**, 401–405 (1948)
67. Szwarc, R.: Positivity of Turán determinants for orthogonal polynomials. In: Harmonic Analysis and Hypergroups (Delhi, 1995), Trends in Mathematics, pp. 165–182. Birkhäuser Boston, Boston, MA (1998)
68. Turán, P.: On the zeros of the polynomials of Legendre. Časopis Pěst. Mat. Fys. **75**, 113–122 (1950)
69. Wiman, A.: Über den fundamentalsatz in der teorie der functionen $E_\alpha(x)$. Acta Math. **29**, 191–201 (1905)
70. Yin, L., Huang, L.G.: Turán type inequalities for generalized Mittag-Leffler functions. J. Math. Inequal. **11**(3), 667–672 (2017)

Chapter 6
Multiplicative Generalized Hardy-Rogers-Type F-Proximal Non-self Mappings and Best Proximity Point Approximation

Nisha Sharma and Hemen Dutta

Abstract In this work, taking into account the multiplicative metric space and Hardy-Rogers-type F-proximal contraction ($F_{\mathcal{PC}}$) and multiplicative generalized almost F-contraction ($F_{\mathcal{AC}}$), we first introduce the multiplicative generalized F-proximal contractions and then multiplicative generalized $F_{\mathcal{PC}}$ of first and second kind with the aim of best proximity theorems in the framework of Complete multiplicative metric spaces ($\mathcal{CMMS}$). The purpose of this paper is to bring the multiplicative behavior of Hardy-Rogers-type F-contraction to the consideration of researchers in the field of research and exemplify its utility. Our work generalizes and improves the results of non-Newtonian calculus-related literature.

Keywords Multiplicative generalized $F_{\mathcal{PC}}$ · $\mathcal{CMMS}$ · Best proximity point, $p-$property

6.1 Introduction

In 1973, Hardy and Rogers [1] studied the principle result which is as follows:

Theorem 6.1 *Let $(\mathscr{M}, \Delta)$ be a metric space and Ψ a self-mapping of $\mathscr{M}$ satisfying the condition for* $\mathfrak{m}, \mathfrak{n} \in \mathscr{M}$

$$\Delta(\Psi\mathfrak{m}, \Psi\mathfrak{n}) \leq \mathfrak{a}_1\Delta(\Psi\mathfrak{m}, \Psi\mathfrak{m}) + \mathfrak{a}_2\Delta(\mathfrak{n}, \Psi\mathfrak{n}) + \mathfrak{a}_3\Delta(\mathfrak{m}, \Psi\mathfrak{n}) + \mathfrak{a}_4\Delta(\mathfrak{n}, \Psi\mathfrak{m}) + \mathfrak{a}_5\Delta(\mathfrak{m}, \mathfrak{n}), \qquad (\mathcal{HR})$$

where $\mathfrak{a}_1, \mathfrak{a}_2, \mathfrak{a}_3, \mathfrak{a}_4, \mathfrak{a}_5 \in \mathbb{R}^+$ *and we set* $\alpha = \mathfrak{a}_1 + \mathfrak{a}_2 + \mathfrak{a}_3 + \mathfrak{a}_4 + \mathfrak{a}_5$. *Then*

N. Sharma (✉)
Department of Mathematics, Pt. J. L. N. Government College, Faridabad, Haryana, India
e-mail: bhardwaj.nnishaa@gmail.com

H. Dutta
Department of Mathematics, Guahati University, Guwahati, India

B. C. Tripathy et al. (eds.), *Operators, Inequalities and Approximation*, Industrial and Applied Mathematics, https://doi.org/10.1007/978-981-97-3238-8_6

($\mathscr{C}_1$) $\mathscr{M}$ is complete and $\alpha < 1$, Ψ has a unique fixed point.
($\mathscr{C}_2$) If $(\mathcal{HR})$ is modified to the condition $\mathfrak{m} \neq \mathfrak{n}$, then it implies

$$\Delta(\Psi\mathfrak{m}, \Psi\mathfrak{n}) < \mathfrak{a}_1\Delta(\Psi\mathfrak{m}, \Psi\mathfrak{m}) + \mathfrak{a}_2\Delta(\mathfrak{n}, \Psi\mathfrak{n}) + \mathfrak{a}_3\Delta(\mathfrak{m}, \Psi\mathfrak{n}) + \mathfrak{a}_4\Delta(\mathfrak{n}, \Psi\mathfrak{m}) + \mathfrak{a}_5\Delta(\mathfrak{m}, \mathfrak{n}), \qquad (\mathcal{HR}')$$

and in this case we assume $\mathscr{M}$ is compact, Ψ is continuous, and $\alpha = 1$, then Ψ has a unique fixed point.

Reich [2] obtained a similar conclusion to that in $\mathscr{C}_1$ in the case where $\alpha = \mathfrak{a}_1 + \mathfrak{a}_2 + \mathfrak{a}_5$. Reich's result, in turn, generalizes Kannan's [3] fixed point theorem, in which $\alpha = \mathfrak{a}_1 + \mathfrak{a}_2$. The observation in part $\mathscr{C}_2$ is a limiting version of the theorem in part $\mathscr{C}_1$, and Edelstein achieved this result in [4] for the case $\alpha = \mathfrak{a}_5 = 1$. Wardowski [5] established a novel concept of contraction and proved a fixed point theorem that generalizes the Banach contraction principle in a way that differs from earlier results. There are numerous extensions to the well-known Banach contraction principle [6], which asserts that every self-mapping Ψ is defined on a complete metric space $(\mathscr{M}, \Delta)$ satisfying

$$\forall_{\mathfrak{m},\mathfrak{n}\in\mathscr{M}}(\Delta(\Psi\mathfrak{m}, \Psi\mathfrak{n}) \leq \mathfrak{c}\Delta(\mathfrak{m}, \mathfrak{n})),$$

where $\mathfrak{c} \in (0, 1)$ has unique fixed point $\forall\ \mathfrak{m} \in \mathscr{M}$.

During the last decade, the following question has been a prominent focus of research in applied mathematics and nonlinear functional analysis (see [9–17]).

Problem. For $\emptyset \neq \mathscr{M}_1,\ \mathscr{M}_2 \subseteq \mathscr{M},\ \overset{\circ}{\Psi} : \mathscr{M}_1 \to \mathscr{M}_2$ is a non-self mapping and $\Delta(\mathscr{M}_1, \mathscr{M}_2) = \inf\{\Delta(\mathfrak{m}, \mathfrak{n}) : \mathfrak{m} \in \mathscr{M}_1,\ \mathfrak{n} \in \mathscr{M}_2\}$, is there existence of a point $\mathfrak{m}_0$ in the metric space $(\mathscr{M}, \Delta)$ such that $\Delta(\mathfrak{m}_0, \overset{\circ}{\Psi}\mathfrak{m}_0) = \Delta(\mathscr{M}_1, \mathscr{M}_2)$?

If there is the existence of such $\mathfrak{m}_0$, it is known as *Best proximity point*. We attempt to find the least conditions on non-self mapping $\overset{\circ}{\Psi}$ for unique best proximity point. This analysis is more extensive than metric fixed point theory. The mappings under consideration are not self-mappings in the sense that fixed point theory seems to be. Fixed point theory employs various mathematical fields of study, including topology and geometry, to prove the existence of $\overset{\circ}{\Psi}\mathfrak{m} = \mathfrak{m}$ solutions within certain restrictions. However, if $\overset{\circ}{\Psi}$ is not a self-mapping, it seems there are no solutions to the $\overset{\circ}{\Psi}\mathfrak{m} = \mathfrak{m}$ equation, and in this case, it is difficult to find an element $\mathfrak{m}$ that really is nearest to $\overset{\circ}{\Psi}\mathfrak{m}$ in any way. We recall that $\mathfrak{m}$ is a best proximity point of a non-self mapping $\overset{\circ}{\Psi} : \mathscr{M}_1 \to \mathscr{M}_2$, if $\mathfrak{m}$ corresponds to $\Delta(\mathfrak{m}, \overset{\circ}{\Psi}\mathfrak{m}) = \Delta(\mathscr{M}_1, \mathscr{M}_2)$, where $\Delta(\mathscr{M}_1, \mathscr{M}_2) = \inf \Delta(\mathfrak{m}, \mathfrak{n}) : \mathfrak{m} \in \mathscr{M}_1,\ \ \mathfrak{n} \in \mathscr{M}_2$. Moreover, if $\mathscr{M}_1 = \mathscr{M}_2$, then the mapping $\overset{\circ}{\Psi}$ is a self-mapping and so the best proximity point theorem becomes to a fixed point result which can be solved by Banach's fixed point theorem. Thus the best proximity point theorem is one of the interesting problems in fixed point theory (refer, [18–23]). Thagafi Shahzad [24] investigates many contraction variants for the existence of the best proximity point (also refer, [25–29]). For relatively nonexpansive maps, Eldred et al. [30] established the best proximity point theorems whereas Basha [13] has studied the best proximity point theory for contractive non-self mappings. Some popular best proximity point theorems can be found in [30, 31].

6.2 Preliminaries

Fan [18] established one of the most appealing theorems, the best approximation theorem, in 1969 as follows:

Theorem 6.2 ([18]) *Let $\mathcal{N}_s$ a subset of a normed space $\mathcal{N}$ and $\mathring{\Psi} : \mathcal{N}_s \to \mathcal{N}$ be a continuous non-self mapping. Then $\exists\ \mathfrak{m} \in \mathcal{N}_s$ such that*

$$||\mathfrak{m} - \mathring{\Psi}\mathfrak{m}|| = \Delta(\mathcal{N}_s, \mathring{\Psi}\mathfrak{m}) = \inf\{||\mathring{\Psi}\mathfrak{m} - \mathfrak{m}_u|| : \mathfrak{m}_u \in \mathcal{N}_s\}.$$

Following that, a number of mathematicians used Fan's theorem in a number of different ways (see [35–38]) and made references to it. Best proximity point theorems for non-self nonexpansive mappings, non-self Kannan-type mappings, and non-self Chatterjea-type mappings were proved by Basha et al. [38] in 2013. Ungchittrakool [39] subsequently proved several best proximity point theorems in complete metric spaces for generalized non-self Kannan-type mappings, non-self Chatterjea-type mappings, and Lipschitzian mappings.

Definition 6.1 ([39]) Let $\mathcal{M}_1$ and $\mathcal{M}_2$ be nonempty subsets of a metric space $(\mathcal{M}, \Delta)$ and let $\mathring{\Psi}_S : \mathcal{M}_2 \to \mathcal{M}_1$ be a mapping. A mapping $\mathring{\Psi}_T : \mathcal{M}_1 \to \mathcal{M}_2$ is said to be a generalized non-self Kannan and non-self Chatterjea mappings with respect to the mapping $\mathring{\Psi}_S$ if $\exists$ non-negative constants $\mathfrak{a}_1,\ \mathfrak{a}_2,\ \mathfrak{a}_3$ such that $\mathfrak{a}_1 + 2\mathfrak{a}_2 + 2\mathfrak{a}_3 < 1$ and

$$\begin{aligned}\Delta(\mathring{\Psi}_T\mathfrak{m}, \mathring{\Psi}_T\mathfrak{n}) \leq \mathfrak{a}_1\Delta(\mathfrak{m}, \mathfrak{n}) &+ \mathfrak{a}_2(\Delta(\mathfrak{m}, \mathring{\Psi}_T\mathring{\Psi}_T\mathfrak{m}) + \Delta(\mathfrak{n}, \mathring{\Psi}_S\mathring{\Psi}_T\mathfrak{n})) \\ &+ \mathfrak{a}_3(\Delta(\mathfrak{m}, \mathring{\Psi}_S\mathring{\Psi}_T\mathfrak{n}) + \Delta(\mathfrak{n}, \mathring{\Psi}_S\mathring{\Psi}_T\mathfrak{m}))\end{aligned}$$

for all $\mathfrak{m},\ \mathfrak{n} \in \mathcal{M}_1$.

Wardowski's [5] work offered an example that exemplifies the validity of their observations, along with numerical data that validates the provided example. In their analysis, they introduced a new type of contraction called $\boldsymbol{F}$-contraction and proved a new fixed point theorem concerning $\boldsymbol{F}$-contraction in their article, which they proved using a mapping $\Psi : \mathbb{R}^+ \to \mathbb{R}$. $\boldsymbol{F}$-contraction is a remarkable generalized Banach's contraction and is defined as follows:

Definition 6.2 ([5]) Let $\boldsymbol{\Theta}$ be the collection of functions $\boldsymbol{F} : \mathbb{R}^+ \to \mathbb{R}$ be a mapping satisfying the following:

($\mathscr{F}_{w_1}$) $\boldsymbol{F}$ is strictly increasing, i.e., $\forall$ $\mathfrak{m}$, $\mathfrak{n} \in \mathbb{R}^+$ such that, $\mathfrak{m} < \mathfrak{n} \implies \boldsymbol{F}(\mathfrak{m}) < F(\mathfrak{n})$;
($\mathscr{F}_{w_1}$) For each sequence $\{\alpha_\Delta\}_{\Delta \in \mathbb{N}}$, $\lim_{\Delta \to \infty} \alpha_\Delta = 0 \iff F(\alpha_\Delta)$ diverges to $-\infty$;
($\mathscr{F}_{w_1}$) There exists $\kappa \in (0, 1)$ such that $\lim_{\alpha \to 0^+} \alpha^\kappa (F(\alpha)) = 0$.

A mapping $\Psi : \mathscr{M} \to \mathscr{M}$ is said to be an $\boldsymbol{F}$-contraction if $\exists$ positive τ such that

$$\forall_{\mathfrak{m},\mathfrak{n} \in \mathscr{M}} (\Delta(\Psi\mathfrak{m}, \Psi\mathfrak{n}) > 0 \implies F(\Delta(\Psi\mathfrak{m}, \Psi\mathfrak{n})) \leq F(\Delta(\mathfrak{m}, \mathfrak{n})) - \tau),$$

where $\Psi\mathfrak{m} \neq \Psi\mathfrak{n}$.

They obtain contractions of the type known from the literature for the concrete mappings Ψ, as well as the Banach contraction. The article included examples of $\boldsymbol{F}_{\mathcal{PC}}$, one of which proved that the generalization result is more significant. Note that, in their study the nature of $\boldsymbol{F}_{\mathcal{PC}}$ is articulated through theoretical considerations validated by computational data. The definition for multiplicative metric spaces ($\mathcal{MMS}$) [7, 8] is defined as follows:

Let $\mathscr{M}$ be a nonempty set. Multiplicative metric [7, 8] is a mapping $\Delta : \mathscr{M} \times \mathscr{M} \to \mathbb{R}^+$ satisfying the following conditions:

($\mathscr{M}_1$) $\Delta(\mathfrak{m}, \mathfrak{n}) > 1$ $\forall$ $\mathfrak{m}$, $\mathfrak{n} \in \mathscr{M}$ and $\Delta(\mathfrak{m}, \mathfrak{n}) = 1$ iff $\mathfrak{m} = \mathfrak{n}$;
($\mathscr{M}_2$) $\Delta(\mathfrak{m}, \mathfrak{n}) = \Delta(\mathfrak{n}, \mathfrak{m})$ $\forall$ $\mathfrak{m}$, $\mathfrak{n} \in \mathscr{M}$;
($\mathscr{M}_3$) multiplicative triangle inequality
$\Delta(\mathfrak{m}, \mathfrak{p}) \leq \Delta(\mathfrak{m}, \mathfrak{n}).\Delta(\mathfrak{n}, \mathfrak{p})$ $\forall$ $\mathfrak{m}$, $\mathfrak{n} \in \mathscr{M}$.

The $\boldsymbol{F}$-contraction in the framework of multiplicative metric spaces is defined as follows:

Definition 6.3 Let $\boldsymbol{\Theta}$ be the collection of functions $\boldsymbol{F} : \mathbb{R}^+ \to \mathbb{R}$ be a mapping satisfying the following:

($\mathscr{F}_1$) $\boldsymbol{F}$ is strictly increasing;
($\mathscr{F}_2$) For each sequence $\{\alpha_\Delta\}_{\Delta \in \mathbb{N}}$, $\lim_{\Delta \to \infty} \alpha_\Delta = 0 \iff F(\alpha_\Delta)$ diverges to 0;
($\mathscr{F}_3$) There exists $\kappa \in (0, 1)$ such that $\lim_{\alpha \to 0^+} (F(\alpha))^{\alpha^\kappa} = 1$.

A mapping $\Psi : \mathscr{M} \to \mathscr{M}$ is said to be an $\boldsymbol{F}_{\mathcal{PC}}$ if $\exists$ $\tau > 1$ such that

$$\forall_{\mathfrak{m},\mathfrak{n} \in \mathscr{M}} (\Delta(\Psi\mathfrak{m}, \Psi\mathfrak{n}) > 1 \implies \tau \times F(\Delta(\Psi\mathfrak{m}, \Psi\mathfrak{n})) \leq F(\Delta(\mathfrak{m}, \mathfrak{n}))).$$

Example 6.1 Let the functions $\boldsymbol{F}_1$, $\boldsymbol{F}_2$, $\boldsymbol{F}_3$, $\boldsymbol{F}_4$, $\boldsymbol{F}_5 : \mathbb{R}^+ \to \mathbb{R}$ defined as follows:

($\mathscr{E}_1$) $\boldsymbol{F}_1(\eta) = \log(1 + \eta)$ for $\eta \in (0, \infty)$; for $\alpha_n = \frac{1}{n}$ and $\kappa = \frac{1}{2}$. It can easily be shown that the conditions $\mathscr{F}_1$, $\mathscr{F}_2$ and condition $\mathscr{F}_3$ are satisfied and hence $\boldsymbol{F}_1 \in \boldsymbol{\Theta}$.

($\mathscr{E}_2$) $\boldsymbol{F}_2(\eta) = \eta \log(\eta)$ for $\eta \in (0, \infty)$; for $\alpha_n = \frac{1}{\sqrt{n}}$ and $\kappa = \frac{1}{2}$. It can easily be shown that the conditions $\mathscr{F}_1$, $\mathscr{F}_2$ and condition $\mathscr{F}_3$ are satisfied and hence $\boldsymbol{F}_2 \in \boldsymbol{\Theta}$.

($\mathscr{E}_3$) $\boldsymbol{F}_3(\eta) = \eta \log(1 + \eta + \eta^2)$ for $\eta \in (0, \infty)$; for $\alpha_n = \frac{1}{n}$ and $\kappa = \frac{1}{2}$. It can easily be shown that the conditions $\mathscr{F}_1$, $\mathscr{F}_1$ and condition $\mathscr{F}_3$ are satisfied and hence $\boldsymbol{F}_3 \in \boldsymbol{\Theta}$.

We will show that in the framework of multiplicative metric spaces, the condition ($\mathscr{F}_2$) can be replaced by an equivalent but a more simple one. To understand the topological properties in the framework of Multiplicative metric space it is requested to refer [7]. First, we need to prove the following two elementary lemmas.

Lemma 6.1 *Let $(\mathscr{M}, \Delta)$ be and multiplicative metric space and a sequence of real numbers $\{\omega_\eta\}_{\eta=1}^{\infty}$ which is bounded and all its convergent subsequences have the same limit ϑ, then $\{\omega_\eta\}_{\eta=1}^{\infty}$ is convergent and $\lim\limits_{\eta \to \infty} \omega_\eta = \vartheta$.*

Proof For two subsequences $\{\omega_{\eta_\kappa}\}_{\kappa \in \mathbb{N}}$, $\{\omega_{\eta_\upsilon}\}_{\upsilon \in \mathbb{N}}$ of $\{\omega_\eta\}$ such that

$$\lim_{\kappa \to \infty} \omega_{\eta_\kappa} = \liminf \omega_\eta \text{ and } \lim_{\upsilon \to \infty} \omega_{\eta_\upsilon} = \limsup \omega_\eta.$$

Therefore we infer that $\liminf \omega_\eta = \limsup \omega_\eta = \vartheta$, i.e., $\lim\limits_{\eta \to \infty} \omega_\eta = \vartheta$.

Lemma 6.2 *Let $(\mathscr{M}, \Delta)$ be an multiplicative metric space and increasing function $\boldsymbol{F} : \mathbb{R}^+ \to \mathbb{R}$ and a sequence $\{\omega_\eta\}_{\eta=1}^{\infty}$ positive real numbers. The following assertions are true:*

(i) if $\boldsymbol{F}(\omega_\eta) \xrightarrow{\eta} 0$, then $\omega_\eta \xrightarrow{\eta} 0$;
(ii) if $\inf \boldsymbol{F} = 0$ and $\omega_\eta \xrightarrow{\eta} 0$, then $\boldsymbol{F}(\omega_\eta) \xrightarrow{\eta} 0$.

Proof (i) It is to observe that $\{\omega_\eta\}_{\eta \in \mathbb{N}}$ is bounded. In case the sequence is unbounded above, a subsequence $\{\omega_{\eta_\kappa}\}_{\kappa \to \eta}$ such that $\omega_{\eta_\kappa} \xrightarrow{\kappa} \infty$. Then, $\forall\ \epsilon > 1$ and for any $\kappa_\epsilon \geq \kappa\epsilon\ \exists\ \kappa_\epsilon \in \mathbb{N}$ such that $\omega_{\eta_\kappa} \geq \epsilon$. Therefore $\boldsymbol{F}(\epsilon) \leq \boldsymbol{F}(\omega_{\eta_\kappa})$, i.e., $\boldsymbol{F}(\epsilon) \leq \lim\limits_{\kappa \to \infty} \boldsymbol{F}(\omega_{\eta_\kappa}) = 0$, which is peculiar. It confirms the boundedness of $\{\omega_\eta\}_{\eta \in \mathbb{N}}$. Therefore there exists a convergent subsequence of $\{\omega_\eta\}$ say $\{\omega_{\eta_\upsilon}\}_{\upsilon \in \mathbb{N}}$. $\jmath = \lim\limits_{\upsilon} \omega_{\eta_\upsilon}$. Clearly $\jmath \geq 0$. Let $\jmath > 0$ and $0 < \epsilon < \jmath$. Then $\exists\ \kappa_\epsilon \in \mathbb{N}$ in such a way that $\omega_{\eta_\upsilon} \in (\frac{\jmath}{\epsilon}, \jmath\epsilon)$ $\forall\ \upsilon \geq \upsilon_\epsilon$. Also, the function $\boldsymbol{F}$ is increasing, have $\boldsymbol{F}(\frac{\jmath}{\epsilon}) \leq \lim\limits_{\upsilon} \boldsymbol{F}(\omega_{\eta_\upsilon}) = 0$ which is contradicting $\boldsymbol{F}(\frac{\jmath}{\epsilon}) \in \mathbb{R}$, which results that $\jmath = 0$. Now using Lemma 6.1, let $\inf \boldsymbol{F} = 0$ and ω_η converges to 0 as $\eta \to \infty$. Let $\epsilon > 1$, for $\jmath > 0$ in such a way $\boldsymbol{F}(\jmath) < \frac{1}{\epsilon}$. Also, there exists $\eta_\jmath \in \mathbb{N}$ in such a way that $\omega_\eta < \jmath\ \forall\ \eta \geq \eta_\jmath$. Therefore, for $\eta \geq \eta_\jmath$, $\boldsymbol{F}(\omega_\eta) \leq \boldsymbol{F}(\jmath) < \frac{1}{\epsilon}$. Thus $\lim\limits_{\eta \to \infty} F(\omega_\eta) = 0$.

Several studies have analyzed fixed point theorems for this $\boldsymbol{F}$-contraction since its introduction. Cosentino and Vetro [40], in particular, introduced Roger-Hardy-type $\boldsymbol{F}$-contraction and proved a fixed point theorem in complete metric spaces in 2014.

Definition 6.4 ([40]) Let $(\mathscr{M}, \Delta)$ be a metric space. Then $\mathring{\Psi} : \mathscr{M} \to \mathscr{M}$ is called $\boldsymbol{F}$-contraction of Roger-Hardy-type if $\exists\ \tau > 0$, $\boldsymbol{F} \in \boldsymbol{\theta}$ such that

$$\Delta(\mathring{\Psi}\mathfrak{m}, \mathring{\Psi}\mathfrak{n}) > 0 \implies \tau + \boldsymbol{F}(\Delta(\mathring{\Psi}\mathfrak{m}, \mathring{\Psi}\mathfrak{n})) \leq F(\mathfrak{a}_1\Delta(\mathfrak{m}, \mathfrak{n}) + \mathfrak{a}_2\Delta(\mathfrak{m}, \mathring{\Psi}\mathfrak{m}) + \mathfrak{a}_3\Delta(\mathfrak{n}, \mathring{\Psi}\mathfrak{n}) + \mathfrak{a}_4\Delta(\mathfrak{m}, \mathring{\Psi}\mathfrak{n}) + \mathfrak{a}_5\Delta(\mathfrak{n}, \mathring{\Psi}\mathfrak{m}))$$

$\forall\ \mathfrak{m},\ \mathfrak{n} \in \mathscr{M}$, where $\mathfrak{a}_1 + \mathfrak{a}_2 + \mathfrak{a}_3 + 2\mathfrak{a}_4 = 1$, $\mathfrak{a}_3 = 1$ and $\mathfrak{a}_5 \geq 0$.

Let $(\mathscr{M}, \Delta)$ be a multiplicative metric space. $\mathscr{M}_1,\ \mathscr{M}_2 \subseteq \mathscr{M}$ and $\mathscr{M}_1,\ \mathscr{M}_2 \neq \emptyset$ then the following notations are substantial for subsequent use.

$$\begin{aligned} \Delta(\mathscr{M}_1, \mathscr{M}_2) &= \inf\{\Delta(\mathfrak{m}, \mathfrak{n}) : \mathfrak{m} \in \mathscr{M}_1 \text{ and } \mathfrak{n} \in \mathscr{M}_2\} \\ \mathscr{M}_1^0 &= \{\mathfrak{m} \in \mathscr{M}_1 : \Delta(\mathfrak{m}, \mathfrak{n}) = \Delta(\mathscr{M}_1, \mathscr{M}_2 \text{ for some } \mathfrak{n} \in \mathscr{M}_2)\} \\ \mathscr{M}_2^0 &= \{\mathfrak{n} \in \mathscr{M}_2 : \Delta(\mathfrak{m}, \mathfrak{n}) = \Delta(\mathscr{M}_1, \mathscr{M}_2 \text{ for some } \mathfrak{m} \in \mathscr{M}_1)\}. \end{aligned}$$

For the subsets $\mathscr{M}_1$ and $\mathscr{M}_2$ of normed space, $\Delta(\mathscr{M}_1, \mathscr{M}_2) > 0$ then the sets $\mathscr{M}_1^0$ and $\mathscr{M}_2^0$ are contained in the boundaries of $\mathscr{M}_1$ and $\mathscr{M}_2$, respectively.

Definition 6.5 Let $(\mathscr{M}, \Delta)$ be a multiplicative metric space. $\mathscr{M}_1,\ \mathscr{M}_2 \neq \emptyset$ and $\mathscr{M}_1,\ \mathscr{M}_2 \subseteq \mathscr{M}$. An element $\mathfrak{m} \in \mathscr{M}$ is known as the best proximity point of a non-self mapping $\mathring{\Psi} : \mathscr{M}_1 \to \mathscr{M}_2$ if $\Delta(\mathfrak{m}, \mathring{\Psi}\mathfrak{m}) = \Delta(\mathscr{M}_1, \mathscr{M}_2)$. Obviously, the best proximity point is reduced to be a fixed point if $\Delta(\mathscr{M}_1, \mathscr{M}_2) = 1$.

Definition 6.6 Let $(\mathscr{M}_1, \Delta)$ and $(\mathscr{M}_2, \Delta_\rho)$ be two multiplicative metric spaces. A mapping $\mathring{\Psi} : \mathscr{M}_1 \to \mathscr{M}_2$ is called a non-self nonexpansive mapping if $\forall\ \mathfrak{m}_1,\ \mathfrak{m}_2 \in \mathscr{M}_2$, $\Delta_\rho(\mathring{\Psi}\mathfrak{m}_1, \mathring{\Psi}\mathfrak{m}_2) \leq \Delta(\mathfrak{m}_1, \mathfrak{m}_2)$.

Remark 6.1 For different values for Δ in Definition 6.6, we have

($\mathscr{R}_1$) A mapping is known as non-self Lipschitzian, if for $\Delta \geq 0$, mapping $\mathring{\Psi} : \mathscr{M}_1 \to \mathscr{M}_2$.

($\mathscr{R}_2$) A mapping $\mathring{\Psi} : \mathscr{M}_1 \to \mathscr{M}_2$ is called a non-self contractive mapping, if $0 \leq \Delta < 1$.

Definition 6.7 Let $(\mathscr{M}, \Delta)$ be a multiplicative metric space and $\mathscr{M}_1,\ \mathscr{M}_2 \neq \emptyset$ and $\mathscr{M}_1,\ \mathscr{M}_2 \subseteq \mathscr{M}$. Let mapping $\mathring{\Psi}_t : \mathscr{M}_1 \to \mathscr{M}_2$ and $\mathring{\Psi}_s : \mathscr{M}_2 \to \mathscr{M}_1$, the pair $(\mathring{\Psi}_s, \mathring{\Psi}_t)$ is known as multiplicative weak Δ−cyclic contraction if $\exists\ 0 \leq \Delta < \frac{1}{2}$ such that

$$\Delta(\overset{\circ}{\Psi}_t \mathfrak{m}_1, \overset{\circ}{\Psi}_s \overset{\circ}{\Psi}_t \mathfrak{m}_1) \leq (\Delta(\mathfrak{m}_1, \overset{\circ}{\Psi}_t \mathfrak{m}_1) \times \Delta(\overset{\circ}{\Psi}_t \mathfrak{m}_1, \overset{\circ}{\Psi}_s \overset{\circ}{\Psi}_t \mathfrak{m}_1))^{\Delta} \times (\Delta(\mathscr{M}_1, \mathscr{M}_2))^{(1-2\Delta)}$$

$\forall\ \mathfrak{m}_1 \in \mathscr{M}_1$

Definition 6.8 Let $(\mathscr{M}, \Delta)$ be a multiplicative metric space and $\mathscr{M}_1,\ \mathscr{M}_2 \neq \emptyset$ and $\mathscr{M}_1,\ \mathscr{M}_2 \subseteq \mathscr{M}$. Let mapping $\overset{\circ}{\Psi}_t : \mathscr{M}_1 \to \mathscr{M}_2$ and $\overset{\circ}{\Psi}_s : \mathscr{M}_2 \to \mathscr{M}_1$, the pair $(\overset{\circ}{\Psi}_s, \overset{\circ}{\Psi}_t)$ is known as multiplicative $\Delta-$cyclic contraction if $\exists\ 0 \leq \Delta < \frac{1}{2}$ such that

$$\Delta(\overset{\circ}{\Psi}_t \mathfrak{m}_1, \overset{\circ}{\Psi}_s \mathfrak{m}_2) \leq (\Delta(\mathfrak{m}_1, \overset{\circ}{\Psi}_t \mathfrak{m}_1) \times \Delta(\mathfrak{m}_2, \overset{\circ}{\Psi}_s \mathfrak{m}_2))^{\Delta} \times (\Delta(\mathscr{M}_1, \mathscr{M}_2))^{(1-2\Delta)}$$

$\forall\ \mathfrak{m}_1 \in \mathscr{M}_1$ and $\mathfrak{m}_2 \in \mathscr{M}_2$

Definition 6.9 ([15]) Let $(\mathscr{M}, \Delta)$ be a metric space and $(\mathscr{M}_1, \mathscr{M}_2)$ be a pair of non-void subsets of $(\mathscr{M}, \Delta)$ with $\mathscr{M}_1^0 = \emptyset$. If for every $\mathfrak{m}', \mathfrak{m}'' \in \mathscr{M}_1$ and every $\mathfrak{n}'$, $\mathfrak{n}'' \in \mathscr{M}_2$

$$\left.\begin{array}{l} \Delta(\mathfrak{m}', \mathfrak{n}') = \Delta(\mathscr{M}_1, \mathscr{M}_2) \\ \Delta(\mathfrak{m}'', \mathfrak{n}'') = \Delta(\mathscr{M}_1, \mathscr{M}_2) \end{array}\right\} \implies \Delta(\mathfrak{m}', \mathfrak{m}'') = \Delta(\mathfrak{n}', \mathfrak{n}'').$$

Then, the pair $(\mathscr{M}_1, \mathscr{M}_2)$ is said to satisfy $p-$property

Definition 6.10 ([13]) $\mathscr{M}_1$ is said to be approximately compact with respect to $\mathscr{M}_2$ if every sequence $\{\mathfrak{m}_\eta\}_{\eta=1}^{\infty}$ of $\mathscr{M}_1$ satisfying the condition that $\Delta(\mathfrak{n}, \mathfrak{m}_\eta) \to \Delta(\mathfrak{n}, \mathscr{M}_1)$ for some $\mathfrak{n}$ in $\mathscr{M}_2$ has a convergent subsequence.

The multiplicative version of the definition of *proximal contraction of the first kind* given by Basha [13] is as follows:

Definition 6.11 A mapping $\Psi : \mathscr{M}_1 \to \mathscr{M}_2$ said to be a multiplicative proximal contraction of first kind if $\exists$ a $\alpha \in (0, 1)$ such that, $\forall\ \mathfrak{p}_1, \mathfrak{p}_2, \mathfrak{q}_1, \mathfrak{q}_2 \in \mathscr{M}_1$

$$\left.\begin{array}{l} \Delta(\mathfrak{p}_1, \Psi \mathfrak{q}_1) = \Delta(\mathscr{M}_1, \mathscr{M}_2) \\ \Delta(\mathfrak{p}_2, \Psi \mathfrak{q}_2) = \Delta(\mathscr{M}_1, \mathscr{M}_2) \end{array}\right\} \implies \Delta(\mathfrak{p}_1, \mathfrak{p}_2) \leq \Delta^{\alpha}(\mathfrak{q}_1, \mathfrak{q}_2).$$

6.3 Best Proximity Point Theorems for Multiplicative Generalized $\boldsymbol{F}_{\mathcal{PC}}$

In this sector, taking into account the Hardy-Rogers-type $\boldsymbol{F}_{\mathcal{PC}}$ [1] and multiplicative calculus, the purpose of this paper is to bring multiplicative generalized $\boldsymbol{F}_{\mathcal{PC}}$ mapping in the framework of $\mathcal{CMMS}$. We begin by describing multiplicative metric space in this research. Furthermore, we proposed the concept of multiplicative generalized $\boldsymbol{F}_{\mathcal{PC}}$ mapping and established several best proximity point theorems in $\mathcal{CMMS}$.

Definition 6.12 A mapping $\overset{\circ}{\Psi} : \mathscr{M}_1 \to \mathscr{M}_2$ is said to be a multiplicative generalized $\boldsymbol{F}_{\mathcal{PC}}$ of the first kind if $\exists\ \boldsymbol{F} \in \boldsymbol{\Theta}$ and $\mathfrak{a}_1,\ \mathfrak{a}_2,\ \mathfrak{a}_3,\ \mathfrak{a}_4 > 0,\ \tau > 1$ with $\mathfrak{a}_1 + \mathfrak{a}_2 + \mathfrak{a}_3 + 2\mathfrak{a}_4 = 1$, $\mathfrak{a}_3 \neq 1$ such that the conditions

$$\left.\begin{aligned} \Delta(\mathfrak{m}^{'}, \overset{\circ}{\Psi}\mathfrak{n}^{'}) = \Delta(\mathscr{M}_1, \mathscr{M}_2) \\ \Delta(\mathfrak{m}^{''}, \overset{\circ}{\Psi}\mathfrak{n}^{''}) = \Delta(\mathscr{M}_1, \mathscr{M}_2) \end{aligned}\right\} \implies \tau \times \boldsymbol{F}(\Delta(\mathfrak{n}^{'}, \mathfrak{n}^{''}))$$

and

$$\begin{aligned} \tau \times \boldsymbol{F}(\Delta(\mathfrak{n}^{'}, \mathfrak{n}^{''})) \leq \boldsymbol{F}(\Delta^{\mathfrak{a}_1}(\mathfrak{n}^{'}, \mathfrak{n}^{''}) \times \Delta^{\mathfrak{a}_2}(\mathfrak{m}^{'}, \mathfrak{n}^{'}) \times \Delta^{\mathfrak{a}_3}(\mathfrak{m}^{''}, \mathfrak{n}^{''}) \\ \times (\Delta(\mathfrak{n}^{'}, \mathfrak{m}^{''}) \times \Delta(\mathfrak{n}^{''}, \mathfrak{m}^{'}))^{\mathfrak{a}_4}) \end{aligned}$$

$\forall\ \mathfrak{m}^{'},\ \mathfrak{m}^{''},\ \mathfrak{n}^{'},\ \mathfrak{n}^{''} \in \mathscr{M}_1$ and $\mathfrak{m}^{'} \neq \mathfrak{m}^{''}$.

Definition 6.13 A mapping $\overset{\circ}{\Psi} : \mathscr{M}_1 \to \mathscr{M}_2$ is said to be a multiplicative generalized $\boldsymbol{F}_{\mathcal{PC}}$ of the second kind if $\exists\ \boldsymbol{F} \in \boldsymbol{\Theta}$ and $\mathfrak{a}_1,\ \mathfrak{a}_2,\ \mathfrak{a}_3,\ \mathfrak{a}_4 > 0,\ \tau > 1$ with $\mathfrak{a}_1 + \mathfrak{a}_2 + \mathfrak{a}_3 + 2\mathfrak{a}_4 = 1$, $\mathfrak{a}_3 \neq 1$ such that the conditions

$$\left.\begin{aligned} \Delta(\mathfrak{m}^{'}, \overset{\circ}{\Psi}\mathfrak{n}^{'}) = \Delta(\mathscr{M}_1, \mathscr{M}_2) \\ \Delta(\mathfrak{m}^{''}, \overset{\circ}{\Psi}\mathfrak{n}^{''}) = \Delta(\mathscr{M}_1, \mathscr{M}_2) \end{aligned}\right\} \implies \tau \times \boldsymbol{F}(\Delta(\overset{\circ}{\Psi}\mathfrak{n}^{'}, \overset{\circ}{\Psi}\mathfrak{n}^{''}))$$

and

$$\begin{aligned} \tau \times \boldsymbol{F}(\Delta(\overset{\circ}{\Psi}\mathfrak{n}^{'}, \overset{\circ}{\Psi}\mathfrak{n}^{''})) \leq \boldsymbol{F}(\Delta^{\mathfrak{a}_1}(\overset{\circ}{\Psi}\mathfrak{n}^{'}, \overset{\circ}{\Psi}\mathfrak{n}^{''}) \times \Delta^{\mathfrak{a}_2}(\overset{\circ}{\Psi}\mathfrak{m}^{'}, \overset{\circ}{\Psi}\mathfrak{n}^{'}) \times \Delta^{\mathfrak{a}_3}(\overset{\circ}{\Psi}\mathfrak{m}^{''}, \overset{\circ}{\Psi}\mathfrak{n}^{''}) \\ \times (\Delta(\overset{\circ}{\Psi}\mathfrak{n}^{'}, \overset{\circ}{\Psi}\mathfrak{m}^{''}) \times \Delta(\overset{\circ}{\Psi}\mathfrak{n}^{''}, \overset{\circ}{\Psi}\mathfrak{m}^{'}))^{\mathfrak{a}_4}) \end{aligned}$$

$\forall\ \mathfrak{m}^{'},\ \mathfrak{m}^{''},\ \mathfrak{n}^{'},\ \mathfrak{n}^{''} \in \mathscr{M}_1$ and $\overset{\circ}{\Psi}\mathfrak{m}^{'} \neq \overset{\circ}{\Psi}\mathfrak{m}^{''}$.

Theorem 6.3 *Let $\mathscr{M}$ be a $\mathcal{CMMS}$ and the pair $(\mathscr{M}_1, \mathscr{M}_2) \neq \emptyset$ closed subsets of $(\mathscr{M}, \Delta)$. If $\mathscr{M}_2$ is approximately compact w.r.t $\mathscr{M}_1$ and $\overset{\circ}{\Psi} : \mathscr{M}_1 \to \mathscr{M}_2$ satisfying*

($\mathscr{C}_1$) *$\overset{\circ}{\Psi}(\mathscr{M}_1^0) \subseteq \mathscr{M}_2^0$ and $(\mathscr{M}_1, \mathscr{M}_2)$ satisfies the $p-$property;*

($\mathscr{C}_2$) *$\overset{\circ}{\Psi}$ be a multiplicative generalized $\boldsymbol{F}_{\mathcal{PC}}$ of the first kind.*

Then, there exists $\mathfrak{m} \in \mathscr{M}$ such that $\Delta(\mathfrak{m}, \overset{\circ}{\Psi}\mathfrak{m}) = \Delta(\mathscr{M}_1, \mathscr{M}_2)$. Furthermore, for arbitrarily chosen $\mathfrak{m}_0 \in \mathscr{M}_1^0$, the sequence $\{\mathfrak{m}_\eta\}_{\eta=1}^{\infty}$ defined as follows

$$\Delta(\mathfrak{m}_{\eta+1}, \overset{\circ}{\Psi}\mathfrak{m}_\eta) = \Delta(\mathscr{M}_1, \mathscr{M}_2)$$

converges to $\mathfrak{m}$, which is a best proximity point.

Proof Let $\mathfrak{m}_0 \in \mathscr{M}_1$, since $\overset{\circ}{\Psi}(\mathscr{M}_1^0) \subseteq \mathscr{M}_2^0$, therefore $\exists\ \mathfrak{m}_1 \in \mathscr{M}_1^0$ such that

$$\Delta(\mathfrak{m}_1, \overset{\circ}{\Psi}\mathfrak{m}_0) = \Delta(\mathscr{M}_1, \mathscr{M}_2).$$

Also, $\overset{\circ}{\Psi}\mathfrak{m}_1 \in \overset{\circ}{\Psi}(\mathscr{M}_1^0) \subseteq \mathscr{M}_2^0$, it furthermore signifies that an element exists say $\mathfrak{m}_2 \in \mathscr{M}_1^0$ in order for

$$\Delta(\mathfrak{m}_2, \overset{\circ}{\Psi}\mathfrak{m}_1) = \Delta(\mathscr{M}_1, \mathscr{M}_2).$$

Inductively, we can choose an element $\mathfrak{m}_{\eta+1} \in \mathscr{M}_1^0$ by continuing in this way, we have

$$\Delta(\mathfrak{m}_{\eta+1}, \overset{\circ}{\Psi}\mathfrak{m}_\eta) = \Delta(\mathscr{M}_1, \mathscr{M}_2), \tag{6.1}$$

$\forall\ \eta \in \mathbb{N} \cup \{0\}$. Owing to the hypothesis that $\overset{\circ}{\Psi}(\mathscr{M}_1^0) \subset \mathscr{M}_2^0$, we can derive

$$\Delta(\mathfrak{m}_\eta, \mathfrak{m}_{\eta+1}) = \Delta(\overset{\circ}{\Psi}\mathfrak{m}_{\eta-1}, \overset{\circ}{\Psi}\mathfrak{m}_\eta),$$

when combining Eq. (6.1) with the $p-$property. If $\Delta(\mathfrak{m}_{\eta_0}, \mathfrak{m}_{\eta_0+1}) = 1$ for some $\eta_0 \geq 0$, we have

$$\Delta(\overset{\circ}{\Psi}\mathfrak{m}_{\eta_0-1}, \overset{\circ}{\Psi}\mathfrak{m}_{\eta_0}) = 1,$$

therefore $\overset{\circ}{\Psi}\mathfrak{m}_{\eta_0-1} = \overset{\circ}{\Psi}\mathfrak{m}_{\eta_0}$. It results that $\Delta(\mathscr{M}_1, \mathscr{M}_2) = \Delta(\mathfrak{m}_{\eta_0}, \overset{\circ}{\Psi}\mathfrak{m}_{\eta_0})$. As a result, the conclusion is obvious. Let for arbitrary $\eta \geq 0$, $\Delta(\mathfrak{m}_\eta, \mathfrak{m}_{\eta+1}) > 1$. Since $\overset{\circ}{\Psi}$ is first kind multiplicative generalized $\boldsymbol{F}_{\mathcal{PC}}$, we have

$$\begin{aligned}
\tau \times \boldsymbol{F}(\Delta(\mathfrak{m}_\eta, \mathfrak{m}_{\eta+1})) &\leq \boldsymbol{F}(\Delta^{\mathfrak{a}_1}(\mathfrak{m}_{\eta-1}, \mathfrak{m}_\eta) \times \Delta^{\mathfrak{a}_2}(\mathfrak{m}_{\eta-1}, \mathfrak{m}_\eta) \times \Delta^{\mathfrak{a}_3}(\mathfrak{m}_\eta, \mathfrak{m}_{\eta+1}) \\
&\quad \times (\Delta(\mathfrak{m}_{\eta-1}, \mathfrak{m}_{\eta+1}) \times \Delta(\mathfrak{m}_\eta, \mathfrak{m}_{\eta+1}))^{\mathfrak{a}_4} \\
&\leq \boldsymbol{F}(\Delta^{\mathfrak{a}_1}(\mathfrak{m}_{\eta-1}, \mathfrak{m}_\eta) \times \Delta^{\mathfrak{a}_2}(\mathfrak{m}_{\eta-1}, \mathfrak{m}_\eta) \times \Delta^{\mathfrak{a}_3}(\mathfrak{m}_\eta, \mathfrak{m}_{\eta+1}) \\
&\quad \times \Delta^{\mathfrak{a}_4}(\mathfrak{m}_{\eta-1}, \mathfrak{m}_\eta) \times \Delta^{\mathfrak{a}_4}(\mathfrak{m}_\eta, \mathfrak{m}_{\eta+1}) \\
&= \boldsymbol{F}(\Delta^{\mathfrak{a}_1+\mathfrak{a}_2+\mathfrak{a}_4}(\mathfrak{m}_{\eta-1}, \mathfrak{m}_\eta) \times \Delta^{\mathfrak{a}_3+\mathfrak{a}_4}(\mathfrak{m}_\eta, \mathfrak{m}_{\eta+1})).
\end{aligned}$$

It is given that $\boldsymbol{F}$ is strictly increasing, we can conclude that

$$\Delta(\mathfrak{m}_\eta, \mathfrak{m}_{\eta+1}) \leq \Delta^{\mathfrak{a}_1+\mathfrak{a}_2+\mathfrak{a}_4}(\mathfrak{m}_{\eta-1}, \mathfrak{m}_\eta) \times \Delta^{\mathfrak{a}_3+\mathfrak{a}_4}(\mathfrak{m}_\eta, \mathfrak{m}_{\eta+1}).$$

Therefore

$$\Delta(\mathfrak{m}_\eta, \mathfrak{m}_{\eta+1}) \leq (\Delta(\mathfrak{m}_\eta, \mathfrak{m}_{\eta-1}))^{\frac{\mathfrak{a}_1+\mathfrak{a}_2+\mathfrak{a}_4}{1-\mathfrak{a}_3-\mathfrak{a}_4}},$$

$\forall\ \eta \in \mathbb{N} \cup \{0\}$. It is given that $\mathfrak{a}_1 + \mathfrak{a}_2 + \mathfrak{a}_3 + 2\mathfrak{a}_4 = 1$ and $\mathfrak{a}_3 \neq 1$, which results that $\mathfrak{a}_3 + \mathfrak{a}_4 < 1$ and hence

$$\Delta(\mathfrak{m}_\eta, \mathfrak{m}_{\eta+1}) = (\Delta(\mathfrak{m}_\eta, \mathfrak{m}_{\eta-1}))^{\frac{\mathfrak{a}_1+\mathfrak{a}_2+\mathfrak{a}_4}{1-\mathfrak{a}_3-\mathfrak{a}_4}} = \Delta(\mathfrak{m}_\eta, \mathfrak{m}_{\eta-1}).$$

As a result, we have $\tau \times \boldsymbol{F}(\Delta(\mathfrak{m}_\eta, \mathfrak{m}_{\eta+1})) \leq \boldsymbol{F}(\Delta(\mathfrak{m}_\eta, \mathfrak{m}_{\eta-1}))$, which shows that

$$\begin{aligned} \boldsymbol{F}(\Delta(\mathfrak{m}_\eta, \mathfrak{m}_{\eta+1})) &\leq \tau^{-1}\boldsymbol{F}(\Delta(\mathfrak{m}_\eta, \mathfrak{m}_{\eta-1})) \\ &\leq \tau^{-2}\boldsymbol{F}(\Delta(\mathfrak{m}_{\eta-1}, \mathfrak{m}_{\eta-2})) \\ &\leq \tau^{-3}\boldsymbol{F}(\Delta(\mathfrak{m}_{\eta-2}, \mathfrak{m}_{\eta-3})) \\ &\leq \ldots \\ &\leq \tau^{-n}\boldsymbol{F}(\Delta(\mathfrak{m}_0, \mathfrak{m}_1)). \end{aligned} \tag{6.2}$$

Let $\omega_\eta = \Delta(\mathfrak{m}_\eta, \mathfrak{m}_{\eta+1})$, by (6.2) $\lim_{\eta\to\infty} \boldsymbol{F}(\omega_\eta) = 0$, using $(\mathscr{F}_2)$, we have

$$\lim_{\eta\to\infty} \omega_\eta = 0.$$

Consider $\kappa \in (0, 1)$ as $\lim_{\eta\to\infty} (\boldsymbol{F}(\omega_\eta))^{\omega_\eta^\kappa} = 0$, refer (6.2), we have

$$\frac{\boldsymbol{F}(\omega_\eta)^{\omega_\eta^\kappa}}{\boldsymbol{F}(\omega_0)^{\omega_\eta^\kappa}} \leq \tau^{\eta\omega_\eta^\kappa} \leq 1. \tag{6.3}$$

For $\kappa \to \infty$ in (6.3), we have

$$\lim_{\eta\to\infty} \eta\omega_\eta^\kappa = 0.$$

Hence $\lim_{\eta\to\infty} \eta\omega_\eta^{(\frac{1}{\kappa})}$, which confirms the convergence of series ω_η. In addition, it implies that $\{\mathfrak{m}_\eta\}$ is a Cauchy sequence. The given multiplicative metric space $(\mathscr{M}, \Delta)$ is complete, it converges to some point $\mathfrak{m}$ of $\mathscr{M}_1$. Furthermore,

$$\begin{aligned} \Delta(\mathfrak{m}, \mathscr{M}_2) \leq \Delta(\mathfrak{m}, \mathring{\Psi}\mathfrak{m}_\eta) &\leq \Delta(\mathfrak{m}_\eta, \mathfrak{m}_{\eta+1}) \times \Delta(\mathfrak{m}_{\eta+1}, \mathring{\Psi}\mathfrak{m}_\eta) \\ &= \Delta(\mathfrak{m}, \mathfrak{m}_{\eta+1}) \times \Delta(\mathscr{M}_1, \mathscr{M}_2) \\ &\leq \Delta(\mathfrak{m}, \mathfrak{m}_{\eta+1}) \times \Delta(\mathfrak{m}, \mathscr{M}_2). \end{aligned}$$

It results that $\Delta(\mathfrak{m}, \mathring{\Psi}\mathfrak{m}_\eta) \to \Delta(\mathfrak{m}, \mathscr{M}_2)$. Considering the fact that $\mathscr{M}_2$ is approximately compact w.r.t. $\mathscr{M}_1$, there exists a subsequence $\{\mathring{\Psi}\mathfrak{m}_{\eta_\Delta}\}$ of $\{\mathring{\Psi}\mathfrak{m}_\eta\}$ which converges to some element $\mathring{\mathfrak{m}}$ in $\mathscr{M}_2$. So, we have

$$\Delta(\mathfrak{m}, \mathring{\mathfrak{m}}) = \lim_{\eta\to\infty} \Delta(\mathfrak{m}_{\eta_\Delta+1}, \mathring{\Psi}\mathfrak{m}_{\eta_\Delta}) = \Delta(\mathscr{M}_1, \mathscr{M}_2). \tag{6.4}$$

Therefore, $\mathfrak{m}$ is necessarily is an element of $\mathscr{M}_1^0$. As, we know that $\overset{\circ}{\Psi}(\mathscr{M}_1^0) \subseteq \mathscr{M}_2^0$, it results that

$$\Delta(\omega, \overset{\circ}{\Psi}\mathfrak{m}) = \Delta(\mathscr{M}_1, \mathscr{M}_2),$$

for some $\omega \in \mathscr{M}_1$. Combining Eq. (6.4) with $p-$ property, we have $\Delta(\mathfrak{m}_{\eta_\Delta+1}, \omega) = \Delta(\overset{\circ}{\Psi}\mathfrak{m}_{\eta_\Delta+1}, \mathfrak{m})$. If $\Delta(\omega, \mathfrak{m}_{\eta_0+1}) = 1$ for arbitrary η_0, then $\Delta(\overset{\circ}{\Psi}\mathfrak{m}_{\eta_0}, \overset{\circ}{\Psi}\mathfrak{m}) = 1$. Therefore $\overset{\circ}{\Psi}\mathfrak{m}_{\eta_0} = \overset{\circ}{\Psi}\mathfrak{m}$, hence $\Delta(\mathscr{M}_1, \mathscr{M}_2) = \Delta(\mathfrak{m}, \overset{\circ}{\Psi}\mathfrak{m})$. As a result, the conclusion is obvious. Now for $\eta \geq 0$, $\Delta(\omega, \mathfrak{m}_{\eta+1}) > 1$. Also, it is given that the mapping $\overset{\circ}{\Psi}$ is first kind $\boldsymbol{F}_{\mathcal{P}C}$, therefore

$$\begin{aligned}\tau \times \boldsymbol{F}(\omega, \mathfrak{m}_{\eta+1})) \leq \boldsymbol{F}(&\Delta^{\mathfrak{a}_1}(\alpha, \alpha_\eta) \times \Delta^{\mathfrak{a}_2}(\omega, \mathfrak{m}) \times \Delta^{\mathfrak{a}_3}(\mathfrak{m}_\eta, \mathfrak{m}_{\eta+1}) \\ &\times (\Delta(\mathfrak{m}_\eta, \mathfrak{m}_{\eta+1}) \times \Delta(\mathfrak{m}_\eta, \omega))^{\mathfrak{a}_4}.\end{aligned}$$

For $\eta \to \infty$,

$$\Delta(\omega, \mathfrak{m}) \leq \Delta(\omega, \mathfrak{m})^{(\mathfrak{a}_2+\mathfrak{a}_4)}.$$

Furthermore, this implies that $\mathfrak{m}$ and ω are necessarily identical. As a consequence,

$$\Delta(\mathfrak{m}, \overset{\circ}{\Psi}\mathfrak{m}) = \Delta(\omega, \overset{\circ}{\Psi}\mathfrak{m}) = \Delta(\mathscr{M}_1, \mathscr{M}_2).$$

To prove the uniqueness, let for the mapping $\overset{\circ}{\Psi}$, $\mathfrak{m}_\zeta$ is distinct best proximity point such that

$$\Delta(\mathfrak{m}_\zeta, \overset{\circ}{\Psi}\mathfrak{m}_\zeta) = \Delta(\mathscr{M}_1, \mathscr{M}_2.)$$

Also, $\tau \times \boldsymbol{F}(\Delta(\mathfrak{m}, \mathfrak{m}_\zeta)) \leq \boldsymbol{F}((\Delta(\mathfrak{m}, \mathfrak{m}_\zeta)^{\mathfrak{a}_1+2\mathfrak{a}_4}))$. Note that, the mapping $\boldsymbol{F}$ is strictly increasing, we have

$$\Delta(\mathfrak{m}, \mathfrak{m}_\zeta) \leq (\Delta(\mathfrak{m}, \mathfrak{m}_\zeta))^{\mathfrak{a}_1+2\mathfrak{a}_4},$$

which proves that $\mathfrak{m} = \mathfrak{m}_\zeta$.

The following best proximity theorem is for second kind multiplicative generalized $\boldsymbol{F}_{\mathcal{P}C}$.

Theorem 6.4 *Let $\mathscr{M}$ be a $\mathcal{CMMS}$ and $(\mathscr{M}_1, \mathscr{M}_2) \neq \emptyset$ closed subsets of $\mathscr{M}$. If $\mathscr{M}_1$ is approximately compact w.r.t $\mathscr{M}_2$ and $\overset{\circ}{\Psi} : \mathscr{M}_1 \to \mathscr{M}_2$ satisfying*

($\mathscr{C}_1$) *$\overset{\circ}{\Psi}(\mathscr{M}_1^0) \subseteq \mathscr{M}_2^0$ and $(\mathscr{M}_1, \mathscr{M}_2)$ satisfies the $p-$property;*

($\mathscr{C}_2$) *$\overset{\circ}{\Psi}$ be a continuous multiplicative generalized $\boldsymbol{F}_{\mathcal{P}C}$ of the second kind.*

Then, there exists $\mathfrak{m} \in \mathscr{M}$ such that $\Delta(\mathfrak{m}, \overset{\circ}{\Psi}\mathfrak{m}) = \Delta(\mathscr{M}_1, \mathscr{M}_2)$. Furthermore, for arbitrarily chosen $\mathfrak{m}_0 \in \mathscr{M}_1^0$, the sequence $\{\mathfrak{m}_\eta\}_{\eta=1}^\infty$ defined as follows:

$$\Delta(\mathfrak{m}_{\eta+1}, \overset{\circ}{\Psi}\mathfrak{m}_{\eta}) = \Delta(\mathscr{M}_1, \mathscr{M}_2)$$

converges to $\mathfrak{m}$, *which is unique best proximity point.*

Proof Using Theorem 6.3, a sequence $\{\mathfrak{m}_{\eta}\}$ can be defined in $\mathscr{M}_1^0$ such that

$$\Delta(\mathfrak{m}_{\eta+1}, \overset{\circ}{\Psi}\mathfrak{m}_{\eta}) = \Delta(\mathscr{M}_1, \mathscr{M}_2), \tag{6.5}$$

$\forall\, \eta \in \mathbb{N} \cup \{0\}$. Owing to the property $p-$ combined with Eq. (6.5), we can derive

$$\Delta(\mathfrak{m}_{\eta}, \mathfrak{m}_{\eta+1}) = \Delta(\overset{\circ}{\Psi}\mathfrak{m}_{\eta-1}, \overset{\circ}{\Psi}\mathfrak{m}_{\eta}).$$

If $\Delta(\mathfrak{m}_{\eta_0+1}, \mathfrak{m}_{\eta_0+2}) = 1$ for some η_0, we have

$$\Delta(\overset{\circ}{\Psi}\mathfrak{m}_{\eta_0}, \overset{\circ}{\Psi}\mathfrak{m}_{\eta_0+1}) = 1.$$

Therefore, $\overset{\circ}{\Psi}\mathfrak{m}_{\eta_0} = \overset{\circ}{\Psi}\mathfrak{m}_{\eta_0+1}$, it results that $\Delta(\mathscr{M}_1, \mathscr{M}_2) = \Delta(\overset{\circ}{\Psi}\mathfrak{m}_{\eta_0+1}, \overset{\circ}{\Psi}\mathfrak{m}_{\eta_0+1})$. As a result, the conclusion is obvious that for any $\eta \leq 0$, $\Delta(\mathfrak{m}_{\eta}, \mathfrak{m}_{\eta+1}) > 1$. Since $\overset{\circ}{\Psi}$ is second kind multiplicative generalized $\boldsymbol{F}_{\mathcal{PC}}$, we have

$$\begin{aligned}
\tau \times \boldsymbol{F}(\Delta(\overset{\circ}{\Psi}\mathfrak{m}_{\eta}, \overset{\circ}{\Psi}\mathfrak{m}_{\eta+1})) &\leq \boldsymbol{F}(\Delta^{\mathfrak{a}_1}(\overset{\circ}{\Psi}\mathfrak{m}_{\eta-1}, \overset{\circ}{\Psi}\mathfrak{m}_{\eta}) \times \Delta^{\mathfrak{a}_2}(\overset{\circ}{\Psi}\mathfrak{m}_{\eta-1}, \overset{\circ}{\Psi}\mathfrak{m}_{\eta}) \times \Delta^{\mathfrak{a}_3}(\overset{\circ}{\Psi}\mathfrak{m}_{\eta}, \overset{\circ}{\Psi}\mathfrak{m}_{\eta+1}) \\
&\quad \times \Delta^{\mathfrak{a}_4}(\overset{\circ}{\Psi}\mathfrak{m}_{\eta-1}, \overset{\circ}{\Psi}\mathfrak{m}_{\eta+1}) \times \Delta^{\mathfrak{a}_4}(\overset{\circ}{\Psi}\mathfrak{m}_{\eta}, \mathfrak{m}_{\eta+1}) \\
&\leq \boldsymbol{F}(\Delta^{\mathfrak{a}_1}(\overset{\circ}{\Psi}\mathfrak{m}_{\eta-1}, \overset{\circ}{\Psi}\mathfrak{m}_{\eta}) \times \Delta^{\mathfrak{a}_2}(\overset{\circ}{\Psi}\mathfrak{m}_{\eta-1}, \overset{\circ}{\Psi}\mathfrak{m}_{\eta}) \times \Delta^{\mathfrak{a}\overset{\circ}{\Psi}_3}(\mathfrak{m}_{\eta}, \overset{\circ}{\Psi}\mathfrak{m}_{\eta+1}) \\
&\quad \times (\Delta(\overset{\circ}{\Psi}\mathfrak{m}_{\eta-1}, \overset{\circ}{\Psi}\mathfrak{m}_{\eta}) \times \Delta^{\mathfrak{a}_4}(\overset{\circ}{\Psi}\mathfrak{m}_{\eta}, \overset{\circ}{\Psi}\mathfrak{m}_{\eta+1}))^{\mathfrak{a}_4} \\
&= \boldsymbol{F}(\Delta^{\mathfrak{a}_1+\mathfrak{a}_2+\mathfrak{a}_4}(\overset{\circ}{\Psi}\mathfrak{m}_{\eta-1}, \overset{\circ}{\Psi}\mathfrak{m}_{\eta}) \times \Delta^{\mathfrak{a}_3+\mathfrak{a}_4}(\overset{\circ}{\Psi}\mathfrak{m}_{\eta}, \overset{\circ}{\Psi}\mathfrak{m}_{\eta+1})).
\end{aligned}$$

It is given that $\boldsymbol{F}$ is strictly increasing, we can conclude that

$$\Delta(\overset{\circ}{\Psi}\mathfrak{m}_{\eta}, \overset{\circ}{\Psi}\mathfrak{m}_{\eta+1}) = \Delta^{\mathfrak{a}_1+\mathfrak{a}_2+\mathfrak{a}_4}(\overset{\circ}{\Psi}\mathfrak{m}_{\eta-1}, \overset{\circ}{\Psi}\mathfrak{m}_{\eta}) \times \Delta^{\mathfrak{a}_3+\mathfrak{a}_4}(\overset{\circ}{\Psi}\mathfrak{m}_{\eta}, \overset{\circ}{\Psi}\mathfrak{m}_{\eta+1}),$$

which results

$$\Delta(\overset{\circ}{\Psi}\mathfrak{m}_{\eta}, \overset{\circ}{\Psi}\mathfrak{m}_{\eta+1}) = (\Delta(\overset{\circ}{\Psi}\mathfrak{m}_{\eta}, \overset{\circ}{\Psi}\mathfrak{m}_{\eta-1}))^{\frac{\mathfrak{a}_1+\mathfrak{a}_2+\mathfrak{a}_4}{1-\mathfrak{a}_3-\mathfrak{a}_4}},$$

$\forall\, \eta \in \mathbb{N} \cup \{0\}$. Also, $\mathfrak{a}_1 + \mathfrak{a}_2 + \mathfrak{a}_3 + 2\mathfrak{a}_4 = 1$ and $\mathfrak{a}_3 \neq 1$, which results that $\mathfrak{a}_3 + \mathfrak{a}_4 < 1$ and hence

$$\Delta(\overset{\circ}{\Psi}\mathfrak{m}_{\eta}, \overset{\circ}{\Psi}\mathfrak{m}_{\eta+1}) = (\Delta(\overset{\circ}{\Psi}\mathfrak{m}_{\eta}, \overset{\circ}{\Psi}\mathfrak{m}_{\eta-1}))^{\frac{\mathfrak{a}_1+\mathfrak{a}_2+\mathfrak{a}_4}{1-\mathfrak{a}_3-\mathfrak{a}_4}} = \Delta(\overset{\circ}{\Psi}\mathfrak{m}_{\eta}, \overset{\circ}{\Psi}\mathfrak{m}_{\eta-1}).$$

As a result, we have

$$\tau \times \boldsymbol{F}(\Delta(\overset{\circ}{\Psi}\mathfrak{m}_\eta, \overset{\circ}{\Psi}\mathfrak{m}_{\eta+1})) \leq \boldsymbol{F}(\Delta(\overset{\circ}{\Psi}\mathfrak{m}_\eta, \overset{\circ}{\Psi}\mathfrak{m}_{\eta-1})),$$

which implies that

$$\begin{aligned}\boldsymbol{F}(\Delta(\overset{\circ}{\Psi}\mathfrak{m}_\eta, \overset{\circ}{\Psi}\mathfrak{m}_{\eta+1})) &\leq \tau^{-1}\boldsymbol{F}(\Delta(\overset{\circ}{\Psi}\mathfrak{m}_\eta, \overset{\circ}{\Psi}\mathfrak{m}_{\eta-1}))\\ &\leq \tau^{-2}\boldsymbol{F}(\Delta(\overset{\circ}{\Psi}\mathfrak{m}_{\eta-1}, \overset{\circ}{\Psi}\mathfrak{m}_{\eta-2}))\\ &\leq \tau^{-3}\boldsymbol{F}(\Delta(\overset{\circ}{\Psi}\mathfrak{m}_{\eta-2}, \overset{\circ}{\Psi}\mathfrak{m}_{\eta-3}))\\ &\leq \ldots\\ &\leq \tau^{-n}\boldsymbol{F}(\Delta(\overset{\circ}{\Psi}\mathfrak{m}_0, \overset{\circ}{\Psi}\mathfrak{m}_1)). \end{aligned} \tag{6.6}$$

Let $\varepsilon_\eta = \Delta(\overset{\circ}{\Psi}\mathfrak{m}_\eta, \overset{\circ}{\Psi}\mathfrak{m}_{\eta+1})$. Using (6.6), we have $\lim_{\eta\to\infty} \boldsymbol{F}(\varepsilon_\eta) = -\infty$. Now using the condition $(\mathscr{F}_2)$, we obtain $\lim_{\eta\to\infty} \varepsilon_\eta = 0$. Considering $\kappa \in (0, 1)$, $\lim_{\eta\to\infty} (\boldsymbol{F}(\varepsilon_\eta))^{\varepsilon_\eta^\kappa} = 1$ and (6.6), we have

$$\frac{\boldsymbol{F}(\varepsilon_\eta)^{\varepsilon_\eta^\kappa}}{\boldsymbol{F}(\varepsilon_0)^{\varepsilon_\eta^\kappa}} \leq \tau^{\eta\varepsilon_\eta^\kappa} \leq 0.$$

For $\kappa \to \infty$ in (6.7), we have $\lim_{\eta\to\infty} \eta\varepsilon_\eta^\kappa = 0$ and hence $\lim_{\eta\to\infty} \eta\varepsilon_\eta^{(\frac{1}{\kappa})} = 0$, which confirms the convergence of series ε_η. Furthermore, it implies that $\{\varepsilon_\eta\}$ is a Cauchy sequence and since the space $(\mathscr{M}, \Delta)$ is complete, it converges to some point ω of $\mathscr{M}_1$. Furthermore,

$$\begin{aligned}\Delta(\overset{\circ}{\mathfrak{m}}, \mathscr{M}_1) \leq \Delta(\overset{\circ}{\mathfrak{m}}, \mathfrak{m}_{\eta+1}) &\leq \Delta(\overset{\circ}{\mathfrak{m}}_\eta, \overset{\circ}{\Psi}\mathfrak{m}_\eta) \times \Delta(\overset{\circ}{\Psi}\mathfrak{m}_\eta, \mathfrak{m}_{\eta+1})\\ &= \Delta(\overset{\circ}{\mathfrak{m}}, \overset{\circ}{\Psi}\mathfrak{m}_\eta) \times \Delta(\mathscr{M}_1, \mathscr{M}_2)\\ &\leq \Delta(\overset{\circ}{\mathfrak{m}}, \overset{\circ}{\Psi}\mathfrak{m}_\eta) \times \Delta(\overset{\circ}{\mathfrak{m}}, \mathscr{M}_1).\end{aligned}$$

Hence $\Delta(\overset{\circ}{\mathfrak{m}}, \mathfrak{m}_\eta) \to \Delta(\overset{\circ}{\mathfrak{m}}, \mathscr{M}_1)$. Considering the fact that $\mathscr{M}_1$ is approximately compact w.r.t. $\mathscr{M}_2$, $\exists$ the subsequence $\{\mathfrak{m}_{\eta_\Delta}\}$ of $\{\mathfrak{m}_\eta\}$ converges to some element $\mathfrak{m} \in \mathscr{M}_1$. So, the following

$$\Delta(\mathfrak{m}, \overset{\circ}{\Psi}\mathfrak{m}) = lim_{\eta\to\infty}\Delta(\mathfrak{m}_{\eta_\Delta+1}, \overset{\circ}{\Psi}\mathfrak{m}_{\eta_\Delta}) = \Delta(\mathscr{M}_1, \mathscr{M}_2) \tag{6.7}$$

is true. To prove the uniqueness, let for the mapping $\overset{\circ}{\Psi}$, $\mathfrak{m}_\zeta \in \mathscr{M}_1$ is distinct best proximity point such that

$$\Delta(\mathfrak{m}_\zeta, \overset{\circ}{\Psi}\mathfrak{m}_\zeta) = \Delta(\mathscr{M}_1, \mathscr{M}_2)$$

Also,

$$\tau \times F(\Delta(\mathring{\Psi}\mathfrak{m}, \mathring{\Psi}\mathfrak{m}_\zeta)) \leq F((\Delta(\mathring{\Psi}\mathfrak{m}, \mathring{\Psi}\mathfrak{m}_\zeta))^{\mathfrak{a}_1+2\mathfrak{a}_4}).$$

Also, the mapping F is strictly increasing, we have

$$\Delta(\mathring{\Psi}\mathfrak{m}, \mathfrak{m}_\zeta) \leq (\Delta(\mathring{\Psi}\mathfrak{m}, \mathfrak{m}_\zeta))^{\mathfrak{a}_1+2\mathfrak{a}_4},$$

which proves that $\mathfrak{m} = \mathfrak{m}_\zeta$.

In the next theorem, we use the mapping $\mathring{\Psi}$, which is non-self multiplicative generalized $F_{\mathcal{PC}}$ of the first kind and second kind. Note that, approximately compactness assumption is not considered for both domain as well codomain.

Theorem 6.5 *Let $\mathcal{M}$ be a $\mathcal{CMMS}$ and $(\mathcal{M}_1, \mathcal{M}_2) \neq \emptyset$ closed subsets of $(\mathcal{M}, \Delta)$. Let $\mathring{\Psi} : \mathcal{M}_1 \to \mathcal{M}_2$ satisfying*

($\mathscr{C}_1$) $\mathring{\Psi}(\mathcal{M}_1^0) \subseteq \mathcal{M}_2^0$ *and $(\mathcal{M}_1, \mathcal{M}_2)$ satisfies the $p-$property;*

($\mathscr{C}_2$) $\mathring{\Psi}$ *be a multiplicative generalized $F_{\mathcal{PC}}$ of first as well of second kind.*

Then there exists $\mathfrak{m} \in \mathcal{M}$ such that $\Delta(\mathfrak{m}, \mathring{\Psi}\mathfrak{m}) = \Delta(\mathcal{M}_1, \mathcal{M}_2)$. Furthermore, for arbitrarily chosen $\mathfrak{m}_0 \in \mathcal{M}_1^0$, the sequence $\{\mathfrak{m}_\eta\}_{\eta=1}^\infty$ defined as follows:

$$\Delta(\mathfrak{m}_{\eta+1}, \mathring{\Psi}\mathfrak{m}_\eta) = \Delta(\mathcal{M}_1, \mathcal{M}_2)$$

converges to $\mathfrak{m}$.

Proof According to Theorem 6.3, a sequence $\{\mathfrak{m}_\eta\}$ can be defined in $\mathcal{M}_1^0$ such that

$$\Delta(\mathfrak{m}_{\eta+1}, \mathring{\Psi}\mathfrak{m}_\eta) = \Delta(\mathcal{M}_1, \mathcal{M}_2),$$

$\forall\ \eta \in \mathbb{N} \cup \{0\}$. From Theorem 6.3, we obtain that the sequence $\{\mathfrak{m}_\eta\}$ is Cauchy and converges to $\mathfrak{m} \in \mathcal{M}_1$. Also from Theorem 6.4, it is also obvious that $\{\mathring{\Psi}\mathfrak{m}_\eta\}$ is Cauchy sequence and converges to $\mathring{\mathfrak{m}}_\eta \in \mathcal{M}_2$ hence the following holds: $\eta \in \mathbb{N} \cup \{0\}$.

$$\Delta(\mathfrak{m}, \mathring{\mathfrak{m}}) = \lim_{\eta\to\infty} \Delta(\mathfrak{m}_{\eta+1}, \mathring{\Psi}) = \Delta(\mathcal{M}_1, \mathcal{M}_2). \tag{6.8}$$

Therefore, $\mathfrak{m} \in \mathcal{M}_1^0$. Consequently, $\mathcal{M}_1^0$ contains $\mathring{\Psi}$ and $\Delta(\omega, \mathring{\Psi}\mathfrak{m}) = \Delta(\mathcal{M}_1, \mathcal{M}_2)$ for $\omega \in \mathcal{M}_1$. By the property p combined with Eq. (6.8) for every non-negative integer η. For arbitrarily chosen η_0, $\Delta(\omega, \mathfrak{m}_{\eta_0+1}) = 1$, which further results $\Delta(\mathring{\Psi}\mathfrak{m}_{\eta_0}, \mathring{\Psi}\mathfrak{m}) = 1$. Therefore $\mathring{\Psi}\mathfrak{m}_{\eta_0} = \mathring{\Psi}\mathfrak{m}$ and hence $\Delta(\mathcal{M}_1, \mathcal{M}_2) = \Delta(\mathfrak{m}, \mathring{\Psi}\mathfrak{m})$.

As a result, the conclusion is obvious that $\Delta(\omega, \mathfrak{m}_{\eta+1}) > 0$ for any non-negative η. By the definition of first kind multiplicative generalized $\boldsymbol{F}_{\mathcal{PC}}$ and $\boldsymbol{F}$ is strictly increasing, therefore the following holds:

$$\begin{aligned}\tau \times \boldsymbol{F}(\Delta(\omega, \mathfrak{m}_{\eta+1})) \leq \boldsymbol{F}(&\Delta^{\mathfrak{a}_1}(\mathfrak{m}, \mathfrak{m}_\eta) \times \Delta^{\mathfrak{a}_2}(\omega, \mathfrak{m}) \times \Delta^{\mathfrak{a}_3}(\mathfrak{m}_\eta, \mathfrak{m}_{\eta+1}) \\ &\times (\Delta(\mathfrak{m}, \mathfrak{m}_{\eta+1}) \times \Delta(\mathfrak{m}_\eta, \omega))^{\mathfrak{a}_4}.\end{aligned}$$

Therefore,

$$\Delta(\omega, \mathfrak{m}_{\eta+1}) \leq \Delta^{\mathfrak{a}_1}(\mathfrak{m}, \mathfrak{m}_\eta) \times \Delta^{\mathfrak{a}_2}(\omega, \mathfrak{m}) \times \Delta^{\mathfrak{a}_3}(\mathfrak{m}_\eta, \mathfrak{m}_{\eta+1}) \times (\Delta(\mathfrak{m}, \mathfrak{m}_\eta) \times \Delta(\mathfrak{m}_{\eta+1}, \omega))^{\mathfrak{a}_4}.$$

For $\eta \to \infty$, $\Delta(\omega, \mathfrak{m}) \leq \Delta(\omega, \mathfrak{m})^{\mathfrak{a}_2+\mathfrak{a}_4}$, which results that $\mathfrak{m}$ and ω must be same and hence $\Delta(\mathfrak{m}, \mathring{\Psi}\mathfrak{m}) = \Delta(\omega, \mathring{\Psi}\mathfrak{m}) = \Delta(\mathscr{M}_1, \mathscr{M}_2)$. Following Theorems 6.3 and 6.4, the uniqueness of the best proximity point is obvious.

6.4 Best Proximity Point Approximation for Multiplicative Khammahawong and Kumam's *F*-Contraction Mapping

Budhia et al. [32] proved various fixed point results based on contraction mappings of F-Suzuki type, in their findings new concepts ($\alpha - type$ almost $-$ $\boldsymbol{F}$contraction and $\alpha - type - \boldsymbol{F}$-Suzuki contraction) and applied the same to a nonlinear differential equation. Note that, in 2019, Salamatbakhsh and Haghi [34] approximate some best proximity point results of generalized almost-$\boldsymbol{F}$-contraction mappings in metric spaces, which generalize various existing results. In this section, we study the multiplicative version of their findings.

Singh et al. [41] proved the theorem for the existence of a fixed point for Roger-Hardy-type $\boldsymbol{F}$-contraction in complete G-metric spaces in 2016. Motivated and inspired by the preceding research, we investigate the best proximity point theorem for the contraction mapping studied by Khammahawong and Kumam in 2017 in the framework of multiplicative metric spaces and defined as follows:

Definition 6.14 Let $(\mathscr{M}, \Delta)$ be a multiplicative metric space. $\mathscr{M}_1$, $\mathscr{M}_2 \neq \emptyset$ and $\mathscr{M}_1, \mathscr{M}_2 \subseteq \mathscr{M}$. A mapping $\mathring{\Psi}_t : \mathscr{M}_1 \to \mathscr{M}_2$ is called multiplicative Khammahawong & Kumam $\boldsymbol{F}$ if $\exists\ \tau > 0$ and $\boldsymbol{F} \in \Theta$ such that $\Delta(\mathring{\Psi}_t\mathfrak{m}_1, \mathring{\Psi}_t\mathfrak{m}_2) > 0$ implies

$$\tau \times \boldsymbol{F}(\Delta(\mathring{\Psi}_t\mathfrak{m}_1, \mathring{\Psi}_t\mathfrak{m}_2)) \leq \boldsymbol{F}\begin{pmatrix}\Delta^{\mathfrak{a}_1}(\mathfrak{m}_1, \mathfrak{m}_2) \times \Delta^{\mathfrak{a}_2}(\mathfrak{m}_1, \mathring{\Psi}_t\mathfrak{m}_1)\times \\ \Delta^{\mathfrak{a}_3}(\mathfrak{m}_2, \mathring{\Psi}_t\mathfrak{m}_2) \times \Delta^{\mathfrak{a}_4}(\mathfrak{m}_1, \mathring{\Psi}_t\mathfrak{m}_2)\times \\ \Delta^{\mathfrak{a}_5}(\mathfrak{m}_2, \mathring{\Psi}_t\mathfrak{m}_1)\end{pmatrix} \qquad (\mathscr{M}_k)$$

$\forall\, \mathfrak{m}_1.\, \mathfrak{m}_2 \in \mathscr{M}_1$ and $\mathfrak{a}_1 + \mathfrak{a}_2 + \mathfrak{a}_3 + \mathfrak{a}_4 + \mathfrak{a}_5 < 1$.

Definition 6.15 Let $(\mathscr{M}, \Delta)$ be a multiplicative metric space. $\mathscr{M}_1,\ \mathscr{M}_2 \neq \emptyset$ and $\mathscr{M}_1, \mathscr{M}_2 \subseteq \mathscr{M}$. A mapping $\mathring{\Psi}_s : \mathscr{M}_2 \to \mathscr{M}_1$. A mapping $\mathring{\Psi}_t : \mathscr{M}_1 \to \mathscr{M}_2$ is called multiplicative Khammahawong & Kumam $\boldsymbol{F}$-contraction mapping w.r.t. $\mathring{\Psi}_s$ if $\exists$ $\tau > 0$ and $\boldsymbol{F} \in \Theta$ such that $\Delta(\mathring{\Psi}_t \mathfrak{m}_1, \mathring{\Psi}_t \mathfrak{m}_2) > 0$ implies

$$\tau \times \boldsymbol{F}(\Delta(\mathring{\Psi}_t \mathfrak{m}_1, \mathring{\Psi}_t \mathfrak{m}_2)) \leq \boldsymbol{F}\begin{pmatrix} \Delta^{\mathfrak{a}_1}(\mathfrak{m}_1, \mathfrak{m}_2) \times \Delta^{\mathfrak{a}_2}(\mathfrak{m}_1, \mathring{\Psi}_s \mathring{\Psi}_t \mathfrak{m}_1) \times \\ \Delta^{\mathfrak{a}_3}(\mathfrak{m}_2, \mathring{\Psi}_s \mathring{\Psi}_t \mathfrak{m}_2) \times \Delta^{\mathfrak{a}_4}(\mathfrak{m}_1, \mathring{\Psi}_s \mathring{\Psi}_t \mathfrak{m}_2) \times \\ \Delta^{\mathfrak{a}_5}(\mathfrak{m}_2, \mathring{\Psi}_s \mathring{\Psi}_t \mathfrak{m}_1) \end{pmatrix} \tag{$\mathscr{M}k_s$}$$

$\forall\, \mathfrak{m}_1.\, \mathfrak{m}_2 \in \mathscr{M}_1$ and $\mathfrak{a}_1 + \mathfrak{a}_2 + \mathfrak{a}_3 + \mathfrak{a}_4 + \mathfrak{a}_5 < 1$ and the following conditions also hold:

($\mathscr{C}_1$) $\dfrac{\mathfrak{a}_1 + \mathfrak{a}_2 + \mathfrak{a}_4}{1 - \mathfrak{a}_3 - \mathfrak{a}_4} \in [0, 1)$;
($\mathscr{C}_2$) $\mathfrak{a}_3 + \mathfrak{a}_4 < 1$.

Theorem 6.6 *Let $(\mathscr{M}, \Delta)$ be a multiplicative metric space and $\mathscr{M}_1,\ \mathscr{M}_2 \neq \emptyset$ are subsets of $\mathscr{M}$. Let a nonexpansive mapping $\mathring{\Psi}_s : \mathscr{M}_2 \to \mathscr{M}_1$ and multiplicative Khammahawong & Kumam $\boldsymbol{F}$-contraction mapping $\mathring{\Psi}_t : \mathscr{M}_1 \to \mathscr{M}_2$ and the pair $(\mathring{\Psi}_s, \mathring{\Psi}_t)$ is a weak $\Delta-$cyclic contraction. Then $\exists\ \mathfrak{m}_1 \in \mathscr{M}_1$ and $\mathfrak{m}_2 \in \mathscr{M}_2$ such that $\Delta(\mathfrak{m}_1, \mathring{\Psi}_t \mathfrak{m}_1) = \Delta(\mathscr{M}_1, \mathscr{M}_2)$, $\Delta(\mathfrak{m}_2, \mathring{\Psi}_s \mathfrak{m}_2) = \Delta(\mathscr{M}_1, \mathscr{M}_2)$ and $\Delta(\mathfrak{m}_1, \mathfrak{m}_2) = \Delta(\mathscr{M}_1, \mathscr{M}_2)$. For arbitrary element $\mathfrak{m}_0 \in \mathscr{M}_1$, $\mathfrak{m}_{2\eta+1} = \mathring{\Psi}_t \mathfrak{m}_{2\eta}$ and $\mathfrak{m}_{2\eta} = \mathring{\Psi}_s \mathfrak{m}_{2\eta-1}$. The sequence converges to best proximity point of $\mathring{\Psi}_t$ and the sequence $\{\mathfrak{m}_{2\eta}\}$ converges to best proximity point of $\mathring{\Psi}_s$.*

Proof Let a fixed element $\mathfrak{m}_0 \in \mathscr{M}_1$, then a sequences $\mathfrak{m}_{2\eta}$ and $\mathfrak{m}_{2\eta+1}$ can be generated such that $\mathfrak{m}_{2\eta+1} = \mathring{\Psi}_s \mathfrak{m}_{2\eta-1}\ \forall\, \eta \geq 1$ and $\mathfrak{m}_{2\eta} = \mathring{\Psi}_t \mathfrak{m}_{2\eta}\ \forall\, \eta \geq 0$, respectively. Let

$$\Delta(\mathfrak{m}_{2\eta}, \mathfrak{m}_{2\eta+2} = \Delta(\mathring{\Psi}_s \mathfrak{m}_{2\eta-1}, \mathring{\Psi}_s \mathfrak{m}_{2\eta+1}). \tag{6.9}$$

Since the mapping $\mathring{\Psi}_s$ is nonexpansive, we have

$$\Delta(\mathring{\Psi}_s \mathfrak{m}_{2\eta-1}, \mathring{\Psi}_s \mathfrak{m}_{2\eta+1}) = \Delta(\mathfrak{m}_{2\eta-1}, \mathfrak{m}_{2\eta+1}) = \Delta(\mathring{\Psi}_t \mathfrak{m}_{2\eta-2}, \mathring{\Psi}_s \mathfrak{m}_{2\eta}). \tag{6.10}$$

Also the mapping $\mathring{\Psi}_t$ is multiplicative Khammahawong and Kumam's $\boldsymbol{F}$-contraction, we have

$$\boldsymbol{F}(\Delta(\mathring{\Psi}_t \mathfrak{m}_{2\eta-2}, \mathring{\Psi}_t \mathfrak{m}_{2\eta})) \leq \tau^{-1} \boldsymbol{F}(\Delta^{\mathfrak{a}_1}(\mathfrak{m}_{2\eta-2}, \mathfrak{m}_{2\eta}) \times \Delta^{\mathfrak{a}_2}(\mathfrak{m}_{2\eta-2}, \mathring{\Psi}_s \mathring{\Psi}_t \mathfrak{m}_{2\eta-2}) \times \Delta^{\mathfrak{a}_3}(\mathfrak{m}_{2\eta}, \mathring{\Psi}_s \mathring{\Psi}_t \mathfrak{m}_{2\eta})$$

$$\times \Delta^{\mathfrak{a}_4}(\mathfrak{m}_{2\eta-2}, \mathring{\Psi}_s\mathring{\Psi}_t\mathfrak{m}_{2\eta}) \times \Delta^{\mathfrak{a}_5}(\mathfrak{m}_{2\eta}, \mathring{\Psi}_s\mathring{\Psi}_t\mathfrak{m}_{2\eta-2}))$$
$$\leq \tau^{-1}\boldsymbol{F}(\Delta^{\mathfrak{a}_1}(\mathfrak{m}_{2\eta-2}, \mathfrak{m}_{2\eta}) \times \Delta^{\mathfrak{a}_2}(\mathfrak{m}_{2\eta-2}, \mathfrak{m}_{2\eta}) \times \Delta^{\mathfrak{a}_3}(\mathfrak{m}_{2\eta}, \mathfrak{m}_{2\eta+2})$$
$$\times \Delta^{\mathfrak{a}_4}(\mathfrak{m}_{2\eta-2}, \mathfrak{m}_{2\eta+2}) \times \Delta^{\mathfrak{a}_5}(\mathfrak{m}_{2\eta}, \mathfrak{m}_{2\eta}))$$
$$\leq \tau^{-1}\boldsymbol{F}(\Delta^{\mathfrak{a}_1}(\mathfrak{m}_{2\eta-2}, \mathfrak{m}_{2\eta}) \times \Delta^{\mathfrak{a}_2}(\mathfrak{m}_{2\eta-2}, \mathfrak{m}_{2\eta}) \times \Delta^{\mathfrak{a}_3}(\mathfrak{m}_{2\eta}, \mathfrak{m}_{2\eta+2})$$
$$\times \Delta^{\mathfrak{a}_4}(\mathfrak{m}_{2\eta-2}, \mathfrak{m}_{2\eta}) \times \Delta^{\mathfrak{a}_4}(\mathfrak{m}_{2\eta}, \mathfrak{m}_{2\eta+2})). \tag{6.11}$$

Hence, Eq. (6.11) combined with the property $\mathscr{F}_1$ results that

$$\Delta(\mathring{\Psi}_t\mathfrak{m}_{2\eta-2}, \mathring{\Psi}_t\mathfrak{m}_{2\eta}) \leq (\Delta(\mathfrak{m}_{2\eta-2}, \mathfrak{m}_{2\eta})^{(\mathfrak{a}_1+\mathfrak{a}_2+\mathfrak{a}_4)} \times \Delta(\mathfrak{m}_{2\eta}, \mathfrak{m}_{2\eta+2}))^{\mathfrak{a}_3+\mathfrak{a}_4}. \tag{6.12}$$

Using Eqs. (6.10), (6.11), and (6.12),

$$\Delta(\mathfrak{m}_{2\eta}, \mathfrak{m}_{2\eta+2}) \leq (\Delta(\mathfrak{m}_{2\eta-2}, \mathfrak{m}_{2\eta}))^{\mathfrak{a}_1+\mathfrak{a}_2+\mathfrak{a}_4} \times \Delta(\mathfrak{m}_{2\eta}, \mathfrak{m}_{2\eta+2}))^{\mathfrak{a}_3+\mathfrak{a}_4}. \tag{6.13}$$

Let $\Delta(\mathfrak{m}_{2\eta}, \mathfrak{m}_{2\eta+2}) = \ell_\eta$. Now, from condition $\mathscr{C}_1$ and Eq. (6.13), we have

$$\ell_\eta \leq (\ell_{\eta-1})^\kappa \leq \ell_{\eta-1}, \text{ where } \kappa = \frac{\mathfrak{a}_1+\mathfrak{a}_2+\mathfrak{a}_4}{1-\mathfrak{a}_3-\mathfrak{a}_4} < 1. \tag{6.14}$$

Therefore $\boldsymbol{F}(\ell_\eta) \leq \tau^{-1}\boldsymbol{F}(\ell_{\eta-1})$. Inductively, we have

$$\boldsymbol{F}(\ell_\eta) \leq \tau^{-1}\boldsymbol{F}(\ell_{\eta-1}) \leq \ldots \leq \tau^{-n}\boldsymbol{F}(\ell_0).$$

Since $\tau > 1$ $\lim_{\eta\to\infty} \boldsymbol{F}(\ell_\eta) = 0$. Now from the condition $\mathscr{F}_2$ and Lemma 6.2, we have $\lim_{\eta\to\infty} \ell_\eta = 0$. Using Eq. (6.14) and the condition that $\ell_\eta \leq (\ell_0)^{\kappa^\eta}$ or $\Delta(\mathfrak{m}_{2\eta}, \mathfrak{m}_{2\eta+2}) \leq (\Delta(\mathfrak{m}_0, \mathfrak{m}_2))^{\kappa^\eta}$. Hence the sequence $\{\mathfrak{m}_{2\eta}\}$ is Cauchy by the condition $\mathscr{C}_2$ in $\mathscr{M}_1$. By the completeness of space, the sequence $\{\mathfrak{m}_{2\eta}\}$ converges to some point of $\mathfrak{m}$ of $\mathscr{M}_1$. Now, let

$$\Delta(\mathfrak{m}_{2\eta+1}, \mathfrak{m}_{2n+3}) = \Delta(\mathring{\Psi}_t\mathfrak{m}_{2\eta}, \mathring{\Psi}_t\mathfrak{m}_{2\eta+2}). \tag{6.15}$$

It is also given that the mapping $\mathring{\Psi}_t$ is multiplicative Khammahawong & Kumam $\boldsymbol{F}$-contraction and hence we have

$$\boldsymbol{F}(\Delta(\mathring{\Psi}_t\mathfrak{m}_{2\eta}, \mathring{\Psi}_t\mathfrak{m}_{2\eta+2})) \leq \tau^{-1}\boldsymbol{F}(\Delta^{\mathfrak{a}_1}(\mathfrak{m}_{2\eta}, \mathfrak{m}_{2\eta+2}) \times \Delta^{\mathfrak{a}_2}(\mathfrak{m}_{2\eta}, \mathring{\Psi}_s\mathring{\Psi}_t\mathfrak{m}_{2\eta})$$
$$\times \Delta^{\mathfrak{a}_3}(\mathfrak{m}_{2\eta+2}, \mathring{\Psi}_s\mathring{\Psi}_t\mathfrak{m}_{2\eta+2}) \times \Delta^{\mathfrak{a}_4}(\mathfrak{m}_{2\eta}, \mathring{\Psi}_s\mathring{\Psi}_t\mathfrak{m}_{2\eta+2})$$
$$\times \Delta^{\mathfrak{a}_5}(\mathfrak{m}_{2\eta+2}, \mathring{\Psi}_s\mathring{\Psi}_t\mathfrak{m}_{2\eta}))$$
$$= \tau^{-1}\boldsymbol{F}(\Delta^{\mathfrak{a}_1}(\mathring{\Psi}_s\mathfrak{m}_{2\eta-1}, \mathring{\Psi}_s\mathfrak{m}_{2\eta+1}) \times \Delta^{\mathfrak{a}_2}(\mathring{\Psi}_s\mathfrak{m}_{2\eta-1}, \mathring{\Psi}_s\mathfrak{m}_{2\eta+1})$$
$$\times \Delta^{\mathfrak{a}_3}(\mathring{\Psi}_s\mathfrak{m}_{2\eta+1}, \mathring{\Psi}_s\mathfrak{m}_{2\eta+3}) \times \Delta^{\mathfrak{a}_4}(\mathring{\Psi}_s\mathfrak{m}_{2\eta-1}, \mathring{\Psi}_s\mathfrak{m}_{2\eta+3})$$
$$\times \Delta^{\mathfrak{a}_5}(\mathring{\Psi}_s\mathfrak{m}_{2\eta+1}, \mathring{\Psi}_s\mathfrak{m}_{2\eta+1})).$$

Since the mapping $\mathring{\Psi}_s$ is nonexpansive, we have

$$\begin{aligned}\boldsymbol{F}(\Delta(\mathring{\Psi}_t\mathfrak{m}_{2\eta}, \mathring{\Psi}_t\mathfrak{m}_{2\eta+2})) &< \boldsymbol{F}(\Delta^{\mathfrak{a}_1}(\mathfrak{m}_{2\eta-1}, \mathfrak{m}_{2\eta+1}) \times \Delta^{\mathfrak{a}_2}(\mathfrak{m}_{2\eta-1}, \mathfrak{m}_{2\eta+1})\\ &\times \Delta^{\mathfrak{a}_3}(\mathfrak{m}_{2\eta+1}, \mathfrak{m}_{2\eta+3}) \times \Delta^{\mathfrak{a}_4}(\mathfrak{m}_{2\eta-1}, \mathfrak{m}_{2\eta+1})\\ &\times \Delta^{\mathfrak{a}_4}(\mathfrak{m}_{2\eta+1}, \mathfrak{m}_{2\eta+3})).\end{aligned} \tag{6.16}$$

It implies from Eq. (6.16) and the condition $\mathscr{F}_2$ that

$$\Delta(\mathring{\Psi}_t\mathfrak{m}_{2\eta}, \mathring{\Psi}_t\mathfrak{m}_{2\eta+2}) \leq \Delta^{\mathfrak{a}_1 + \mathfrak{a}_2 + \mathfrak{a}_4}(\mathfrak{m}_{2\eta-1}, \mathfrak{m}_{2\eta+1}) \times \Delta^{\mathfrak{a}_3 + \mathfrak{a}_4}(\mathfrak{m}_{2\eta+1}, \mathfrak{m}_{2\eta+3}). \tag{6.17}$$

Now, from Eqs. (6.15) and (6.17)

$$\Delta(\mathfrak{m}_{2\eta+1}, \mathfrak{m}_{2\eta+3}) \leq (\Delta(\mathfrak{m}_{2\eta-1}, \mathfrak{m}_{2\eta+1}))^{\mathfrak{a}_1 + \mathfrak{a}_2 + \mathfrak{a}_4} \times (\Delta(\mathfrak{m}_{2\eta+1}, \mathfrak{m}_{2\eta+3}))^{\mathfrak{a}_3 + \mathfrak{a}_4}. \tag{6.18}$$

Considering $\tilde{\ell}_\eta = \Delta(\mathfrak{m}_{2\eta+1}, \mathfrak{m}_{2\eta+3})$, Eq. (6.19), and the condition $\mathscr{C}_1$, we have

$$\tilde{\ell}_\eta \leq (\tilde{\ell}_{\eta-1})^\kappa \leq \tilde{\ell}_{\eta-1}, \text{ where } \kappa = \frac{\mathfrak{a}_1 + \mathfrak{a}_2 + \mathfrak{a}_4}{1 - \mathfrak{a}_3 - \mathfrak{a}_4} < 1. \tag{6.19}$$

Therefore $\boldsymbol{F}(\tilde{\ell}_\eta) \leq \tau^{-1}\boldsymbol{F}(\tilde{\ell}_{\eta-1})$. Inductively, we have

$$\boldsymbol{F}(\tilde{\ell}_\eta) \leq \tau^{-1}\boldsymbol{F}(\tilde{\ell}_{\eta-1}) \leq \ldots \leq \tau^{-n}\boldsymbol{F}(\tilde{\ell}_0).$$

Since $\tau > 1$ $\lim_{\eta\to\infty} \boldsymbol{F}(\ell_\eta) = 0$. Now from the condition $\mathscr{F}_2$ and Lemma 6.2, we have $\lim_{\eta\to\infty} \tilde{\ell}_\eta = 0$. Using Eq. (6.14) and the condition that $\tilde{\ell}_\eta \leq (\ell_0)^{\kappa^\eta}$ or $\Delta(\mathfrak{m}_{2\eta+1}, \mathfrak{m}_{2\eta+3}) \leq (\Delta(\mathfrak{m}_1, \mathfrak{m}_3))^{\kappa^\eta}$. Hence the sequence $\{\mathfrak{m}_{2\eta+1}\}$ is Cauchy by the condition $\mathscr{C}_1$ in $\mathscr{M}_2$. By the completeness of space, the sequence $\{\mathfrak{m}_{2\eta}\}$ converges to some point of $\mathfrak{m}_\mu$ of $\mathscr{M}_2$. Consequently, we have

$$\begin{aligned}\Delta(\mathfrak{m}_{2\eta+2}, \mathring{\Psi}_s\mathring{\Psi}_t\mathfrak{m}) &= \Delta(\mathring{\Psi}_s\mathring{\Psi}_t\mathfrak{m}_{2\eta}, \mathring{\Psi}_s\mathring{\Psi}_t\mathfrak{m})\\ &\leq \Delta(\mathring{\Psi}_t\mathfrak{m}_{2\eta}, \mathring{\Psi}_t\mathfrak{m})\\ &= \Delta(\mathfrak{m}_{2\eta+1}, \mathring{\Psi}_t\mathfrak{m}).\end{aligned} \tag{6.20}$$

Since the mapping $\mathring{\Psi}_t$ is multiplicative Khammahawong and Kumam $\boldsymbol{F}$-contraction, it results that

$$\boldsymbol{F}(\Delta(\mathring{\Psi}_t\mathfrak{m}_{2\eta}, \mathring{\Psi}_t\mathfrak{m})) \leq \tau^{-1}\boldsymbol{F}(\Delta^{\mathfrak{a}_1}(\mathfrak{m}_{2\eta}, \mathfrak{m}) \times \Delta^{\mathfrak{a}_2}(\mathfrak{m}_{2\eta}, \mathring{\Psi}_s\mathring{\Psi}_t\mathfrak{m}_{2\eta})$$

$$\begin{aligned}
&\times \Delta^{\mathfrak{a}_3}(\mathfrak{m}, \mathring{\Psi}_s\mathring{\Psi}_t\mathfrak{m}) \times \Delta^{\mathfrak{a}_4}(\mathfrak{m}_{2\eta}, \mathring{\Psi}_s\mathring{\Psi}_t\mathfrak{m})\\
&\times \Delta^{\mathfrak{a}_5}(\mathfrak{m}, \mathring{\Psi}_s\mathring{\Psi}_t\mathfrak{m}_{2\eta}))\\
&< \boldsymbol{F}(\Delta^{\mathfrak{a}_1}(\mathfrak{m}_{2\eta}, \mathfrak{m}) \times \Delta^{\mathfrak{a}_2}(\mathfrak{m}_{2\eta}, \mathring{\Psi}_s\mathring{\Psi}_t\mathfrak{m}_{2\eta})\\
&\times \Delta^{\mathfrak{a}_3}(\mathfrak{m}, \mathring{\Psi}_s\mathring{\Psi}_t\mathfrak{m}) \times \Delta^{\mathfrak{a}_4}(\mathfrak{m}_{2\eta}, \mathring{\Psi}_s\mathring{\Psi}_t\mathfrak{m})\\
&\times \Delta^{\mathfrak{a}_5}(\mathfrak{m}, \mathring{\Psi}_s\mathring{\Psi}_t\mathfrak{m}_{2\eta}))\\
&= \boldsymbol{F}(\Delta^{\mathfrak{a}_1}(\mathfrak{m}_{2\eta}, \mathfrak{m}) \times \Delta^{\mathfrak{a}_2}(\mathfrak{m}_{2\eta}, \mathfrak{m}_{2\eta+2})\\
&\times \Delta^{\mathfrak{a}_3}(\mathfrak{m}, \mathring{\Psi}_s\mathring{\Psi}_t\mathfrak{m}) \times \Delta^{\mathfrak{a}_4}(\mathfrak{m}_{2\eta}, \mathring{\Psi}_s\mathring{\Psi}_t\mathfrak{m})\\
&\times \Delta^{\mathfrak{a}_5}(\mathfrak{m}, \mathfrak{m}_{2\eta+2}).
\end{aligned} \tag{6.21}$$

Now, from Eqs. (6.20) and (6.21) together with the condition $\mathscr{F}_2$, we have

$$\begin{aligned}
\Delta(\mathfrak{m}_{2\eta+2}, \mathring{\Psi}_s\mathring{\Psi}_t\mathfrak{m}) \leq \Delta(\mathfrak{m}_{2\eta+1}, \mathring{\Psi}_t\mathfrak{m}) &\leq \Delta^{\mathfrak{a}_1}(\mathfrak{m}_{2\eta}, \mathfrak{m}) \times \Delta^{\mathfrak{a}_2}(\mathfrak{m}_{2\eta}, \mathfrak{m}_{2\eta+2}) \times \Delta^{\mathfrak{a}_3}(\mathfrak{m}, \mathring{\Psi}_s\mathring{\Psi}_t\mathfrak{m})\\
&\times \Delta^{\mathfrak{a}_4}(\mathfrak{m}_{2\eta}, \mathring{\Psi}_s\mathring{\Psi}_t\mathfrak{m}) \times \Delta^{\mathfrak{a}_5}(\mathfrak{m}, \mathfrak{m}_{2\eta+2}),
\end{aligned} \tag{6.22}$$

which results $\Delta(\mathfrak{m}, \mathring{\Psi}_s\mathring{\Psi}_t\mathfrak{m}_{2\eta}) \leq \Delta(\mathfrak{m}_\mu, \mathring{\Psi}_t\mathfrak{m}) < (\Delta(\mathfrak{m}, \mathring{\Psi}_s\mathring{\Psi}_t\mathfrak{m}))^{\mathfrak{a}_3+\mathfrak{a}_4}$. Using condition $\mathscr{C}_2$, it is obvious that $\Delta(\mathfrak{m}_\mu, \mathring{\Psi}_t\mathfrak{m}) = 1$. Therefore $\mathring{\Psi}_t\mathfrak{m} = \mathfrak{m}_\mu$. Also

$$\Delta(\mathfrak{a}_{2\eta+3}, \mathring{\Psi}_t\mathring{\Psi}_s\mathfrak{m}_\mu) = \Delta(\mathring{\Psi}_t\mathring{\Psi}_s\mathfrak{a}_{2\eta}, \mathring{\Psi}_t\mathring{\Psi}_s\mathfrak{m}_\mu). \tag{6.23}$$

Note that the mapping $\mathring{\Psi}_t$ is multiplicative Khammahawong & Kumam $\boldsymbol{F}$-contraction and $\mathring{\Psi}_s$ is nonexpansive, for $\eta \to \infty$ it yields that

$$\begin{aligned}
\boldsymbol{F}(\Delta(\mathring{\Psi}_s\mathring{\Psi}_t\mathfrak{m}_{2\eta+1}, \mathring{\Psi}_t\mathring{\Psi}_s\mathfrak{m}^*)) \leq \tau^{-1}&\boldsymbol{F}(\Delta^{\mathfrak{a}_1}(\mathring{\Psi}_s\mathfrak{m}_{2\eta+1}, \mathring{\Psi}_s\mathfrak{m}_\mu) \times \Delta^{\mathfrak{a}_2}(\mathring{\Psi}_s\mathfrak{m}_{2\eta+1}, \mathring{\Psi}_s\mathring{\Psi}_t\mathring{\Psi}_s\mathfrak{m}_{2\eta+1})\\
&\times \Delta^{\mathfrak{a}_3}(\mathring{\Psi}_s\mathfrak{m}_\mu, \mathring{\Psi}_s\mathring{\Psi}_t\mathring{\Psi}_s\mathfrak{m}_\mu) \times \Delta^{\mathfrak{a}_4}(\mathring{\Psi}_s\mathfrak{m}_\mu, \mathring{\Psi}_s\mathring{\Psi}_t\mathring{\Psi}_s\mathfrak{m}_\mu)\\
&\times \Delta^{\mathfrak{a}_5}(\mathring{\Psi}_s\mathfrak{m}_\mu, \mathring{\Psi}_s\mathring{\Psi}_t\mathring{\Psi}_s\mathfrak{m}_{2\eta+1}))\\
&< \boldsymbol{F}(\Delta^{\mathfrak{a}_1}(\mathring{\Psi}_s\mathfrak{m}_{2\eta+1}, \mathring{\Psi}_s\mathfrak{m}_\mu) \times \Delta^{\mathfrak{a}_2}(\mathring{\Psi}_s\mathfrak{m}_{2\eta+1}, \mathring{\Psi}_s\mathring{\Psi}_t\mathring{\Psi}_s\mathfrak{m}_{2\eta+1})\\
&\times \Delta^{\mathfrak{a}_3}(\mathring{\Psi}_s\mathfrak{m}_\mu, \mathring{\Psi}_s\mathring{\Psi}_t\mathring{\Psi}_s\mathfrak{m}_\mu) \times \Delta^{\mathfrak{a}_4}(\mathring{\Psi}_s\mathfrak{m}_\mu, \mathring{\Psi}_s\mathring{\Psi}_t\mathring{\Psi}_s\mathfrak{m}_\mu)\\
&\times \Delta^{\mathfrak{a}_5}(\mathring{\Psi}_s\mathfrak{m}_\mu, \mathring{\Psi}_s\mathring{\Psi}_t\mathring{\Psi}_s\mathfrak{m}_{2\eta+1}))\\
&\leq \boldsymbol{F}(\Delta^{\mathfrak{a}_1}(\mathfrak{m}_{2\eta+1}, \mathfrak{m}_\mu) \times \Delta^{\mathfrak{a}_2}(\mathring{\Psi}_s\mathfrak{m}_{2\eta+1}, \mathfrak{m}_{2\eta+4})\\
&\times \Delta^{\mathfrak{a}_3}(\mathfrak{m}_\mu, \mathring{\Psi}_t\mathring{\Psi}_s\mathfrak{m}_\mu) \times \Delta^{\mathfrak{a}_4}(\mathfrak{m}^{2\eta+1}, \mathring{\Psi}_t\mathring{\Psi}_s\mathfrak{m}_\mu)\\
&\times \Delta^{\mathfrak{a}_5}(\mathfrak{m}_\mu, \mathfrak{m}_{2\eta+3}))\\
&= \boldsymbol{F}(\Delta^{\mathfrak{a}_1}(\mathfrak{m}_{2\eta+1}, \mathfrak{m}_\mu) \times \Delta^{\mathfrak{a}_2}(\mathring{\Psi}_s\mathfrak{m}_{2\eta+1}, \mathring{\Psi}_s\mathfrak{m}_{2\eta+3})\\
&\times \Delta^{\mathfrak{a}_3}(\mathfrak{m}_\mu, \mathring{\Psi}_t\mathring{\Psi}_s\mathfrak{m}_\mu) \times \Delta^{\mathfrak{a}_4}(\mathfrak{m}^{2\eta+1}, \mathring{\Psi}_t\mathring{\Psi}_s\mathfrak{m}_\mu) \times \Delta^{\mathfrak{a}_5}(\mathfrak{m}_\mu, \mathfrak{m}_{2\eta+3}))\\
&\leq \boldsymbol{F}(\Delta^{\mathfrak{a}_1}(\mathfrak{m}_{2\eta+1}, \mathfrak{m}_\mu) \times \Delta^{\mathfrak{a}_2}(\mathfrak{m}_{2\eta+1}, \mathfrak{m}_{2\eta+3})
\end{aligned}$$

$$\times \Delta^{\mathfrak{a}_3}(\mathfrak{m}_\mu, \mathring{\Psi}_t\mathring{\Psi}_s\mathfrak{m}_\mu) \times \Delta^{\mathfrak{a}_4}(\mathfrak{m}^{2\eta+1}, \mathring{\Psi}_t\mathring{\Psi}_s\mathfrak{m}_\mu) \times \Delta^{\mathfrak{a}_5}(\mathfrak{m}_\mu, \mathfrak{m}_{2\eta+3})). \tag{6.24}$$

Using (6.23), (6.24), and condition $\mathscr{F}_1$, we have

$$\begin{aligned}\Delta(\mathfrak{m}_{2\eta+3}, \mathring{\Psi}_t\mathring{\Psi}_s\mathfrak{m}^*) &\leq \Delta^{\mathfrak{a}_1}(\mathfrak{m}_{2\eta+1}, \mathfrak{m}_\mu) \times \Delta^{\mathfrak{a}_2}(\mathring{\Psi}_s\mathfrak{m}_{2\eta+1}, \mathfrak{m}_{2\eta+4})\\ &\times \Delta^{\mathfrak{a}_3}(\mathfrak{m}_\mu, \mathring{\Psi}_t\mathring{\Psi}_s\mathfrak{m}_\mu) \times \Delta^{\mathfrak{a}_4}(\mathfrak{m}^{2\eta+1}, \mathring{\Psi}_t\mathring{\Psi}_s\mathfrak{m}_\mu) \times \Delta^{\mathfrak{a}_5}(\mathfrak{m}_\mu, \mathfrak{m}_{2\eta+3})\\ &\leq \Delta^{\mathfrak{a}_1}(\mathfrak{m}_{2\eta+1}, \mathfrak{m}_\mu) \times \Delta^{\mathfrak{a}_2}(\mathfrak{m}_{2\eta+1}, \mathfrak{m}_{2\eta+4})\\ &\times \Delta^{\mathfrak{a}_3}(\mathfrak{m}_\mu, \mathring{\Psi}_t\mathring{\Psi}_s\mathfrak{m}_\mu) \times \Delta^{\mathfrak{a}_4}(\mathfrak{m}^{2\eta+1}, \mathring{\Psi}_t\mathring{\Psi}_s\mathfrak{m}_\mu) \times \Delta^{\mathfrak{a}_5}(\mathfrak{m}_\mu, \mathfrak{m}_{2\eta+3}),\end{aligned} \tag{6.25}$$

for $\eta \to \infty$ it yields that

$$\Delta(\mathfrak{m}_\mu, \mathring{\Psi}_t\mathring{\Psi}_s\mathfrak{m}_\mu) \leq \Delta^{\mathfrak{a}_2}(\mathring{\Psi}_s\mathfrak{m}_\mu, \mathfrak{m}) \times \Delta^{(\mathfrak{m}_3+\mathfrak{m}_4)}(\mathfrak{m}_\mu, \mathring{\Psi}_t\mathring{\Psi}_s y) \leq \Delta^{(\mathfrak{m}_3+\mathfrak{m}_4)}(\mathfrak{m}_\mu, \mathring{\Psi}_t\mathring{\Psi}_s\mathfrak{m}_\mu),$$

which results $\Delta(\mathring{\Psi}_s\mathfrak{m}_\mu) = 1$ which further yields that $\mathring{\Psi}_s = \mathfrak{m}_\mu$. Note that the pair $(\mathring{\Psi}_s, \mathring{\Psi}_t)$ satisfies multiplicative weak $\Delta-$cyclic contraction hence $\exists\ \Delta \in [0, \frac{1}{2})$ such that

$$\begin{aligned}\Delta(\mathfrak{m}, \mathfrak{m}_\mu) &= \Delta(\mathring{\Psi}_t\mathfrak{m}, \mathring{\Psi}_s\mathring{\Psi}_t\mathfrak{m})\\ &\leq (\Delta(\mathfrak{m}, \mathring{\Psi}_t\mathfrak{m}) \times \Delta(\mathring{\Psi}_t\mathfrak{m}, \mathring{\Psi}_s\mathring{\Psi}_t\mathfrak{m}))^{\Delta} \times (\Delta(\mathscr{M}_1, \mathscr{M}_2))^{(1-2\Delta)}\\ &\leq (\Delta(\mathfrak{m}, \mathfrak{m}_\mu) \times \Delta(\mathfrak{m}, \mathfrak{m}_\mu))^{\Delta} \times (\Delta(\mathscr{M}_1, \mathscr{M}_2))^{(1-2\Delta)}\\ &= \Delta^{2\Delta}(\mathfrak{m}, \mathfrak{m}_\mu) \times \Delta^{(1-2\Delta)}(\mathscr{M}_1, \mathscr{M}_2).\end{aligned} \tag{6.26}$$

So, we have $\Delta(\mathfrak{m}, \mathfrak{m}_\mu) = \Delta(\mathscr{M}_1, \mathscr{M}_2)$. Therefore $\Delta(\mathscr{M}_1, \mathscr{M}_2) = \Delta(\mathfrak{m}, \mathfrak{m}_\mu) = \Delta(\mathfrak{m}, \mathring{\Psi}_t\mathfrak{m}) = \Delta(\mathring{\Psi}_t\mathfrak{m}, \mathring{\Psi}_s\mathring{\Psi}_t\mathfrak{m})$ and $\Delta(\mathscr{M}_1, \mathscr{M}_2) = \Delta(\mathfrak{m}, \mathfrak{m}_\mu) = \Delta(\mathfrak{m}_\mu, \mathring{\Psi}_s\mathfrak{m}_\mu)$, which proves that $\mathfrak{m}_1$ is a best proximity point of $\mathring{\Psi}_t$ and $\mathfrak{m}_\mu$ is a best proximity point of $\mathring{\Psi}_s$.

Example 6.2 Let $\mathscr{M} = \mathbb{R} \times \mathbb{R}$ and the metric Δ on $\mathscr{U}_s$ is defined by

$$\Delta((\mathfrak{a}_1, \mathfrak{a}_2), (\mathfrak{b}_1, \mathfrak{b}_2)) = \left|\frac{\mathfrak{a}_1}{\mathfrak{b}_1}\right| \times \left|\frac{\mathfrak{a}_2}{\mathfrak{b}_2}\right|,$$

$\forall\,(\mathfrak{a}_1, \mathfrak{a}_2), (\mathfrak{b}_1, \mathfrak{b}_2) \in \mathbb{R} \times \mathbb{R}$. For the complete multiplicative metric space $(\mathscr{M}, \Delta)$, $\boldsymbol{F} : \mathbb{R}^+ \to \mathbb{R}$ is defined as $\boldsymbol{F}(\varkappa) = \ln(1 + \varkappa)$.

References

1. Hardy, G.E., Rogers, T.D.: A generalization of a fixed point of Reich. Canad. Math. Bull. **16**, 1–6 (1973)
2. Reich, S.: Kannan's fixed point theorem. Bull. Univ. Mat. Italiana **4**, 1–11 (1971)
3. Kannan, R.: Some remarks on fixed points. Bull. Calcutta Math. Soc. **60**, 71–76 (1960)
4. Edelstein, M.: On fixed and periodic points under contractive mappings. J. Lond. Math. Soc. **37**, 74–79 (1962)
5. Wardowski, D.: Fixed points of a new type of contractive mappings in complete metric spaces. Fixed Point Theory Appl. **94**, 1–6 (2012)
6. Banach, S.: Sur les opérations dans les ensembles abstraits et leur applications aux équations intégrales. Fund Math. **3**, 133–181 (1922)
7. Ozavsar, M., Cevikel, A.C., Fixed points of multiplicative contraction mappings on multiplicative metric spaces (2012). arXiv:1205.5131v1 [math.GN]
8. Bashirov, A.E., Kurpınar, E.M., Ozyapıcı, A.: Multiplicative calculus and its applications. J. Math. Anal. App. **337**, 36–48 (2008)
9. Khammahawong, K., Kumam, P.: A best proximity point theorem for Roger-Hardy type generalized F–contractive mappings in complete metric spaces with some examples. RACSAM **112**, 1503–1519 (2018)
10. Sharma, N., Dutta, H.: Generalized multiplicative nonlinear elastic matching distance measure and fixed point approximation. Math. Meth. Appl. Sci. **2021**, 1–17 (2021)
11. Altun, I., Aslantas, M., Sahin, H.: Best proximity point results for p-proximal contractions. Acta Math. Hung. **162**, 393–402 (2020)
12. Aslantas, M., Sahin, H., Altun, I.: Best proximity point theorems for cyclic pcontractions with some consequences and applications. Nonlinear Anal. Modell. Control **26**(1), 113–129 (2021)
13. Basha, S.S.: Best proximity points, optimal solutions. J. Optim. Theory Appl. **151**, 210–216 (2011)
14. Eldered, A.A., Veeramani, P.: Existence and convergence for best proximity points. J. Math. Anal. Appl. **33**, 1001–1006 (2006)
15. Raj, V.S.: Best proximity point theorems for nonself mappings. Fixed Point Theory **14**(2), 447–454 (2013)
16. Sahin, H., Aslantas, M., Altun, I.: Feng-Liu type approach to best proximity point results for multivalued mappings. J. Fixed Point Theory Appl. **22**, 11 (2020)
17. Sultana, A., Vetrivel, V.: On the existence of best proximity points for generalized contractions. Appl. Gen. Topol. **15**(1), 55–63 (2014)
18. Fan, K., Extensions of two fixed point theorems of F.E. Browder. Math. Z. **112**, 234–240 (1969)
19. Prolla, J.B., Fixed point theorems for set valued mappings and existence of best approximations. Numer. Funct.Anal. Optim. **5**, 449–455 (1982–1983)
20. Reich, S.: Approximate selections, best approximations, fixed points and invariant sets. J. Math. Anal. Appl. **62**, 104–113 (1978)
21. Vetrivel, V., Veeramani, P., Bhattacharyya, P.: Some extensions of fans best approximation theorem. Numer. Funct. Anal. Optim. **13**, 397–402 (1992)
22. Sehgal, V.M., Singh, S.P.: A generalization to multifunctions of fans best approximation theorem. Proc. Am. Math. Soc. **102**, 534–537 (1988)
23. Sehgal, V.M., Singh, S.P.: A theorem on best approximations. Numer. Funct. Anal. Optim. **10**, 181–184 (1989)
24. Al-Thagafi, M.A., Shahzad, N.: Convergence and existence results for best proximity points. Nonlinear Anal. **70**, 3665–3671 (2009)
25. Anthony Eldred, A., Veeramani, P.: Existence and convergence of best proximity points. J. Math. Anal. Appl. **323**, 1001–1006 (2006)
26. Di Bari, C., Suzuki, T., Vetro, C.: Best proximity points for cyclic Meir-Keeler contractions. Nonlinear Anal. **69**, 3790–3794 (2008)
27. Karpagam, S., Agrawal, S., Best proximity point theorems for p-cyclic Meir-Keeler contractions. Fixed Point Theory Appl. Art. ID 197308 (2009)

28. Suzuki, T., Kikkawa, M., Vetro, C.: The existence of best proximity points in metric spaces with the property UC. Nonlinear Anal. **71**, 2918–2926 (2009)
29. Vetro, C.: Best proximity points: convergence and existence theorems for $p-$cyclic mappings. Nonlinear Anal. **73**, 2283–2291 (2010)
30. Eldred, A.A., Kirk, W.A., Veeramani, P.: Proximinal normal structure and relatively nonexpansive mappings. Studia Math. **171**, 283–293 (2005)
31. Shahzad, N., Sadiq Basha, S., Jeyaraj, R.: Common best proximity points: global optimal solutions. J. Optim. Theory Appl. **148**, 69–78 (2011)
32. Budhia, L.B., Kumam, P., Martinez-Moreno, J.: Extensions of almost-F and F-Suzuki contractions with graph and some applications to fractional calculus. Fixed Point Theory Appl. **2016**, 1–14 (2016)
33. Salimi, P., Vetro, P.: A Best Proximity Point Theorem for Generalized GeraghtySuzuki Contractions. Bull. Malays. Math. Sci. Soc. **39**, 245–256 (2016)
34. Salamatbakhsh, M., Haghi, R.H.: On best proximity points of generalized almost-$F - -$ contraction point. J. Indones. Math. Soc. **25**, 16–23 (2019)
35. Kirk, W.A., Srinivasan, P.S., Veeramani, P.: Fixed points for mappings satisfying cyclical contractive conditions. Fixed Point Theory **4**, 79–89 (2003)
36. Sadiq Basha, S.: Best proximity point theorems generalizing the contraction principle. Nonlinear Anal. **74**, 5844–5850 (2011)
37. Sadiq Basha, S.: Common best proximity points: global minimization of multi-objective functions. J. Glob. Optim. **54**(2), 367–373 (2012)
38. Sadiq Basha, S., Shahzad, N., Jeyaraj, R.: Best proximity points: approximation and optimization. Optim. Lett. **7**(1), 145–155 (2013)
39. Ungchittrakool, K.: A best proximity point theorem for generalized non-self-Kannan-type and Chatterjea type mappings and Lipschitzian mappings in complete metric spaces. J. Funct. Spaces **2016**, 9321082 (2016)
40. Cosentino, M., Vetroa, P.: Fixed point results for $F - -$contractive mappings of Hardy-Rogers-type. Filomat **28**(4), 715–722 (2014)
41. Singh, D., Joshi, V., Kumam, P., Singh, N.: Fixed point results for generalized $F-$contraction and Roger Hardy type $F-$contractive mappings in G-metric spaces, Revista de la Real Academia de Ciencias Exacts Físicas y Naturales. Serie A. Matemátics **111**(2), 473–487 (2017)

Chapter 7
Best Proximity Point Problems in G-Metric Spaces and Its Applications

Talat Nazir and Mujahid Abbas

Abstract The aim of this chapter is to establish certain new classes of proximal contractive mappings in G-metric spaces. We also obtain the best proximity point theorems for such kind of mappings in G-metric spaces. As a consequence of these results, we deduce certain new best proximity and fixed point results. Moreover, we present an example to illustrate the usability of obtained results. Our results generalize, extend and unify various results in the existing literature.

Keywords Best proximity point · Fixed point · G-F-proximal contractive · Approximately compact · G-metric

7.1 Introduction

Fixed points problems and their solutions are import research area and there are several useful applications of fixed point theory. Best approximity point give us an approximate solution to the fixed point problem $u = fu$, where f is a non-selfmapping f having no fixed point. The study of best approximity point is called best approximation theory. Fan [14] presented an important result which states that if U is a nonempty convex compact subset of a Hausdorff locally convex topological vector space V and $f : U \to V$ is a non-self continuous map, then a point $p \in U$ exists that satisfies $d(p, fp) = \inf\{d(q, fp) : q \in U\}$, where d is a metric on V.

For two nonempty subsets $\mathcal{A}_1$ and $\mathcal{A}_2$ in (X, d) having a map $f : \mathcal{A}_1 \to \mathcal{A}_2$, a fixed point problem defined by a pair $(\mathcal{A}_1, \mathcal{A}_2)$ of sets and mapping f is to find x^* in $\mathcal{A}_1$ that gives $d(x^*, fx^*) = 0$. It is clear that if $\mathcal{A}_1 \cap \mathcal{A}_2 = \emptyset$, then solution set of fixed point problem defined by $(\mathcal{A}_1, \mathcal{A}_2)$ is empty set. An element x^* in $\mathcal{A}_1$ satisfying

T. Nazir (✉)
Department of Mathematical Sciences, University of South Africa, Florida 0003, South Africa
e-mail: dr.talatnazir@gmail.com

M. Abbas
Department of Mechanical Engineering Sciences, University of Johannesburg, Auckland Park 2006, South Africa
e-mail: mujahida@uj.ac.za

B. C. Tripathy et al. (eds.), *Operators, Inequalities and Approximation*, Industrial and Applied Mathematics, https://doi.org/10.1007/978-981-97-3238-8_7

$d(x^*, fx^*) \leq d(a, fx^*)$ for every a in $\mathcal{A}_1$ is the nearest point to $fx^* \in \mathcal{A}_2$ in $\mathcal{A}_1$. Such a point is known as an approximate fixed point of nonself map f.

It's a reasonable demand to settle down with such a point when an operator equation $fx = x$ does not admit a solution. The study of necessary constraints that gives unique approximate fixed point results of the maps f is an important area of research.

Consider $d(\mathcal{A}_1, \mathcal{A}_2) = \inf\{d(a_1, a_2) : a_1 \in \mathcal{A}_1, a_2 \in \mathcal{A}_2\}$ that gives a shorter distance between sets $\mathcal{A}_1$ and $\mathcal{A}_2$. If x^* in $\mathcal{A}_1$ satisfy $d(x^*, fx^*) = d(\mathcal{A}_1, \mathcal{A}_2)$, then x^* is known as best proximity point of f.

Best proximity results deal with sufficient conditions such that the nonlinear minimization problem

$$\min_{a_1 \in \mathcal{A}_1} d(a_1, fa_1) \text{ has a solution.}$$

So the best proximity point problems of a map $f : \mathcal{A}_1 \to \mathcal{A}_2$ defined by a pair $(\mathcal{A}_1, \mathcal{A}_2)$ of sets is to find x^* in $\mathcal{A}_1$ that gives $d(x^*, fx^*) = d(\mathcal{A}_1, \mathcal{A}_2)$.

The best proximity point problem becomes a problem of fixed point in case we have $\mathcal{A}_1 = \mathcal{A}_2$. Hence problem best proximity point is a natural extension and generalization of problem of fixed point. Also, theorems of unique best proximity point have some more general conditions of maps which are more general than theorems related to fixed point results of maps.

The theory of best proximity point has proved to be simple and applicable in solving real-world problems in economics, optimization theory, game theory and analysis, see, for instance, in [6, 8, 15–17, 44], and the references mentioned their in.

Recently, Işık and Aydi [21] obtained solutions of best proximity problems by employing the Z-proximal type contraction and shown it's application. Işık et al. [19] established results of best proximity point of multivalued maps having Z_σ-proximal type contraction. Further useful results are appeared in [20, 22, 39].

On the other hand, in 1992, Dhage [10] proposed notion of a D-metric in an attempt to get parallel results to those for metric spaces, but in some more general form. Dhage presented topological structure of D-metric together with several fixed point results in [11–13]. These works have been the basis for a substantial number of results by other authors.

Unfortunately, most of results having the fundamental topological properties of D-metric are not correct and nullify the validity of many results obtained in D-metric.

To get better come under the basic flaws in Dhage's results of D-metric [10], Mustafa and Sims [27] presented a closer relevant extension of metric space structure. Afterwards, Mustafa and Sims [28] defined a G-metric with having some modifications. Then Mustafa et al. [29–34] developed many useful results of fixed points of various contractive maps in G-metric. Further interesting results in this direction are obtained in [1, 2, 4, 5, 7, 9, 18, 23, 24, 26, 35–38, 40–43].

This chapter deals with best proximity point results in G-metric. We establish new proximal F-contractive maps in G-metric spaces. We also establish best proximity point results of these maps in G-metric. Further, we obtain some fixed point and best proximity point results as special cases of our main results. We gave an example that gives the validity of main results presented therein.

7.2 G-Metric Spaces

Mustafa and Sims [27] gave the definition of G-metric as an extension and generalization of metric space.

Definition 7.1 ([27]) For a nonempty set X, suppose that a map $G : X \times X \times X \to \mathbb{R}^+$ satisfies:

($\mathfrak{g}_1$) $G(\lambda, \eta, \gamma) = 0$ if and only if $\lambda = \eta = \gamma$,
($\mathfrak{g}_2$) $0 < G(\gamma, \eta, \eta)$ for all $\eta, \gamma \in X$, with $\gamma \neq \eta$,
($\mathfrak{g}_3$) $G(\lambda, \lambda, \gamma) \leq G(\lambda, \eta, \gamma)$ for all $\lambda, \eta, \gamma \in X$, with $\eta \neq \gamma$,
($\mathfrak{g}_4$) $G(\lambda, \eta, \gamma) = G(\eta, \lambda, \gamma) = G(\gamma, \lambda, \eta) = \ldots$ (symmetry in all triplets) and
($\mathfrak{g}_5$) $G(\gamma, \lambda, \eta) \leq G(\lambda, \varsigma, \varsigma) + G(\varsigma, \eta, \gamma)$ for all $\lambda, \eta, \gamma, \varsigma \in X$.

Then G is known as G-metric on X and (X, G) is known as G-metric space.

Definition 7.2 ([27]) In a G-metric on X, if $G(\lambda, \lambda, \gamma) = G(\lambda, \gamma, \gamma)$ holds for all $\lambda, \gamma \in X$, then G-metric is known as symmetric G-metric.

Proposition 7.1 ([27]) *In a G-metric on X, we can define d_G as*

$$d_G(\lambda, \eta) = G(\lambda, \eta, \eta) + G(\lambda, \lambda, \eta), \ \forall \lambda, \eta \in X. \tag{7.1}$$

This d_G is a metric on X.

And then in case of symmetric G-metric,

$$d_G(\lambda, \eta) = 2G(\lambda, \eta, \eta), \ \forall \lambda, \eta \in X. \tag{7.2}$$

And for case of non symmetric G-metric, it satisfies the inequality

$$\frac{3}{2}G(\lambda, \eta, \eta) \leq d_G(\lambda, \eta) \leq 3G(\lambda, \eta, \eta), \ \forall \lambda, \eta \in X. \tag{7.3}$$

Example 7.1 ([27]) For any interval $X = [\mu, \nu]$ in $\mathbb{R}^+$ with $\mu \leq \nu$, $G(\lambda, \eta, \gamma) = \max\{|\lambda - \eta|, |\eta - \gamma|, |\lambda - \gamma|\}$ is a G-metric on X.

Example 7.2 For a discrete subset $X = \{\kappa, \upsilon\}$ of $\mathbb{R}$, we define $G : X \times X \times X \to \mathbb{R}^+$ by

$(\lambda, \delta, \gamma)$	$G(\lambda, \delta, \gamma)$
$(\kappa, \kappa, \kappa), (\upsilon, \upsilon, \upsilon)$	0
$(\kappa, \kappa, \upsilon), (\kappa, \upsilon, \kappa), (\upsilon, \kappa, \kappa)$	0.5
$(\kappa, \upsilon, \upsilon), (\upsilon, \kappa, \upsilon), (\upsilon, \upsilon, \kappa)$	1

Here, $G(\upsilon, \kappa, \kappa) \neq G(\upsilon, \upsilon, \kappa)$ for distinct υ, κ in X. Therefore, G is a non symmetric G-metric.

Proposition 7.2 *In G-metric on X and* $k > 0$,

$$(E_1) \quad G_1(\lambda, \eta, \gamma) = \min\{k, G(\lambda, \eta, \gamma)\},$$

$$(E_2) \quad G_2(\lambda, \eta, \gamma) = \frac{G(\lambda, \eta, \gamma)}{1 + G(\lambda, \eta, \gamma)},$$

are also G-metrics on X. Further, if $X = \cup_{i=1}^{n} A_i$ *is any partition of X, then*

$$(E_3) \quad G_3(\lambda, \eta, \gamma) = \begin{cases} G(\lambda, \eta, \gamma), & \text{if for some } i, \text{ we have } \lambda, \eta, \gamma \in A_i \\ k + G(\lambda, \eta, \gamma), & \text{otherwise} \end{cases}$$

are also G-metric on X.

Now, we consider some properties that hold in G-metric.

Proposition 7.3 *In G-metric, for* $\lambda, \eta, \gamma, \varsigma \in X$, *the following are true:*

(i) *if* $G(\lambda, \eta, \gamma) = 0$, *then* $\lambda = \eta = \gamma$,
(ii) $G(\lambda, \eta, \gamma) \leq G(\lambda, \lambda, \gamma) + G(\lambda, \lambda, \eta)$,
(iii) $G(\lambda, \eta, \eta) \leq 2G(\lambda, \lambda, \eta)$,
(iv) $G(\eta, \lambda, \gamma) \leq G(\gamma, \lambda, \varsigma) + G(\varsigma, \eta, \gamma)$,
(v) $G(\lambda, \eta, \gamma) \leq \frac{2}{3}[G(\lambda, \varsigma, \gamma) + G(\lambda, \eta, \varsigma) + G(\varsigma, \eta, \gamma)]$,
(vi) $G(\lambda, \eta, \gamma) \leq G(\varsigma, \eta, \varsigma) + G(\lambda, \varsigma, \varsigma) + G(\varsigma, \varsigma, \gamma)$,
(vii) $|G(\lambda, \eta, \gamma) - G(\lambda, \eta, \varsigma)| \leq \max\{G(\varsigma, \lambda, \lambda), G(\lambda, \varsigma, \varsigma)\}$,
(viii) $|G(\lambda, \eta, \gamma) - G(\lambda, \eta, \varsigma)| \leq G(\lambda, \varsigma, \gamma)$,
(ix) $|G(\lambda, \eta, \gamma) - G(\eta, \gamma, \gamma)| \leq \max\{G(\lambda, \gamma, \gamma), G(\gamma, \lambda, \lambda)\}$,
(x) $|G(\lambda, \eta, \eta) - G(\eta, \lambda, \lambda)| \leq \max\{G(\lambda, \eta, \eta), G(\lambda, \lambda, \eta)\}$.

Definition 7.3 In G-metric on X, for u_0 in X with $\epsilon > 0$, the G-ball having centre u_0 with radius ϵ is

$$\mathcal{B}_G(u_0, \epsilon) = \{\upsilon \in X : G(u_0, \upsilon, \upsilon) < \epsilon\}.$$

Proposition 7.4 *In G-metric on X, for* u_0 *in X with* $\epsilon > 0$, *we have*

(1) $\upsilon, \mu \in \mathcal{B}_G(u_0, \epsilon)$ *provided that* $G(u_0, \upsilon, \mu) < \epsilon$;
(2) *we have* $\epsilon^* > 0$ *such that* $\mathcal{B}_G(\mu, \epsilon^*) \subseteq \mathcal{B}_G(u_0, \epsilon)$ *provided that* $\mu \in B_G(x_0, r)$.

Proposition 7.5 *In G-metric on X, for* $u_0 \in X$ *with* $\epsilon > 0$, *we have*

$$\mathcal{B}_G(u_0, \frac{\epsilon}{3}) \subseteq \mathcal{B}_{d_G}(u_0, \epsilon) \subseteq \mathcal{B}_G(u_0, \epsilon).$$

Consequently, the G-metric topology $\tau(G)$ on X consistent with the metric topology arising from d_G. Therefore, while 'isometrically' is separate, every G-metric space and metric space is topologically equivalent.

Definition 7.4 In G-metric on X, a sequence $\{\varkappa_n\}$ in X is G-convergent to $\varkappa$ whenever it converges to $\varkappa$ under the topology $\tau(G)$ of G-metric.

Proposition 7.6 *In G-metric on X, these statements are equivalent:*

(a) *sequence* $\{\varkappa_l\}$ *is G-convergent to* $\varkappa$,
(b) $G(\varkappa_l, \varkappa_l, \varkappa) \to 0$ *when* $l \to \infty$,
(c) $G(\varkappa_l, \varkappa, \varkappa) \to 0$ *when* $l \to \infty$,
(d) $G(\varkappa_l, \varkappa_n, \varkappa) \to 0$ *when* $l, n \to \infty$.

Definition 7.5 Let (X, G), (X', G') be G-metric spaces. A mapping $f : X \to X'$ is G-continuous at x_0 in X if $f^{-1}(B_{G'}(f(x_0), r)) \in \tau(G)$, for all $r > 0$. We say f is G-continuous whenever f is G-continuous at all points of X, that is, continuous as function from X with the $\tau(G)-$ topology to X' with $\tau(G')-$ topology.

Proposition 7.7 *Let* (X, G), (X', G') *be G-metric spaces. Map* $f : X \to X'$ *is G-continuous at* u_0 *in X if and only if f is G-sequentially continuous at* u_0*; that is, if* $\{\varkappa_n\}$ *is G-convergent to* $\varkappa$ *in* (X, G), *then it implies* $\{f(\varkappa_n)\}$ *is G-convergent to* $f(\varkappa)$ *in* (X', G').

Proposition 7.8 *In G-metric on X, if* $\{x_n\}$, $\{y_m\}$ *and* $\{z_l\}$ *are sequences in X having* $x_n \to x$, $y_m \to y$ *and* $z_l \to z$, *then* $G(x_n, y_m, z_l) \to G(x, y, z)$ *when* $n, m, l \to \infty$.

Proposition 7.9 *In G-metric on X, if a sequence* $\{\varkappa_n\}$ *in X having* $\{\varkappa_n\}$ *converges to* $\varkappa$ *and* $\{\varkappa_n\}$ *converges to* $\varkappa^*$ *for some* $\varkappa$ *and* $\varkappa^*$ *in X, then* $\varkappa = \varkappa^*$.

Definition 7.6 In G-metric on X, a sequence $\{\varkappa_n\}$ is

(i) a *G-Cauchy* if, for any $\epsilon > 0$, we can find positive integer k having for all $l, n, m \geq k$, $G(\varkappa_l, \varkappa_n, x_m) < \epsilon$;
(ii) a *G-convergent* if, for any $\epsilon > 0$, we have $\varkappa$ in X and can find a positive integer k having for all $l, n \geq k$, $G(\varkappa_l, \varkappa_n) < \epsilon$.

Proposition 7.10 *In G-metric on X, these two statements are equivalent:*

(a) *The sequence* $\{\varkappa_n\}$ *is G-Cauchy.*
(b) $G(\varkappa_l, \varkappa_n, \varkappa_m) \to 0$ *as* $l, n, m \to \infty$.

Corollary 7.1 *In a G-metric on X, if a sequence* $\{\varkappa_n\}$ *in X is G-convergent, then* $\{\varkappa_n\}$ *is G-Cauchy.*

Corollary 7.2 *In a G-metric on X, if a G-Cauchy sequence has G-convergent subsequence, then this sequence itself is G-convergent.*

Definition 7.7 In a G-metric space X, if every G-Cauchy sequence $\{\varkappa_n\}$ in X is G-convergent to some $\varkappa$ in X, then (X, G) is known as G-complete.

Proposition 7.11 (X, G) *is a G-complete metric space if and only if* (X, d_G) *is a complete metric space.*

Consistent with Wardowski [45], we consider the following definitions.

Let F be the set of all maps $F : \mathbb{R}^+ \to \mathbb{R}$ for which the following are true:

($\mathcal{F}_1$) for all $\alpha, \beta \in \mathbb{R}_+$ having $\alpha < \beta$ implies that $F(\alpha) < F(\beta)$. (strictly increasing);

($\mathcal{F}_2$) for every sequence $\{\alpha_n\}$ in $\mathbb{R}^+$, $\lim_{n\to\infty} \alpha_n = 0$ and $\lim_{n\to\infty} F(\alpha_n) = -\infty$ are equivalent.

($\mathcal{F}_3$) $\lim_{\alpha\to 0^+} \alpha^h F(\alpha) = 0$ for $h \in (0, 1)$.

Definition 7.8 ([45]) In a metric space (X, d) and $F \in F$, a map $f : X \to X$ is known as F-contraction on X if there is $\tau > 0$ having

$$d(f\mu, f\upsilon) > 0 \text{ implies that } \tau + F(d(f\mu, f\upsilon)) \leq F(d(\mu, \upsilon)),$$

for all $\mu, \upsilon \in X$.

Note that every F-contraction is continuous (see [45]).

Wardowski [45] obtained the following result.

Theorem 7.1 *In a complete metric space (X, d), if a map $f : X \to X$ is F-contraction, then a unique $\varkappa$ in X exists that has $\varkappa = f\varkappa$. Moreover, for any $\varkappa_0$ in X, sequence $\{x_n\}$ defined by $x_{n+1} = f(x_n)$ converges to $\varkappa$.*

7.3 Best Proximity Points in G-Metric Spaces

In a G-metric space (X, G), let $\mathcal{A}_1$ and $\mathcal{A}_2$ be two nonempty subsets of X. We consider two sets

$$\begin{aligned} \mathcal{A}_0 &= \{a_1 \in \mathcal{A}_1 : d_G(a_1, a_2) = d_G(\mathcal{A}_1, \mathcal{A}_2) \text{ for some } a_2 \in \mathcal{A}_2\}, \\ \mathcal{B}_0 &= \{a_2 \in \mathcal{A}_2 : d_G(a_1, a_2) = d_G(\mathcal{A}_1, \mathcal{A}_2) \text{ for some } a_1 \in \mathcal{A}_1\}, \end{aligned} \tag{7.4}$$

where $d_G(x, y) = \inf\{d_G(a_1, a_2) : a_1 \in \mathcal{A}_1, a_2 \in \mathcal{A}_2\}$.

Definition 7.9 In a G-metric space (X, G), let $\mathcal{A}_1$ and $\mathcal{A}_2$ be two nonempty subsets of X. Set A_2 is called approximately compact w.r.t. $\mathcal{A}_1$ if every sequence $\{u_n\}$ in $\mathcal{A}_2$, satisfy condition that $d_G(a_1, u_n) \to d_G(a_1, \mathcal{A}_2)$ for some a_1 in $\mathcal{A}_1$, then $\{u_n\}$ has a convergent subsequence.

Definition 7.10 In a G-metric space (X, G), let $\mathcal{A}_1$ and $\mathcal{A}_2$ be two nonempty subsets of a X. A nonself map $T : \mathcal{A}_1 \to \mathcal{A}_2$ is known as G-$\mathcal{F}$-proximal contractive whenever a $\tau > 0$ exists, such that for

$$\begin{aligned} d_G(x, y) &= d_G(\mathcal{A}_1, \mathcal{A}_2) \\ d_G(\mu^*, T\mu) &= d_G(\mathcal{A}_1, \mathcal{A}_2) \\ d_G(\upsilon, Ty) &= d_G(\mathcal{A}_1, \mathcal{A}_2) \end{aligned}$$

with $G(\mu, \mu^*, \upsilon) > 0$ implies

$$\tau + F\left(G\left(\mu, \mu^*, \upsilon\right)\right) \leq F\left(G\left(x, \mu, y\right)\right) \tag{7.5}$$

hold for $F \in F$.

Theorem 7.2 *In a G-metric space (X, G), let $\mathcal{A}_1$ and $\mathcal{A}_2$ be two nonempty subsets of X such that $(\mathcal{A}_1, G)$ is a G-complete metric space. Suppose that $\mathcal{A}_0$ is nonempty set and $\mathcal{A}_2$ is approximately compact w.r.t. $\mathcal{A}_1$ and $T : \mathcal{A}_1 \to \mathcal{A}_2$ is a G-$\mathcal{F}$-proximal contractive map having $T(\mathcal{A}_0) \subseteq \mathcal{B}_0$. Then a unique $\upsilon \in \mathcal{A}_1$ exists that satisfies $d_G(\upsilon, T\upsilon) = d_G(\mathcal{A}_1, \mathcal{A}_2)$.*

Proof Let $\varkappa_0 \in \mathcal{A}_0$ be an arbitrary point. By taking $T\varkappa_0 \in T(\mathcal{A}_0) \subseteq \mathcal{B}_0$, we obtain that $\varkappa_1 \in \mathcal{A}_0$ that has $d_G(\varkappa_1, T\varkappa_0) = d_G(\mathcal{A}_1, \mathcal{A}_2)$. Again, as $T\varkappa_1 \in T(\mathcal{A}_0)$, we can find a point $\varkappa_2 \in \mathcal{A}_0$ that has $d_G(\varkappa_2, T\varkappa_1) = d_G(\mathcal{A}_1, \mathcal{A}_2)$. Thus, we get a sequence $\{\varkappa_n\}$ in $\mathcal{A}_0$ that has

$$d_G(\varkappa_{n+1}, T\varkappa_n) = d_G(\mathcal{A}_1, \mathcal{A}_2) \text{ for all } n \in \mathbb{N} \cup \{0\}. \tag{7.6}$$

And so

$$\begin{aligned} d_G(\mu, Tx) &= d_G(\mathcal{A}_1, \mathcal{A}_2), \\ d_G(\mu^*, T\mu) &= d_G(\mathcal{A}_1, \mathcal{A}_2), \\ d_G(\upsilon, Ty) &= d_G(\mathcal{A}_1, \mathcal{A}_2), \end{aligned} \tag{7.7}$$

where $x = \varkappa_{n-1}$, $\mu = \varkappa_n$, $\mu^* = \varkappa_{n+1}$, $y = \varkappa_n$ and $\upsilon = \varkappa_{n+1}$. So, by (7.5) along with $G(\varkappa_n, \varkappa_{n+1}, \varkappa_{n+1}) > 0$, it gives

$$\tau + F(G(\varkappa_n, \varkappa_{n+1}, \varkappa_{n+1})) \leq F(G(\varkappa_{n-1}, \varkappa_n, \varkappa_n)),$$

and we get $G(\varkappa_n, \varkappa_{n+1}, \varkappa_{n+1}) < G(\varkappa_{n-1}, \varkappa_n, \varkappa_n)$. Thus $\{G(\varkappa_n, \varkappa_{n+1}, \varkappa_{n+1})\}$ becomes a decreasing sequence in $\mathbb{R}^+$ that converge to an element η in $\mathbb{R}^+$. Thus

$$F(G(\varkappa_n, \varkappa_{n+1}, \varkappa_{n+1})) \leq F(G(\varkappa_{n-1}, \varkappa_n, \varkappa_n)) - \tau$$

for all $n \in \mathbb{N} \cup \{0\}$. So

$$\begin{aligned} F(G(\varkappa_n, \varkappa_{n+1}, \varkappa_{n+1})) &\leq F(G(\varkappa_{n-1}, \varkappa_n, \varkappa_n)) - \tau \\ &\leq F(G(\varkappa_{n-2}, \varkappa_{n-1}, \varkappa_{n-1})) - 2\tau \\ &\leq \cdots \\ &\leq F(G(\varkappa_0, \varkappa_1, \varkappa_1)) - n\tau \end{aligned}$$

and $\lim_{n\to\infty} F(G(\varkappa_n, \varkappa_{n+1}, \varkappa_{n+1})) = -\infty$ and by using $(\mathcal{F}_2)$

$$\lim_{n\to\infty} G(\varkappa_{n+1}, \varkappa_{n+1}, \varkappa_n) = 0.$$

Using $(\mathcal{F}_3)$, we have $h \in (0, 1)$ having

$$\lim_{n\to\infty} [G(\varkappa_n, \varkappa_{n+1}, \varkappa_{n+1})]^h F(G(\varkappa_n, \varkappa_{n+1}, \varkappa_{n+1})) = 0.$$

Thus

$$\begin{aligned} &\left[G(\varkappa_n, \varkappa_{n+1}, \varkappa_{n+1})\right]^h F(G(\varkappa_n, \varkappa_{n+1}, \varkappa_{n+1})) \\ &-\left[G(\varkappa_n, \varkappa_{n+1}, \varkappa_{n+1})\right]^h F(G(\varkappa_0, \varkappa_1, \varkappa_1)) \\ &\leq -n\tau\left[G(\varkappa_n, \varkappa_{n+1}, \varkappa_{n+1})\right]^h \leq 0. \end{aligned}$$

By $n \to \infty$ we obtain

$$\lim_{n\to\infty} n[G(\varkappa_n, \varkappa_{n+1}, \varkappa_{n+1})]^h = 0.$$

Since $\lim_{n\to\infty} n^{\frac{1}{h}} G(\varkappa_n, \varkappa_{n+1}, \varkappa_{n+1}) = 0$, there exists $n_1 \in \mathbb{N} \cup \{0\}$ having

$$n^{\frac{1}{h}} G(\varkappa_n, \varkappa_{n+1}, \varkappa_{n+1}) \leq 1$$

for all $n \geq n_1$. So

$$G(\varkappa_n, \varkappa_{n+1}, \varkappa_{n+1}) \leq \frac{1}{n^{1/h}}$$

for all $n \geq n_1$. For $m, n \in \mathbb{N} \cup \{0\}$ with $m > n \geq n_1$, implies

$$\begin{aligned} &G(\varkappa_n, \varkappa_m, \varkappa_m) \\ &\leq G(\varkappa_n, \varkappa_{n+1}, \varkappa_{n+1}) + G(\varkappa_{n+1}, \varkappa_{n+2}, \varkappa_{n+2}) + \cdots + G(\varkappa_{m-1}, \varkappa_m, \varkappa_m) \\ &\leq \frac{1}{n^{1/h}} + \frac{1}{(n+1)^{1/h}} + \cdots + \frac{1}{(m-1)^{1/h}} \\ &\leq \sum_{i=n}^{\infty} \frac{1}{i^{1/h}}. \end{aligned}$$

As the series $\sum_{i=1}^{\infty} \frac{1}{i^{1/h}}$ is convergent, we obtain $G(\varkappa_n, \varkappa_m, \varkappa_m) \to 0$ as $n, m \to \infty$. So $\{\varkappa_n\}$ is a G-Cauchy sequence in X. By G-completeness of $(\mathcal{A}_1, G)$, we have an z in $\mathcal{A}_1$ that has $\varkappa_n \to u$ as $n \to \infty$. Also, for all $n \geq 0$,

$$\begin{aligned} d_G(z, \mathcal{A}_2) &\leq d_G(z, T\varkappa_n) \\ &\leq d_G(z, \varkappa_{n+1}) + d_G(\varkappa_{n+1}, T\varkappa_n) \\ &= d_G(z, \varkappa_{n+1}) + d(\mathcal{A}_1, \mathcal{A}_2). \end{aligned} \tag{7.8}$$

By limiting $n \to \infty$ implies

$$\lim_{n\to\infty} d_G(z, T\varkappa_n) = d_G(z, \mathcal{A}_2) = d_G(\mathcal{A}_1, \mathcal{A}_2). \tag{7.9}$$

As $\mathcal{A}_2$ is approximately compact w.r.t. $\mathcal{A}_1$, $\{T\varkappa_n\}$ has a $\{T\varkappa_{n_k}\}$ that converge to an element y^* in $\mathcal{A}_2$. So

$$d_G(z, y^*) = \lim_{n\to\infty} d_G(\varkappa_{n_{k+1}}, T\varkappa_{n_k}) = d_G(\mathcal{A}_1, \mathcal{A}_2), \tag{7.10}$$

which implies $z \in \mathcal{A}_0$. As $Tz \in T(\mathcal{A}_0) \subseteq \mathcal{B}_0$, we must have ω in $\mathcal{A}_0$ that has $d_G(\omega, Tz) = d_G(\mathcal{A}_1, \mathcal{A}_2)$.

By (7.5) with $x = \varkappa_n$, $\mu = \varkappa_{n+1}$, $\mu^* = \varkappa_{n+2}$, $y = z$, and $\upsilon = \omega$, having $G(z, z, \omega) > 0$, it implies

$$\tau + F(G(\varkappa_{n+1}, \varkappa_{n+2}, \omega)) \leq F(G(\varkappa_n, \varkappa_{n+1}, z)),$$

and so

$$G(\varkappa_{n+1}, \varkappa_{n+2}, \omega) < G(\varkappa_n, \varkappa_{n+1}, z)$$

and limiting $n \to \infty$ gives

$$G(z, z, \omega) \leq 0,$$

a contradiction. Therefore $G(z, z, \omega) = 0$ and $\omega = z$. So $d_G(z, Tz) = d_G(\mathcal{A}_1, \mathcal{A}_2)$. Thus T has the best proximity point. For uniqueness, let $\omega_1 \neq \omega_2$ that has $d_G(\omega_1, T\omega_1) = d_G(\mathcal{A}_1, \mathcal{A}_2)$ and $d_G(\omega_2, T\omega_2) = d_G(\mathcal{A}_1, \mathcal{A}_2)$. Using (1) having $x = \mu = \mu^* = \omega_1$ and $y = \upsilon = \omega_2$, with $G(\omega_1, \omega_1, \omega_2) > 0$, we get

$$\tau + F(G(\omega_1, \omega_1, \omega_2)) \leq F(G(\omega_1, \omega_1, \omega_2)),$$

that gives $G(\omega_1, \omega_1, \omega_2) < G(\omega_1, \omega_1, \omega_2)$, a contradiction. Hence $\omega_1 = \omega_2$.

Example 7.3 Let $X = \mathbb{R}^+$ and $G(\lambda, \eta, \gamma) = \frac{1}{2}(|\lambda - \eta| + |\eta - \gamma| + |\lambda - \gamma|)$ for all λ, η, γ in X. Then $d_G(\eta, \gamma) = |\eta - \gamma|$. Take $\mathcal{A}_1 = \{1, 2, 3, 4, 5\}$ and $\mathcal{A}_2 = \{6, 7, 8, 9, 10\}$. Define $T : \mathcal{A}_1 \to \mathcal{A}_2$ as

$$T(\eta) = \begin{cases} 6 & \text{if } \eta = 5, \\ \eta + 5, & \text{if } \eta \in \{1, 2, 3, 4\}. \end{cases}$$

And consider $F : \mathbb{R}^+ \to \mathbb{R}$ defined as $F(\eta) = \ln(\eta)$.

Then, $d_G(\mathcal{A}_1, \mathcal{A}_2) = 1$, $\mathcal{A}_0 = \{5\}$, $\mathcal{B}_0 = \{6\}$, that has $T\mathcal{A}_0 \subseteq \mathcal{B}_0$. Consider $d_G(\mu, Tx) = d_G(\mathcal{A}_1, \mathcal{A}_2)$ and $d_G(\upsilon, Ty) = d_G(\mathcal{A}_1, \mathcal{A}_2)$. This implies $(\mu, x), (\upsilon, y) \in \{(5, 1), (5, 5)\}$. Note that, if $d_G(\mu^*, T\mu) = d_G(\mathcal{A}_1, \mathcal{A}_2) = 1$, then $\mu^* = 5$. And in case

$$\begin{aligned} d_G(\mu, Tx) &= d_G(\mathcal{A}_1, \mathcal{A}_2), \\ d_G(\mu^*, T\mu) &= d_G(\mathcal{A}_1, \mathcal{A}_2), \\ d_G(\upsilon, Ty) &= d_G(\mathcal{A}_1, \mathcal{A}_2), \end{aligned}$$

we get

$$(\mu, \mu^*, \upsilon, x, y) \in \{(5, 5, 5, 5, 1), (5, 5, 5, 1, 1), (5, 5, 5, 1, 5), (5, 5, 5, 5, 5)\}.$$

As $\mu = \mu^* = \upsilon = 5$ gives $G(\mu, \mu^*, \upsilon) = 0$. Thus,

$$\begin{aligned} d_G(\mu, Tx) &= d_G(\mathcal{A}_1, \mathcal{A}_2), \\ d_G(\mu^*, T\mu) &= d_G(\mathcal{A}_1, \mathcal{A}_2), \\ d_G(\upsilon, Ty) &= d_G(\mathcal{A}_1, \mathcal{A}_2), \end{aligned}$$

such that $G(\mu, \mu^*, \upsilon) > 0$ implies

$$\tau + F\left(G\left(\mu, \mu^*, \upsilon\right)\right) \le F\left(G\left(x, \mu, y\right)\right),$$

holds for $F \in \digamma$. Therefore map T is a G-$\mathcal{F}$-proximal contractive. So all axioms of Theorem 7.2 are satisfied. And a unique $\upsilon = 5 \in \mathcal{A}_1$ exists that satisfies $d_G(\upsilon, T\upsilon) = d_G(\mathcal{A}_1, \mathcal{A}_2)$.

In Theorem 7.2, by taking $F(\lambda) = \ln(\lambda)$, we get this result.

Corollary 7.3 *In a G-metric space (X, G), let $\mathcal{A}_1$ and $\mathcal{A}_2$ be two nonempty subsets of a X such that $(\mathcal{A}_1, G)$ is a G-complete metric space. Suppose that $\mathcal{A}_0$ is nonempty set and $\mathcal{A}_2$ is approximately compact w.r.t. $\mathcal{A}_1$ and $T : \mathcal{A}_1 \to \mathcal{A}_2$ is a map satisfies for $x, y, \mu^*, \upsilon \in \mathcal{A}_1$,*

$$\begin{aligned} d_G(\mu, Tx) &= d_G(\mathcal{A}_1, \mathcal{A}_2), \\ d_G(\mu^*, T\mu) &= d_G(\mathcal{A}_1, \mathcal{A}_2), \\ d_G(\upsilon, Ty) &= d_G(\mathcal{A}_1, \mathcal{A}_2), \end{aligned}$$

such that $G(\mu, \mu^*, \upsilon) > 0$ implies

$$G\left(\mu, \mu^*, \upsilon\right) \le e^{-\tau} G\left(x, \mu, y\right), \tag{7.11}$$

holds, where $\tau > 0$ and $T(\mathcal{A}_0) \subseteq \mathcal{B}_0$. Then a unique $\upsilon \in \mathcal{A}_1$ exists that satisfies $d_G(\upsilon, T\upsilon) = d_G(\mathcal{A}_1, \mathcal{A}_2)$.

Theorem 7.3 *In a G-metric space (X, G), let $\mathcal{A}_1$ and $\mathcal{A}_2$ be two nonempty subsets of a X such that $(\mathcal{A}_1, G)$ is a G-complete metric space. Suppose that $\mathcal{A}_0$ is nonempty set and $\mathcal{A}_2$ is approximately compact w.r.t. $\mathcal{A}_1$ and $T : \mathcal{A}_1 \to \mathcal{A}_2$ is a map satisfies for $x, y, \mu, \mu^*, \mu^{**}, \upsilon, \upsilon^* \in \mathcal{A}_1$,*

$$\begin{gathered} d_G(\mu, Tx) = d_G(\mathcal{A}_1, \mathcal{A}_2), \\ d_G(\mu^*, T\mu) = d_G(\mathcal{A}_1, \mathcal{A}_2), \\ d_G(\mu^{**}, T\mu^*) = d_G(\mathcal{A}_1, \mathcal{A}_2), \\ d_G(\mu^*, T\mu) = d_G(\mathcal{A}_1, \mathcal{A}_2), \\ d_G(\upsilon, Ty) = d_G(\mathcal{A}_1, \mathcal{A}_2), \end{gathered}$$

such that $G(\mu, \upsilon, \upsilon^*) > 0$ *implies*

$$\tau + F\left(G\left(\mu, \upsilon, \upsilon^*\right)\right) \leq F\left(\lambda[G\left(x, \mu, \mu^*\right) + G(y, \upsilon, \upsilon^*) + G\left(x, \mu, \upsilon\right) + G\left(y, \upsilon, \mu^{**}\right)]\right), \tag{7.12}$$

where $0 \leq \lambda < \frac{1}{3}$, $\tau > 0$ *and* $F \in \mathcal{F}$ *and* $T(\mathcal{A}_0) \subseteq \mathcal{B}_0$. *Then a unique* $\upsilon \in \mathcal{A}_1$ *exists that satisfies* $d_G(\upsilon, T\upsilon) = d_G(\mathcal{A}_1, \mathcal{A}_2)$.

Proof Using similar steps as in the Proof of Theorem 7.2, we develop a sequence $\{\varkappa_n\}$ in $\mathcal{A}_0$ that has

$$d_G(\varkappa_{n+1}, T\varkappa_n) = d_G(\mathcal{A}_1, \mathcal{A}_2) \text{ for all } n \in \mathbb{N} \cup \{0\}. \tag{7.13}$$

By (7.12), having $x = \varkappa_{n-1}$, $\mu = \varkappa_n$, $\mu^* = \varkappa_{n+1}$, $\mu^{**} = \varkappa_{n+2}$, $y = \varkappa_n$, $\upsilon = \varkappa_{n+1}$ and $\upsilon^* = \varkappa_{n+2}$, having $G(\varkappa_n, \varkappa_{n+1}, \varkappa_{n+2}) > 0$, we have

$$\begin{aligned} \tau + F\left(G\left(\varkappa_n, \varkappa_{n+1}, \varkappa_{n+2}\right)\right) &\leq F(\lambda[G\left(\varkappa_{n-1}, \varkappa_n, \varkappa_{n+1}\right) + G(\varkappa_n, \varkappa_{n+1}, \varkappa_{n+2}) \\ &\quad + G\left(\varkappa_{n-1}, \varkappa_n, \varkappa_{n+1}\right) + G\left(\varkappa_n, \varkappa_{n+1}, \varkappa_{n+2}\right)]) \\ &= F(2\lambda[G\left(\varkappa_{n-1}, \varkappa_n, \varkappa_{n+1}\right) + G(\varkappa_n, \varkappa_{n+1}, \varkappa_{n+2})]). \end{aligned}$$

Now, suppose that $G(\varkappa_{n-1}, \varkappa_n, \varkappa_{n+1}) < G(\varkappa_n, \varkappa_{n+1}, \varkappa_{n+2})$, then

$$\begin{aligned} \tau + F\left(G\left(\varkappa_n, \varkappa_{n+1}, \varkappa_{n+2}\right)\right) &\leq F(2\lambda[G\left(\varkappa_{n-1}, \varkappa_n, \varkappa_{n+1}\right) + G(\varkappa_n, \varkappa_{n+1}, \varkappa_{n+2})]) \\ &< F(3\lambda G(\varkappa_n, \varkappa_{n+1}, \varkappa_{n+2})), \end{aligned}$$

a contradiction. Hence $G(\varkappa_{n-1}, \varkappa_n, \varkappa_{n+1}) \geq G(\varkappa_n, \varkappa_{n+1}, \varkappa_{n+2})$ and so

$$\begin{aligned} \tau + F\left(G\left(\varkappa_n, \varkappa_{n+1}, \varkappa_{n+2}\right)\right) &\leq F(2\lambda[G\left(\varkappa_{n-1}, \varkappa_n, \varkappa_{n+1}\right) + G(\varkappa_n, \varkappa_{n+1}, \varkappa_{n+2})]) \\ &\leq F(3\lambda G(\varkappa_{n-1}, \varkappa_n, \varkappa_{n+1})) \\ &\leq F(G(\varkappa_{n-1}, \varkappa_n, \varkappa_{n+1})), \end{aligned}$$

that is,

$$F\left(G\left(\varkappa_n, \varkappa_{n+1}, \varkappa_{n+2}\right)\right) \leq F\left(G\left(\varkappa_{n-1}, \varkappa_n, \varkappa_{n+1}\right)\right) - \tau$$

for all $n \geq 0$. Now

$$\begin{aligned} F(G(\varkappa_n,\varkappa_{n+1},\varkappa_{n+2})) &\le F(G(\varkappa_{n-1},\varkappa_n,\varkappa_{n+1}))-\tau \\ &\le F(G(\varkappa_{n-2},\varkappa_{n-1},\varkappa_n))-2\tau \\ &\le \cdots \\ &\le F(G(\varkappa_0,\varkappa_1,\varkappa_2))-n\tau \end{aligned}$$

implies $\lim_{n\to\infty} F(G(\varkappa_n,\varkappa_{n+1},\varkappa_{n+2}))=-\infty$ and by $(\mathcal{F}_2)$ implies

$$\lim_{n\to\infty} G(\varkappa_n,\varkappa_{n+1},\varkappa_{n+2})=0.$$

Now by $(\mathcal{F}_3)$, we have $h\in(0,1)$ having

$$\lim_{n\to\infty}[G(\varkappa_n,\varkappa_{n+1},\varkappa_{n+2})]^h F(G(\varkappa_n,\varkappa_{n+1},\varkappa_{n+2}))=0.$$

Thus

$$\begin{aligned} &\left[G(\varkappa_n,\varkappa_{n+1},\varkappa_{n+2})\right]^h F(G(\varkappa_n,\varkappa_{n+1},\varkappa_{n+2})) \\ &-\left[G(\varkappa_n,\varkappa_{n+1},\varkappa_{n+2})\right]^h F(G(\varkappa_0,\varkappa_1,\varkappa_2)) \\ &\le -n\tau\left[G(\varkappa_n,\varkappa_{n+1},\varkappa_{n+2})\right]^h \le 0. \end{aligned}$$

By $n\to\infty$ we obtain

$$\lim_{n\to\infty} n[G(\varkappa_n,\varkappa_{n+1},\varkappa_{n+2})]^h=0.$$

As $\lim_{n\to\infty} n^{\frac{1}{h}} G(\varkappa_n,\varkappa_{n+1},\varkappa_{n+2})=0$, there exists $n_1\in\mathbb{N}\cup\{0\}$ such that

$$n^{\frac{1}{h}} G(\varkappa_n,\varkappa_{n+1},\varkappa_{n+2})\le 1$$

for all $n\ge n_1$. Therefore,

$$G(\varkappa_n,\varkappa_{n+1},\varkappa_{n+2})\le \frac{1}{n^{1/h}}$$

for all $n\ge n_1$. For $m,n\in\mathbb{N}\cup\{0\}$ with $m>n\ge n_1$,

$$
\begin{aligned}
& G(\varkappa_n, \varkappa_m, \varkappa_m) \\
& \leq G(\varkappa_n, \varkappa_{n+1}, \varkappa_{n+1}) + G(\varkappa_{n+1}, \varkappa_{n+2}, \varkappa_{n+2}) + \cdots + G(\varkappa_{m-1}, \varkappa_m, \varkappa_m) \\
& \leq G(\varkappa_n, \varkappa_{n+1}, \varkappa_{n+2}) + G(\varkappa_{n+1}, \varkappa_{n+2}, \varkappa_{n+3}) + \cdots + G(\varkappa_{m-1}, \varkappa_m, \varkappa_{m+1}) \\
& \leq \frac{1}{n^{1/h}} + \frac{1}{(n+1)^{1/h}} + \cdots + \frac{1}{(m-1)^{1/h}} \\
& \leq \sum_{i=n}^{\infty} \frac{1}{i^{1/h}}.
\end{aligned}
$$

As the series $\sum_{i=1}^{\infty} \frac{1}{i^{1/h}}$ is convergent, we get $\{\varkappa_n\}$ is a G-Cauchy sequence in X. By G-completeness of $(\mathcal{A}_1, G)$, we have an z in $\mathcal{A}_1$ that has $\varkappa_n \to u$ as $n \to \infty$. Also, for all $n \geq 0$,

$$
\begin{aligned}
d_G(z, \mathcal{A}_2) & \leq d_G(z, T\varkappa_n) \\
& \leq d_G(z, \varkappa_{n+1}) + d_G(\varkappa_{n+1}, T\varkappa_n) \\
& = d_G(z, \varkappa_{n+1}) + d(\mathcal{A}_1, \mathcal{A}_2).
\end{aligned} \tag{7.14}
$$

By limiting $n \to \infty$ implies

$$
\lim_{n\to\infty} d_G(z, T\varkappa_n) = d_G(z, \mathcal{A}_2) = d_G(\mathcal{A}_1, \mathcal{A}_2). \tag{7.15}
$$

As $\mathcal{A}_2$ is approximately compact w.r.t. $\mathcal{A}_1$, $\{T\varkappa_n\}$ has a $\{T\varkappa_{n_k}\}$ that converge to an element y^* in $\mathcal{A}_2$. So

$$
d_G(z, y^*) = \lim_{n\to\infty} d_G(\varkappa_{n_{k+1}}, T\varkappa_{n_k}) = d_G(\mathcal{A}_1, \mathcal{A}_2), \tag{7.16}
$$

which implies $z \in \mathcal{A}_0$. As $Tz \in T(\mathcal{A}_0) \subseteq \mathcal{B}_0$, we must have ω in $\mathcal{A}_0$ that has $d_G(\omega, Tz) = d_G(\mathcal{A}_1, \mathcal{A}_2)$. Also $T\omega \in T(\mathcal{A}_0) \subseteq \mathcal{B}_0$, we must have ω^* in $\mathcal{A}_0$ that has $d_G(\omega^*, T\omega^*) = d_G(\mathcal{A}_1, \mathcal{A}_2)$.

By (7.12) having $x = \varkappa_{n-1}$, $\mu = \varkappa_n$, $\mu^* = \varkappa_{n+1}$, $\mu^{**} = \varkappa_{n+2}$, $y = z$, $v = \omega$ and $v^* = \omega^*$, implies

$$
\begin{aligned}
\tau + F\left(G\left(\varkappa_n, \omega, \omega^*\right)\right) & \leq F(\lambda[G\left(\varkappa_{n-1}, \varkappa_n, \varkappa_{n+1}\right) + G(z, \omega, \omega^*) + G\left(\varkappa_{n-1}, \varkappa_n, \omega\right) \\
& \quad + G\left(z, \omega, \varkappa_{n+2}\right)])
\end{aligned}
$$

and limiting $n \to \infty$ gives

$$
\begin{aligned}
\tau + F\left(G\left(z, \omega, \omega^*\right)\right) & \leq F(\lambda[G\left(z, z, z\right) + G(z, \omega, \omega^*) + G(z, z, \omega) + G(z, \omega, z)]) \\
& = F(\lambda[G(z, \omega, \omega^*) + 2G(z, z, \omega)]),
\end{aligned}
$$

that is,

$$G\left(z,\omega,\omega^*\right) < \frac{2\lambda}{1-\lambda} G\left(z,z,\omega\right). \tag{7.17}$$

Let us assume that $\omega \neq \omega^*$. As by $(\mathfrak{g}_3)$, implies $G\left(z,z,\omega\right) \leq G\left(z,w,\omega^*\right)$. And by (7.17) implies

$$\begin{aligned} G\left(z,\omega,\omega^*\right) &< \frac{2\lambda}{1-\lambda} G\left(z,z,\omega\right) \\ &< G\left(z,\omega,\omega^*\right), \end{aligned} \tag{7.18}$$

a contradiction and so $\omega = \omega^*$. Thus $d_G\left(\omega, T\omega\right) = d_G\left(\omega^*, T\omega\right) = d_G\left(\mathcal{A}_1, \mathcal{A}_2\right)$.

For uniqueness, let $\omega_1 \neq \omega_2$ that has $d_G(\omega_1, T\omega_1) = d_G(\mathcal{A}_1, \mathcal{A}_2)$ and $d_G(\omega_2, T\omega_2) = d_G(\mathcal{A}_1, \mathcal{A}_2)$. Using (7.12) having $x = \mu = \mu^* = \mu^{**} = \omega_1$ and $y = v = v^* = \omega_2$, with $G\left(\omega_1, \omega_1, \omega_2\right) > 0$, and

$$\begin{aligned} &\tau + F(G\left(\omega_1, \omega_1, \omega_2\right)) \\ &\leq F(\lambda[G\left(\omega_1, \omega_1, \omega_1\right) + G(\omega_2, \omega_2, \omega_2) + G\left(\omega_1, \omega_1, \omega_2\right) + G\left(\omega_2, \omega_2, \omega_1\right)]) \\ &\leq F(3\lambda G\left(\omega_1, \omega_1, \omega_2\right)) \\ &\leq F(G\left(\omega_1, \omega_1, \omega_2\right)), \end{aligned}$$

that gives $G\left(\omega_1, \omega_1, \omega_2\right) < G\left(\omega_1, \omega_1, \omega_2\right)$, a contradiction. Hence $\omega_1 = \omega_2$.

Theorem 7.4 *In a G-metric space (X, G), let $\mathcal{A}_1$ and $\mathcal{A}_2$ be two nonempty subsets of a X such that $(\mathcal{A}_1, G)$ is a G-complete metric space. Suppose that $\mathcal{A}_0$ is nonempty set and $\mathcal{A}_2$ is approximately compact w.r.t. $\mathcal{A}_1$ and $T : \mathcal{A}_1 \to \mathcal{A}_2$ is a map satisfies for $x, y, \mu, v \in \mathcal{A}_1$,*

$$\begin{aligned} d_G(\mu, Tx) &= d_G(\mathcal{A}_1, \mathcal{A}_2), \\ d_G(v, Ty) &= d_G(\mathcal{A}_1, \mathcal{A}_2), \end{aligned}$$

such that $G\left(\mu, v, v^*\right) > 0$ implies

$$\tau + F(G\left(\mu, v, v\right) \leq F(\max\{G(x,\mu,y), \frac{\sqrt{G(x,y,y)G(x,\mu,\mu)}}{1+G(\mu,v,v)}, G\left(x,y,y\right), G\left(x,\mu,\mu\right)\}), \tag{7.19}$$

where $\tau > 0$ and $F \in F$ and $T(\mathcal{A}_0) \subseteq \mathcal{B}_0$. Then a unique $v \in \mathcal{A}_1$ exists that satisfies $d_G\left(v, Tv\right) = d_G\left(\mathcal{A}_1, \mathcal{A}_2\right)$.

Proof Using similar steps as in the Proof of Theorem 7.2, we develop a sequence $\{\varkappa_n\}$ in $\mathcal{A}_0$ that has

$$d_G(\varkappa_{n+1}, T\varkappa_n) = d_G(\mathcal{A}_1, \mathcal{A}_2) \text{ for all } n \in \mathbb{N} \cup \{0\}. \tag{7.20}$$

From (7.19) with $x = \varkappa_{n-1}$, $\mu = \varkappa_n$, $y = \varkappa_n$, and $\upsilon = \varkappa_{n+1}$, implies

$$\begin{aligned}
&\tau + F\left(G\left(\varkappa_n, \varkappa_{n+1}, \varkappa_{n+1}\right)\right) \\
&\leq F(\max\{G(\varkappa_{n-1}, \varkappa_n, \varkappa_n), \frac{\sqrt{G(\varkappa_{n-1}, \varkappa_n, \varkappa_n)G(\varkappa_{n-1}, \varkappa_n, \varkappa_n)}}{1 + G(\varkappa_n, \varkappa_{n+1}, \varkappa_{n+1})}, \\
&G\left(\varkappa_{n-1}, \varkappa_n, \varkappa_n\right), G\left(\varkappa_{n-1}, \varkappa_n, \varkappa_n\right)\} \\
&\leq F(G(\varkappa_{n-1}, \varkappa_n, \varkappa_n))
\end{aligned} \tag{7.21}$$

for all $n \in \mathbb{N} \cup \{0\}$. So

$$\begin{aligned}
F\left(G\left(\varkappa_n, \varkappa_{n+1}, \varkappa_{n+1}\right)\right) &\leq F\left(G\left(\varkappa_{n-1}, \varkappa_n, \varkappa_n\right)\right) - \tau \\
&\leq F\left(G\left(\varkappa_{n-2}, \varkappa_{n-1}, \varkappa_{n-1}\right)\right) - 2\tau \\
&\leq \cdots \\
&\leq F\left(G\left(\varkappa_0, \varkappa_1, \varkappa_1\right)\right) - n\tau
\end{aligned}$$

implies $\lim_{n\to\infty} F\left(G\left(\varkappa_n, \varkappa_{n+1}, \varkappa_{n+1}\right)\right) = -\infty$ and using $(\mathcal{F}_2)$, it implies

$$\lim_{n\to\infty} G\left(\varkappa_n, \varkappa_{n+1}, \varkappa_{n+1}\right) = 0. \tag{7.22}$$

By $(\mathcal{F}_3)$, we have $h \in (0, 1)$ having

$$\lim_{n\to\infty} [G\left(\varkappa_n, \varkappa_{n+1}, \varkappa_{n+1}\right)]^h F\left(G\left(\varkappa_n, \varkappa_{n+1}, \varkappa_{n+1}\right)\right) = 0.$$

Thus

$$\begin{aligned}
&\left[G\left(\varkappa_n, \varkappa_{n+1}, \varkappa_{n+1}\right)\right]^h F\left(G\left(\varkappa_n, \varkappa_{n+1}, \varkappa_{n+1}\right)\right) \\
&- \left[G\left(\varkappa_n, \varkappa_{n+1}, \varkappa_{n+1}\right)\right]^h F\left(G\left(\varkappa_0, \varkappa_1, \varkappa_1\right)\right) \\
&\leq -n\tau \left[G\left(\varkappa_n, \varkappa_{n+1}, \varkappa_{n+1}\right)\right]^h \leq 0.
\end{aligned}$$

By $n \to \infty$,

$$\lim_{n\to\infty} n[G\left(\varkappa_n, \varkappa_{n+1}, \varkappa_{n+1}\right)]^h = 0.$$

As $\lim_{n\to\infty} n^{\frac{1}{h}} G\left(\varkappa_n, \varkappa_{n+1}, \varkappa_{n+1}\right) = 0$, we have $n_1 \in \mathbb{N} \cup \{0\}$ having

$$n^{\frac{1}{h}} G\left(\varkappa_n, \varkappa_{n+1}, \varkappa_{n+1}\right) \leq 1$$

for all $n \geq n_1$. So

$$G\left(\varkappa_n, \varkappa_{n+1}, \varkappa_{n+1}\right) \leq \frac{1}{n^{1/h}}$$

for all $n \geq n_1$. For $m, n \in \mathbb{N} \cup \{0\}$ with $m > n \geq n_1$, implies

$$\begin{aligned} G(\varkappa_n, \varkappa_m, \varkappa_m) &\leq G(\varkappa_n, \varkappa_{n+1}, \varkappa_{n+1}) + G(\varkappa_{n+1}, \varkappa_{n+2}, \varkappa_{n+2}) + \cdots \\ &\quad + G(\varkappa_{m-1}, \varkappa_m, \varkappa_m) \\ &\leq \frac{1}{n^{1/h}} + \frac{1}{(n+1)^{1/h}} + \cdots + \frac{1}{(m-1)^{1/h}} \\ &\leq \sum_{i=n}^{\infty} \frac{1}{i^{1/h}}. \end{aligned}$$

As the series $\sum_{i=1}^{\infty} \frac{1}{i^{1/h}}$ is convergent, we get $\{\varkappa_n\}$ is a G-Cauchy sequence in X. By G-completeness of $(\mathcal{A}_1, G)$, we have an z in $\mathcal{A}_1$ that has $\varkappa_n \to u$ as $n \to \infty$. Follows step in Proof of Theorem 7.2, we have $\omega \in \mathcal{A}_0$ that has $d_G(\omega, Tz) = d_G(\mathcal{A}_1, \mathcal{A}_2)$.

By (7.19) having $x = \varkappa_{n-1}$, $\mu = \varkappa_n$, $y = z$, and $v = w$, implies

$$\begin{aligned} \tau + F(G(\varkappa_n, \omega, \omega)) &\leq F(\max\{G(\varkappa_{n-1}, \varkappa_n, z), \frac{\sqrt{G(\varkappa_{n-1}, z, z)G(\varkappa_{n-1}, \varkappa_n, \varkappa_n)}}{1 + G(\varkappa_n, \omega, \omega)}, \\ &\quad G(\varkappa_{n-1}, z, z), G(\varkappa_{n-1}, \varkappa_n, \varkappa_n)\} \end{aligned} \tag{7.23}$$

and limiting as $n \to \infty$ implies $\lim_{n,m\to\infty} G(z, \omega, \omega) = 0$, and so $z = \omega$. Thus $d_G(\omega, T\omega) = d_G(\omega^*, T\omega) = d_G(\mathcal{A}_1, \mathcal{A}_2)$.

For uniqueness, let $\omega_1 \neq \omega_2$ that has $d_G(\omega_1, T\omega_1) = d_G(\mathcal{A}_1, \mathcal{A}_2)$ and $d_G(\omega_2, T\omega_2) = d_G(\mathcal{A}_1, \mathcal{A}_2)$. Using (7.19) having $x = \mu = \omega_1$ and $y = v = \omega_2$, with $G(\omega_1, \omega_1, \omega_2) > 0$, we get

$$\begin{aligned} \tau + F(G(\omega_1, \omega_2, \omega_2)) &\leq F(\max\{G(\omega_1, \omega_1, \omega_2), \frac{\sqrt{G(\omega_1, \omega_2, \omega_2)G(\omega_1, \omega_1, \omega_1)}}{1 + G(\omega_1, \omega_2, \omega_2)}, \\ &\quad G(\omega_1, \omega_2, \omega_2), G(\omega_1, \omega_1, \omega_1\} \\ &\leq F(G(\omega_1, \omega_1, \omega_2)), \end{aligned}$$

that gives

$$G(\omega_1, \omega_2, \omega_2) < G(\omega_1, \omega_1, \omega_2). \tag{7.24}$$

Again using (7.19) having $y = \mu = \omega_1$ and $x = v = \omega_2$, we have

$$\begin{aligned} \tau + F(G(\omega_2, \omega_1, \omega_1)) &\leq F(\max\{G(\omega_2, \omega_2, \omega_1), \frac{\sqrt{G(\omega_2, \omega_1, \omega_1)G(\omega_2, \omega_2, \omega_2)}}{1 + G(\omega_2, \omega_1, \omega_1)}, \\ &\quad G(\omega_2, \omega_1, \omega_1), G(\omega_2, \omega_2, \omega_2\} \\ &\leq F(G(\omega_2, \omega_2, \omega_1)), \end{aligned}$$

that gives

$$G(\omega_2, \omega_1, \omega_1) < G(\omega_2, \omega_2, \omega_1). \tag{7.25}$$

By (7.24) and (7.25), we get a contradiction. Hence $\omega_1 = \omega_2$.

Example 7.4 Let $X = \mathbb{R}^+$ and $G(\lambda, \eta, \gamma) = \max\{|\lambda - \eta|, |\eta - \gamma|, |\lambda - \gamma|\}$ for all $\lambda, \eta, \gamma \in X$. Then $d_G(\eta, \gamma) = |\eta - \gamma|$. Take $\mathcal{A}_1 = \{0, 2, 4, 6\}$ and $\mathcal{A}_2 = \{8, 10, 12, 14\}$. Define $T : \mathcal{A}_1 \to \mathcal{A}_2$ as

$$T(\eta) = \begin{cases} \eta + 6 \text{ if } \eta \in \{0, 2, 4\}, \\ 8 \qquad\quad \text{if } \eta = 6. \end{cases}$$

And consider $F : \mathbb{R}^+ \to \mathbb{R}$ defined as $F(\eta) = \ln(\eta) + \eta$. Then clearly $F \in \digamma$. Also, $d_G(\mathcal{A}_1, \mathcal{A}_2) = 2$, $\mathcal{A}_0 = \{6\}$, $\mathcal{B}_0 = \{8\}$ that has $T\mathcal{A}_0 \subseteq \mathcal{B}_0$.

For any μ, υ, x, y in $\mathcal{A}_1$ such that $\mu \neq \upsilon$ with $d_G(\mu, Tx) = d_G(\mathcal{A}_1, \mathcal{A}_2)$ and $d_G(\upsilon, Ty) = d_G(\mathcal{A}_1, \mathcal{A}_2)$, we have $(\mu, x), (\upsilon, y) \in \{(4, 0), (6, 2), (6, 6)\}$.

By taking $\tau = 2$ with $\mu \neq \upsilon$, the following cases arise:

(I) For $(\mu, x) = (4, 0)$, $(\upsilon, y) = (6, 2)$,

$$\begin{aligned} &\tau + F(G(\mu, \upsilon, \upsilon)) \\ &= \tau + \ln(2) + 2 \\ &\leq \ln(4) + 4 \\ &\leq F(G(x, \mu, y)) \\ &\leq F(\max\{G(x, \mu, y), \frac{\sqrt{G(x, y, y)G(x, \mu, \mu)}}{1 + G(\mu, \upsilon, \upsilon)}, G(x, y, y), G(x, \mu, \mu)\}). \end{aligned}$$

(II) In case $(\mu, x) = (4, 0)$, $(\upsilon, y) = (6, 6)$,

$$\begin{aligned} &\tau + F(G(\mu, \upsilon, \upsilon)) \\ &= \tau + \ln(2) + 2 \\ &\leq \ln(6) + 6 \\ &\leq F(G(x, y, y)) \\ &\leq F(\max\{G(x, \mu, y), \frac{\sqrt{G(x, y, y)G(x, \mu, \mu)}}{1 + G(\mu, \upsilon, \upsilon)}, G(x, y, y), G(x, \mu, \mu)\}). \end{aligned}$$

Thus $T : \mathcal{A}_1 \to \mathcal{A}_2$ is a map satisfies for $x, y, \mu, \upsilon \in \mathcal{A}_1$,

$$\begin{aligned} d_G(\mu, Tx) &= d_G(\mathcal{A}_1, \mathcal{A}_2), \\ d_G(\upsilon, Ty) &= d_G(\mathcal{A}_1, \mathcal{A}_2), \end{aligned}$$

such that $G(\mu, \upsilon, \upsilon^*) > 0$ implies

$$\tau + F(G(\mu, \upsilon, \upsilon) \leq F(\max\{G(x, \mu, y), \frac{\sqrt{G(x, y, y)G(x, \mu, \mu)}}{1 + G(\mu, \upsilon, \upsilon)}, G(x, y, y), G(x, \mu, \mu)\}),$$

where $\tau > 0$ and $F \in \digamma$ and $T(\mathcal{A}_0) \subseteq \mathcal{B}_0$. So all axioms of Theorem 7.4 are satisfied. And a unique $\upsilon = 6 \in \mathcal{A}_1$ exists that satisfies $d_G(\upsilon, T\upsilon) = d_G(\mathcal{A}_1, \mathcal{A}_2)$.

Corollary 7.4 *In a G-metric space* (X, G)*, let* $\mathcal{A}_1$ *and* $\mathcal{A}_2$ *be two nonempty subsets of a* X *such that* $(\mathcal{A}_1, G)$ *is a G-complete metric space. Suppose that* $\mathcal{A}_0$ *is nonempty set and* $\mathcal{A}_2$ *is approximately compact w.r.t.* $\mathcal{A}_1$ *and* $T : \mathcal{A}_1 \to \mathcal{A}_2$ *is a map satisfies for* $x, y, \mu, \upsilon \in \mathcal{A}_1$,

$$d_G(\mu, Tx) = d_G(\mathcal{A}_1, \mathcal{A}_2),$$
$$d_G(\upsilon, Ty) = d_G(\mathcal{A}_1, \mathcal{A}_2),$$

such that $G(\mu, \upsilon, \upsilon^*) > 0$ *implies*

$$\tau + F(G(\mu, \upsilon, \upsilon)) \leq F(G(x, \mu, y)), \tag{7.26}$$

where $\tau > 0$ *and* $F \in \mathcal{F}$ *and* $T(\mathcal{A}_0) \subseteq \mathcal{B}_0$*. Then a unique* $\upsilon \in \mathcal{A}_1$ *exists that satisfies* $d_G(\upsilon, T\upsilon) = d_G(\mathcal{A}_1, \mathcal{A}_2)$.

7.4 Application to Fixed Point Results

Now, we show fixed point results as applications of obtained best proximity results. Notice that in case

$$\begin{aligned} d_G(\mu, Tx) &= d_G(\mathcal{A}_1, \mathcal{A}_2), \\ d_G(\mu^*, T\mu) &= d_G(\mathcal{A}_1, \mathcal{A}_2), \\ d_G(\upsilon, Ty) &= d_G(\mathcal{A}_1, \mathcal{A}_2) \end{aligned} \tag{7.27}$$

and $\mathcal{A}_1 = \mathcal{A}_2 = X$, then $\mu = Tx$, $\mu^* = T\mu$ and $\upsilon = Ty$. Thus, $\mu^* = T^2x$. In Theorem 7.2, having $\mathcal{A}_1 = \mathcal{A}_2 = X$, we get this result.

Theorem 7.5 *In a G-complete metric space* (X, G)*, consider a selfmap* $T : X \to X$*. Suppose that a* $\tau > 0$ *exists having for all* $\mathfrak{u}_1, \mathfrak{u}_2 \in X$ *with* $G\left(T\mathfrak{u}_1, T^2\mathfrak{u}_1, T\mathfrak{u}_2\right) > 0$,

$$\tau + F(G\left(T\mathfrak{u}_1, T^2\mathfrak{u}_1, T\mathfrak{u}_2\right)) \leq F(G(\mathfrak{u}_1, T\mathfrak{u}_1, \mathfrak{u}_2)) \tag{7.28}$$

holds, where $F \in \mathcal{F}$*. Then a unique* $\upsilon \in \mathcal{X}$ *exists that satisfies* $\upsilon = T\upsilon$.

Corollary 7.5 *In a G-complete metric space* (X, G)*, consider a selfmap* $T : X \to X$*. Suppose that a* $\tau > 0$ *exists such that for all* $\mathfrak{u}_1, \mathfrak{u}_2 \in X$ *with* $G\left(T\mathfrak{u}_1, T^2\mathfrak{u}_1, T\mathfrak{u}_2\right) > 0$,

$$G\left(T\mathfrak{u}_1, T^2\mathfrak{u}_1, T\mathfrak{u}_2\right)) \leq e^{-\tau} G(\mathfrak{u}_1, T\mathfrak{u}_2, \mathfrak{u}_2) \tag{7.29}$$

holds. Then a unique $\upsilon \in \mathcal{X}$ *exists that satisfies* $\upsilon = T\upsilon$.

Follows from Theorem 7.3, we get this result.

Theorem 7.6 *In a G-complete metric space* (X, G)*, consider a selfmap* $T : X \to X$*. Suppose that a* $\tau > 0$ *exists such that for all* $\mathfrak{u}_1, \mathfrak{u}_2 \in X$ *with* $G\left(T\mathfrak{u}_1, T\mathfrak{u}_2, T^2\mathfrak{u}_2\right) > 0$,

$$\tau + F\left(G\left(T\mathrm{u}_1, T\mathrm{u}_2, T^2\mathrm{u}_2\right)\right) \leq F(\lambda \max\{G\left(\mathrm{u}_1, T\mathrm{u}_1, T^2\mathrm{u}_1\right), G(\mathrm{u}_2, T\mathrm{u}_2, T^2\mathrm{u}_2\right), G\left(\mathrm{u}_1, T\mathrm{u}_1, T\mathrm{u}_2\right), G\left(\mathrm{u}_2, T\mathrm{u}_2, T^3\mathrm{u}_1\right)\} \tag{7.30}$$

holds, where $F \in \mathcal{F}$ and $\lambda \in [0, \frac{1}{3})$. Then a unique $\upsilon \in \mathcal{X}$ exists that satisfies $\upsilon = T\upsilon$.

Follows from Theorem 7.4, we get this result.

Theorem 7.7 *In a G-complete metric space (X, G), consider a selfmap $T : X \to X$. Suppose that a $\tau > 0$ exists such that for all $\mathrm{u}_1, \mathrm{u}_2 \in X$ with $G\left(T\mathrm{u}_1, T\mathrm{u}_2, T\mathrm{u}_2\right) > 0$,*

$$\tau + F\left(G\left(T\mathrm{u}_1, T\mathrm{u}_2, T\mathrm{u}_2\right)\right) \leq F(\max\{G(\mathrm{u}_1, T\mathrm{u}_1, \mathrm{u}_2), \frac{\sqrt{G(\mathrm{u}_1, \mathrm{u}_2, \mathrm{u}_2)G(\mathrm{u}_1, T\mathrm{u}_1, T\mathrm{u}_1)}}{1 + G(T\mathrm{u}_1, T\mathrm{u}_2, T\mathrm{u}_2)}, G\left(\mathrm{u}_1, \mathrm{u}_2, \mathrm{u}_2\right), G\left(\mathrm{u}_1, T\mathrm{u}_1, T\mathrm{u}_1\right)\}, \tag{7.31}$$

holds, where $F \in \mathcal{F}$. Then a unique $\upsilon \in \mathcal{X}$ exists that satisfies $\upsilon = T\upsilon$.

Corollary 7.6 *In a G-complete metric space (X, G), consider a selfmap $T : X \to X$. Suppose that a $\tau > 0$ exists such that for all $\mathrm{u}_1, \mathrm{u}_2 \in X$ with $G\left(T\mathrm{u}_1, T\mathrm{u}_2, T\mathrm{u}_2\right) > 0$,*

$$\tau + F\left(G\left(T\mathrm{u}_1, T\mathrm{u}_2, T\mathrm{u}_2\right)\right) \leq F\left(G(\mathrm{u}_1, T\mathrm{u}_2, \mathrm{u}_2)\right), \tag{7.32}$$

holds, where $F \in \mathcal{F}$. Then a unique $\upsilon \in \mathcal{X}$ exists that satisfies $\upsilon = T\upsilon$.

7.5 Application to Solution of Functional Equations

We are applying our obtained results to obtain the solution of a functional equation appearing in the dynamic programming.

Let B_1 and B_2 be two Banach spaces with $U \subseteq B_1$ and $V \subseteq B_2$. Suppose that

$$\mu \colon U \times V \longrightarrow U, \quad \sigma : U \times V \longrightarrow \mathbb{R}, \quad f \colon U \times V \times \mathbb{R} \longrightarrow \mathbb{R}.$$

If we take U and V as the state and decision spaces, respectively, then the problem of dynamic programming brings down the problem to solve a functional equation:

$$\rho(\nu) = \sup_{z \in V}\{\sigma(\nu, z) + f(\nu, z, \rho(\mu(\nu, z)))\}, \text{ for } \nu \in U. \tag{7.33}$$

We study the existence and uniqueness of bounded solution of functional equation (7.33) arising in dynamic programming in G-metric space.

Let $B(U)$ denotes the set of all bounded real valued functions on U. For an arbitrary $\eta \in B(U)$, define $\|\eta\| = \sup_{\tau \in U} |\eta(\tau)|$. Then $(B(U), \|\cdot\|)$ is a Banach space. Now consider

$$G(\lambda, \eta, \gamma) = \sup_{\tau \in U} |\lambda(\tau) - \eta(\tau)| + \sup_{\tau \in U} |\gamma(\tau) - \eta(\tau)| + \sup_{\tau \in U} |\gamma(\tau) - \lambda(\tau)|,$$

where $\lambda, \eta, \gamma \in B(U)$. Then $(B(U), G)$ is a $G-$ complete metric space.
Assume that
(C_1) : f and σ are bounded and continuous.
(C_2): For $x \in U$, $\eta \in B(U)$ and $b > 0$, take $T : B(U) \to B(U)$ as

$$T\eta(\upsilon) = \sup_{y \in V}\{\sigma(\upsilon, y) + f(\upsilon, y, \eta(\mu(\upsilon, y)))\} \text{ for } \upsilon \in U. \tag{7.34}$$

Moreover, for (υ, y) in $U \times V$, $\tau \in U$ having $f(\upsilon, y, \eta(\tau)) \neq f(\upsilon, y, \xi(\tau))$ implies

$$2|f(\upsilon, y, \eta(\tau)) - f(\upsilon, y, \xi(\tau))| \leq G(\eta(\tau), T\eta(\tau), \xi(\tau)). \tag{7.35}$$

Theorem 7.8 *Assume that the conditions* (C_1)*–*(C_2) *hold. Then, functional equation (7.33) has a unique bounded solution in* $B(U)$*.*

Proof Since $(B(U), G)$ is a G-complete metric space. By (C_1), T is a selfmap of $B(U)$. By (7.34) in (C_2), it follows that for any $\eta, \xi \in B(U)$, choose $x \in U$ and $y_1, y_2 \in V$ such that

$$T\eta < \sigma(\upsilon, y_1) + f(\upsilon, y_1, \eta(\mu(\upsilon, y_1))), \tag{7.36}$$

$$T\xi < \sigma(\upsilon, y_2) + f(\upsilon, y_2, \xi(\mu(\upsilon, y_2))), \tag{7.37}$$

which further implies that

$$T\eta \geq \sigma(\upsilon, y_2) + f(\upsilon, y_2, \eta(\mu(\upsilon, y_2))) \tag{7.38}$$

$$T\xi \geq \sigma(\upsilon, y_1) + f(\upsilon, y_1, \xi(\mu(\upsilon, y_1))). \tag{7.39}$$

From (7.36) and (7.39) together with (7.35) implies

$$\begin{aligned} 2(T\eta(\tau) - T\xi(\tau)) &< 2(f(\upsilon, y_1, \eta(\kappa(x, y_1))) - f(\upsilon, y_1, \xi(\mu(x, y_1)))) \\ &\leq G(\eta(\tau), T\eta(\tau), \xi(\tau)). \end{aligned} \tag{7.40}$$

From (7.37) and (7.38) together with (7.35) implies

$$\begin{aligned} 2(T\xi(\tau) - T\eta(\tau)) &< 2(f(\upsilon, y_2, \xi(\kappa(\upsilon, y_2))) - f(\upsilon, y_2, \eta(\mu(\upsilon, y_2)))) \\ &\leq G(\eta(\tau), T\eta(\tau), \xi(\tau)). \end{aligned} \tag{7.41}$$

From (7.41), we get

$$2|T\eta(\tau) - T\xi(\tau)| < G(\eta(\tau), T\eta(\tau), \xi(\tau)). \tag{7.42}$$

From (7.42),

$$G(T\eta(\tau), T\xi(\tau), T\xi(\tau)) < G(\eta(\tau), T\eta(\tau), \xi(\tau)). \tag{7.43}$$

Applying $F \in \mathcal{F}$ on both sides of (7.43) gives

$$F(G(T\eta(\tau), T\xi(\tau), T\xi(\tau))) < F(G(\eta(\tau), T\eta(\tau), \xi(\tau))). \tag{7.44}$$

Consequently, there exists $\tau > 0$ having for all $\eta(\tau), \xi(\tau) \in B(U)$ that has $T\eta(\tau) \neq T\xi(\tau)$,

$$\tau + F(G(T\eta(\tau), T\xi(\tau), T\xi(\tau))) \leq F(G(\eta(\tau), T\eta(\tau), \xi(\tau))) \tag{7.45}$$

holds. Therefore, all conditions of Corollary 7.6 hold. Thus, a fixed point of T exists, i.e. $\eta^* \in B(U)$, where $\eta^*(\tau)$ is a solution of functional equation (7.33).

7.6 Conclusion

We defined certain new classes of proximal F-contractive maps in the setup of G-metric and established best proximity point results for these kinds of mappings. Moreover, we were able to present examples to illustrate main result and an application related to the existence of unique bounded solution in functional equation appearing in domain of the dynamic programming as well. As G_b-metric spaces are linked in a very natural way with applications of scientific problems (see [25]), so the best proximity point results of this paper can be investigated in connection with G_b metric spaces with some applications.

References

1. Abbas, M., Nazir, T.: Fixed points of T-Hardy Rogers type contraction mapping in G-metric spaces. Afrika Mat. **25**(1), 103–113 (2014)
2. Abbas, M., Nazir, T., Radenovic, S.: Some periodic point results in generalized metric spaces. Appl. Math. Comput. **217**(8), 4094–4099 (2010)
3. Abbas, M., Khan, S.H., Nazir, T.: Common fixed points of R-weakly commuting maps in generalized metric spaces. Fixed Point Theory Appl. **41**, 11 (2011)
4. Abbas, M., Nazir, T., Shatanawi, W., Mustafa, Z.: Fixed and related fixed point theorems for three maps in G-metric spaces. Hacet. J. Math. Stat. **41**(2), 291–306 (2012)
5. Abbas, M., Sintunavarat, W., Kumam, P.: Coupled fixed point of generalized contractive mappings on partially ordered $G-$ metric spaces. Fixed Point Theory Appl. **2012**, 31 (2012)
6. Amini-Harandi, A., Hussain, N., Akbar, F.: Best proximity point results for generalized contractions in metric spaces. Fixed Point Theory Appl. Article ID 164 (2013)
7. Asadi, M., Karapinar, E., Salimi, P.: A new approach to $G-$metric and related fixed point theorems. J. Inequal. Appl. 2013, Article 454 (2013)

8. di Bari, C., Suzuki, T., Vetro, C.: Best proximity points forcyclic Meir-Keeler contractions. Nonlinear Anal. 69(11), 3790–3794 (2008)
9. Choudhury, B.S., Maity, P.: Coupled fixed point results in generalized metric spaces. Math. Comput. Modeling **54**, 73–79 (2011)
10. Dhage, B.C.: Generalized metric space and mapping with fixed point. Bull. Calcutta Math. Soc. **84**, 329–336 (1992)
11. Dhage, B.C.: Generalized metric spaces and topological structure, I. An. Stiint. Univ. Al. I. Cuza Iasi. Mat (N.S) **46**, 3–24 (2000)
12. Dhage, B.C.: On generalized metric spaces and topological structure II. Pure Appl. Math. Sci. **40**(1–2), 37–41 (1994)
13. Dhage, B.C.: On continuity of mappings in D-metric spaces. Bull. Calcutta Math. Soc. **86**, 503–508 (1994)
14. Fan, K.: Extensions of two fixed point theorems of F. E. Browder. Mathematische Zeitschrift **112**(3), 234–240 (1969)
15. Hussain, N., Latif, A., Salimi, P.: Best proximity point results in G-metric spaces. Abstr. Appl. Anal. Article ID 837943, 8 (2014)
16. Hussain, N., Kutbi, M.A., Salimi, P.: Best proximity point results for modified $\alpha - \psi$-proximal rational contractions. Abstr. Appl. Anal. **2013**, Article ID 927457, 14 (2013)
17. Hussain, N., Latif, A., Salimi, P.: Best proximity point results for modified Suzuki $\alpha - \psi$-proximal contractions. Fixed Point Theory Appl. **2014**, Article ID 10 (2014)
18. Hussain, N., Karapinar, E., Salimi, P., Vetro, P.: Fixed point results for G^m-Meir-Keeler contractive and $G - (\alpha, \psi)$-Meir-Keeler contractive mappings. Fixed Point Theory Appl. **2013**, Article ID, 34 (2013)
19. Işık, H., Ali, A., Uddin, F., Asif, A., Arshad, M.: Best proximity points for multivalued mappings satisfying Z_σ-proximal contractions with applications. In: Debnath, P., Konwar, N., Radenović, S. (eds.) Metric Fixed Point Theory, pp. 227–245. Forum for Interdisciplinary Mathematics. Springer, Singapore (2022)
20. Işık, H., Aydi, H., Mlaik, N., Radenović, S.: Best proximity point results for Geraghty type Z-proximal contractions with an application. Axioms **8**(3), 81 (2019)
21. Işık, H., Aydi, H.: Best proximity problems for new types of Z-proximal contractions with an application. Commun. Fac. Sci. Univ. Ank. Ser. A1 Math. Stat. **69**(2), 1405–1417 (2020)
22. Işık, H., Sezen, M.S., Vetro, C.: φ-Best proximity point theorems and applications to variational inequality problems. J. Fixed Point Theory Appl. **19**, 3177–3189 (2017)
23. Jleli, M., Samet, B.: Remarks on G-metric spaces and fixed point theorems. Fixed Point Theory Appl. **2012**, Article ID 210 (2012)
24. Khan, S.H., Abbas, M., Nazir, T.: Fixed point results for generalized Chatterjea type contractive conditions in partially ordered G-metric spaces. Sci. World J. **2014**, Article ID 341751, 8 (2014)
25. Makran, N., Haddouchi, A.E., Marzouki, B.: A generalized common fixed points for multi-valued mappings in G_b-metric spaces with an application. U.P.B. Sci. Bull. Series A **83**(1), 157–168 (2021)
26. Moradlou, F., Salimi, P., Vetro, P.: Some new extensions of Edelstein-Suzuki-type fixed point theorem to $G-$ metric and G-cone metric spaces. Acta Mathematica Scientia B **33**(4), 1049–1058 (2013)
27. Mustafa, Z., Sims, B.: Some remarks concerning D-metric spaces. In: Proceedings of International Conference on Fixed Point Theory and Applications, pp. 189–198. Valencia, Spain (2003)
28. Mustafa, Z., Sims, B.: A new approach to generalized metric spaces. J. Nonlinear Convex Anal. **7**(2), 289–297 (2006)
29. Mustafa, Z., Obiedat, H., Awawdeh, F.: Some fixed point theorem for mapping on complete G-metric spaces. Fixed Point Theory Appl. **2008**, Article ID 189870, 12 (2008)
30. Mustafa, Z., Shatanawi, W., Bataineh, M.: Existence of fixed point results in G-metric spaces. Int. J. Math. Math. Sci. **2009**, Article ID 283028, 10 (2009)
31. Mustafa, Z., Sims, B.: Fixed point theorems for contractive mappings in complete G-metric spaces. Fixed Point Theory Appl. **2009**, Article ID 917175, 10 (2009)

32. Mustafa, Z., Awawdeh, F., Shatanawi, W.: Fixed point theorem for expansive mappings in G-metric spaces. Int. J. Contemp. Math. Sci. **5**, 2463–2472 (2010)
33. Mustafa, Z., Khandagji, M., Shatanawi, W.: Fixed point results on complete G-metric spaces. Stud. Sci. Math. Hung. **48**(3), 304–319 (2011)
34. Mustafa, Z.: A new structure for generalized metric spaces with applications to fixed point theory [Ph.D. thesis], The University of Newcastle, New South Wales, Australia (2005)
35. Mustafa, Z., Parvaneh, V., Abbas, M., Roshan, J.R.: Some coincidence point results for generalized $(\psi, \phi)-$weakly contractive mappings in ordered G-metric spaces. Fixed Point Theory Appl. **2013**, article 326 (2013)
36. Razani, A., Parvaneh, V.: On generalized weakly $G-$ contractive mappings in partially ordered $G-$metric spaces. Abstr. Appl. Anal. **2012**, Article ID 701910, 18 (2012)
37. Radenovic, S., Pantelic, S., Salimi, P., Vujakovic, J.: A note on some tripled coincidence point results in $G-$metric spaces. Int. J. Math. Sci. Eng. Appl. **6**, 23–38 (2012)
38. Saadati, R., Vaezpour, S.M., Vetro, P., Rhoades, B.E.: Fixed point theorems in generalized partially ordered G-metric spaces. Math. Comput. Model. **52**, 797–801 (2010)
39. Saleem, N., Işık, H., Khaleeq, S., Park, C.: Interpolative Ćirić-Reich-Rus-type best proximity point results with applications. AIMS Math. **7**(6), 9731–9747 (2022)
40. Salimi, P., Vetro, P.: A result of Suzuki type in partial $G-$metric spaces. Acta Math. Sci. Ser. B Engl. Ed. **34**(2), article 111 (2014)
41. Samet, B., Vetro, C., Vetro, F.: Remarks on G-metric spaces. Int. J. Anal. **2013**, Article ID 917158, 6 (2013)
42. Shatanawi, W.: Fixed point theory for contractive mappings satisfying Φ-maps in G-metric spaces. Fixed Point Theory Appl. **2010**, Article ID 181650, 9 (2010)
43. Shatanawi, W., Abbas, M., Nazir, T.: Common coupled fixed points results in two generalized metric spaces. Fixied Point Theory Appl. **80**, 13 (2011)
44. Suzuki, T., Kikkawa, M., Vetro, C.: The existence of best proximity points in metric spaces with the property UC. Nonlinear Anal. **71**(7–8), 2918–2926 (2009)
45. Wardowski, D.: Fixed point of a new type of contractive mappings in complete metric spaces. Fixed Point Theory Appl. **94**, 06 (2012)

Chapter 8
On a New Subclass of Bi-Univalent Analytic Functions Characterized by $(\mathcal{P}, \mathcal{Q})$-Lucas Polynomial Coefficients via Sălăgean Differential Operator

Arzu Akgül

Abstract In this study, by using Lucas polynomials, subordination, and Sălăgean differential operator, we investigate the class $\mathcal{Q}_{\Sigma}(m, n; \Theta)$ of bi-univalent functions. Also we examine $(\mathcal{P}, \mathcal{Q})$-Lucas polynomial coefficient estimates and Fekete–Szegö inequalities for functions belonging $\mathcal{Q}_{\Sigma}(m, n; \Theta)$.

Keywords $(\mathcal{P} \cdot \mathcal{Q})$-Lucas polynomials · Differential operator · Coefficient bounds · Bi-univalent functions

Mathematics Subject Classification 30C45

8.1 Main Concepts and Properties of Univalent Functions

Complex analysis is an important topic in mathematics which contains functions of complex numbers. In addition to the theory of numbers, applied mathematics, physics, and also electrical engineering, it offers an incredibly potent instrument with an unanticipatedly wide range of applications. Students across many disciplines continue to be interested in studying complex analysis due to its rapid development in theory and applications. As a result, complex analysis today has an important situation in mathematics curriculum around the globe and is taught at different levels at practically every institution.

Geometric Function Theory, an important branch of Complex Analysis, is a traditional subject. The theory of partial differential equations, nonlinear integrable systems, current mathematical physics, and more traditional fields of physics like fluid dynamics are just a few of the areas where it continues to find new applications.

Let $\mathcal{D}(\xi_0, r)$ be the open unit disc of radius $r > 0$ centred at $\xi_0 \in \mathbb{C}$,

A. Akgül (✉)
Faculty of Arts and Science, Department of Mathematics, Kocaeli University, Umuttepe Campus, Kocaeli, Turkey
e-mail: akgul@kocaeli.edu.tr

B. C. Tripathy et al. (eds.), *Operators, Inequalities and Approximation*, Industrial and Applied Mathematics, https://doi.org/10.1007/978-981-97-3238-8_8

$$\mathcal{D}(\xi_0, r) = \{\xi \in \mathbb{C} : |\xi - \xi_0| < r, r \in \mathbb{R}\},$$

where $\mathbb{C}$ is set of complex numbers and $\mathbb{R}$ is the set of real numbers.

The open disc $\mathcal{D}(0, r)$ and the unit disc $\mathcal{D}_1$ will be shown with $\mathcal{D}_r$ and $\mathcal{D}$.

Definition 8.1 ([2], p. 24) If every pair of points ξ_1, ξ_2 in B can be joined by a polygonal path that lies entirely in B, then an open set B is named to be connected. Line segments that are either horizontal or vertical may be contained by the polygonal path. An open connected set is named a domain (Fig. 8.1).

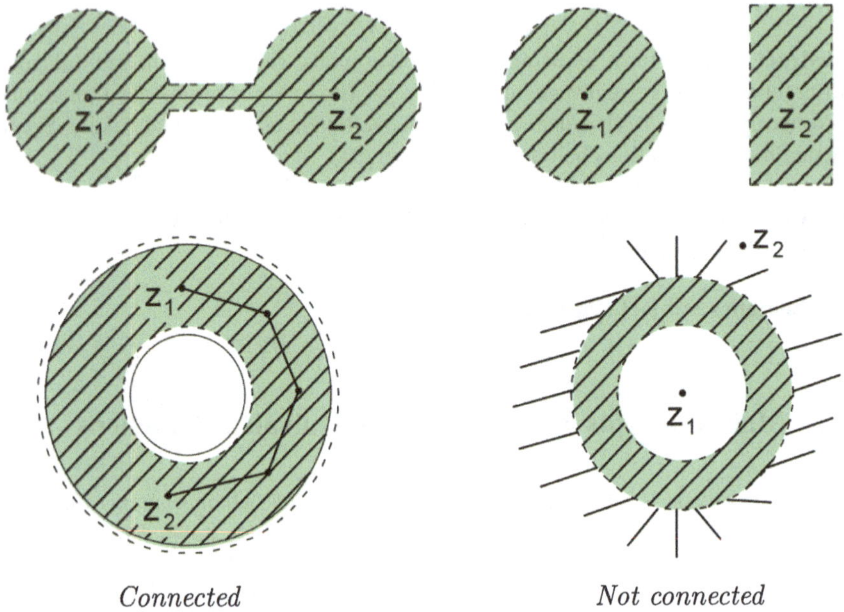

Fig. 8.1 Connected domain and not connected domain

Definition 8.2 ([2], p. 81) A simply connected domain S is a domain satisfying the next property: If γ is any simple closed contour lying in B, then the domain interior to γ lies wholly in B. The fact that the interior of a simple closed curve is simply connected is an important corollary of the Jordan Curve Theorem. If a domain is not simply connected then it is named multiply connected.

In a complex plane, the continuous image of a line segment is known as an *arc*. Strictly speaking, an *arc* must be viewed as a continuous mapping $\xi = \phi(t)$ of an interval $a \leq t \leq b$ into $\mathbb{C}$.

If an *arc* has a length, then it is a *rectifiable arc*; that is, for which the mapping function ϕ is of bounded variation. A *Jordan arc* is an *arc* except self-intersections.

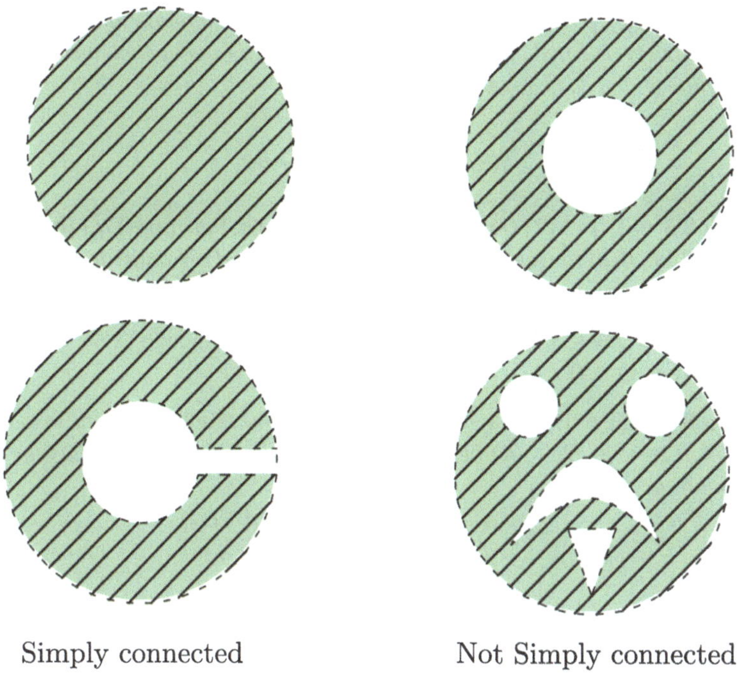

Fig. 8.2 Simply connected domain and not simply connected domain

Examples of simply connected and not simply connected domains are shown in Fig. 8.2. The continuous image of a circle is a *closed curve*, or an *arc* whose end-points coincide. A *simple closed curve* or a *Jordan curve* is a closed *curve* except self-intersections. According to the Jordan theorem, every *Jordan curve* divides the plane into two regions, the interior and the exterior of the curve. The interior of the Jordan curve is named as *Jordan domain* ([25], p. 1).

Now, we remember a special class of functions explained as analytic functions.

Definition 8.3 ([2], p. 37) Assume that $\mathfrak{f}$ be a function established in a neighbourhood of a point ξ_0. Thus, the derivative of $\mathfrak{f}$ at ξ_0 is explained by

$$\frac{d\mathfrak{f}}{d\xi}(\xi_0) = \mathfrak{f}'(\xi_0) = \lim_{\Delta\xi \to 0} \frac{\mathfrak{f}(\xi_0 + \Delta\xi) - \mathfrak{f}(\xi_0)}{\Delta\xi} \tag{8.1}$$

if this limit exists. Hence the function $\mathfrak{f}$ is named to be differentiable at the point ξ_0.

In case of $\mathfrak{f}$ is differentiable in some neighbourhood of $\xi_0 \in B \subset \mathbb{C}$, then $\mathfrak{f}$ is named to be analytic (or regular or holomorphic). If $\mathfrak{f}$ is analytic at all points in domain B, then $\mathfrak{f}$ is analytic in B. One can argue that $\mathfrak{f}$ is an entire function if it is analytic on the entire complex plane.

Theorem 8.4 ([2], p. 111) (Cauchy's Integral Formula) *Suppose that γ is a simple, closed, positively oriented contour. If $\mathfrak{f}$ is analytic in some simply connected domain B containing γ and ξ_0 is any point inside γ, then*

$$\mathfrak{f}(\xi_0) = \frac{1}{2\pi i}\int_\gamma \frac{\mathfrak{f}(\xi)}{\xi - \xi_0} d\xi.$$

Theorem 8.5 ([2], p. 117) (Differentiation of Analytic Functions) *If $\mathfrak{f}$ is analytic in a domain B, then all its derivatives $\mathfrak{f}, \mathfrak{f}', ..., \mathfrak{f}^{(n)}$ exist and are analytic in B.*

A version of this theorem for real functions does not fulfill, for instance, the function $\mathfrak{f}(\Theta) = \Theta^{3/2}$, $\Theta \in \mathbb{R}$ is differentiable for all real Θ, although $\mathfrak{f}'(\Theta) = \frac{3}{2}\Theta^{1/2}$ has no derivative at $\Theta = 0$.

Theorem 8.6 ([2], p. 159) (Taylor's Theorem) *Assume that $\mathfrak{f}$ is analytic in a domain B and let at $\xi_0 \in B$, $\mathcal{D}(\xi_0, R)$ be the largest open disc in B. Then, $\mathfrak{f}$ has the series representation*

$$\mathfrak{f}(\xi) = \sum_{n=0}^{\infty} a_n (\xi - \xi_0)^n, \; a_n = \frac{\mathfrak{f}^{(n)}(\xi_0)}{n!} \tag{8.2}$$

which converges for all $\xi \in \mathcal{D}(\xi_0, R)$. Also, for any $0 \leq r \leq R$ the convergence is uniform on the closed disc $\overline{\mathcal{D}}(\xi_0, r)$.This series is named the Taylor series of $\mathfrak{f}$ at ξ_0.

In a simple way, the Taylor series development is obtained by the Cauchy integral formula for derivatives:

Theorem 8.7 ([2], p. 118) (Cauchy's Integral Formula for Derivatives) *Let γ be a simple, closed, positively oriented contour. If If $\mathfrak{f}$ is analytic inside and on γ and if ξ is any point inside γ, then*

$$\mathfrak{f}(\xi) = \frac{n!}{2\pi i}\int_\gamma \frac{\mathfrak{f}(\xi)}{(\xi - \xi)^{n+1}} d\xi, n = 1, 2, \cdots.$$

The Cauchy integral formula is derived for $n = 0$ from the Cauchy integral theorem:

$$\int_\gamma \mathfrak{f}(\xi) d\xi = 0.$$

Theorem 8.8 ([2], p. 120) (Morera's Theorem) *If $\mathfrak{f}$ is continuous in a domain B and $\int_\gamma \mathfrak{f}(\xi)d\xi = 0$ for every closed contour γ in B, then $\mathfrak{f}$ is analytic.*

Theorem 8.9 ([2], p. 133) (Maximum Modulus Principle) *If $\mathfrak{f}$ is analytic in a domain B and $|\mathfrak{f}(\xi)|$ achieves its maximum value at a point ξ_0 in B, then $\mathfrak{f}$ is constant in B.*

A simple but important corollary of this theorem may be expressed as follows:

Lemma 8.10 ([25], p. 3) (Schwarz Lemma) *Let $\mathfrak{f}$ be analytic in the unit disc $\mathcal{D}$, with $\mathfrak{f}(0) = 0$ and $|\mathfrak{f}(\xi)| < 1$ in $\mathcal{D}$.Then $\left|\mathfrak{f}'(0)\right| \leq 1$ and $|\mathfrak{f}(\xi)| \leq \xi$ in $\mathcal{D}$. Strictly inequality holds in both estimates unless $\mathfrak{f}$ is a rotation of a disc: $\mathfrak{f}(\xi) = e^{i\theta}\xi$.*

The argument principle is usually a convenient tool for the determination of the range and valence of an analytic function. As an easy application of the residue theorem, the principle can be proved.

Argument Principle ([25], ***p*.3**): Assume that $\mathfrak{f}$ is analytic in the closure of a domain $\mathcal{B}$ bounded by a rectifiable Jordan curve γ, and suppose that $\mathfrak{f}(\xi) \neq 0$ on γ.Then the number of zeros of $\mathfrak{f}$ in B, counted according to multiplicity, is equal to $\frac{1}{2\pi}$ times the net change in the argument of $\mathfrak{f}(\xi)$ as ξ traverses γ in the positive direction.

An essential utilized form of this principle is given as follows:

Theorem 8.11 ([25], p. 4) (Rouche's Theorem) *Assume that $\mathfrak{f}$ and g are analytic inside and on a rectifiable Jordan curve γ,with $|\mathfrak{f}(\xi)| < |\mathfrak{g}(\xi)|$ on γ. Thus $\mathfrak{f}$ and* $(\mathfrak{f} + \mathfrak{g})$ *have the same number of zeros, counted according to multiplicity, inside γ.*

Theorem 8.12 ([25], p. 4) (Hurvitz Theorem) *Let $\mathfrak{f}_n$be analytic in a domain B, and assume that $\mathfrak{f}_n(\xi) \rightarrow \mathfrak{f}(\xi)$ as $n \rightarrow \infty$, uniformly on each compact subset of B. Then either $\mathfrak{f}(\xi) = 0$ in B or every zero of $\mathfrak{f}$ is a limit point of a sequence of zeros of the functions $\mathfrak{f}_n$.*

8.1.1 Conformal Mappings

Conformal mappings get the angle-preserving property. Now we recall these mappings.

We will show the function $\eta = \mathfrak{f}(\xi)$ as a mapping from a subset of the ξ—plane (domain of f) to the η-plane (range of $\mathfrak{f}$).

As we know that the transformation

$$\eta = A\xi \tag{8.3}$$

is named a dilation, where $0 \neq A \in \mathbb{C}$, A is constant and $0 \neq \xi \in \mathbb{C}$. By writing $A = ae^{i\alpha}$, $\xi = re^{i\theta}$we get

$$\eta = (ar)e^{i(\alpha+\theta)}. \tag{8.4}$$

It is known that linear mapping (8.3) which is nonconstant is an expansion or contraction and a rotation, followed by a transformation. Thus, applying a linear transformation, we see that the angles between any two intersecting arcs in the ξ-plane and the angle between the images in the η-plane are equal. Transformations that have such angle conservation property are named conformal transformations. In order to solve boundary value problems containing Laplace's equation, these mappings are of vital importance ([2], p. 64).

The Riemann mapping theorem is one of the most fundamental conclusions in the study of univalent functions in one variable. One of the primary distinctions between complex analysis in one variable and higher dimensions is its failure in many variables.

Theorem 8.13 ([2], p. 260) (Riemann Mapping Theorem) *Every simply connected domain B which is a proper subset of $\mathbb{C}$ can be mapped conformally onto the unit disc.*

Also, if $\xi_0 \in B$, there is a unique conformal map of B onto $\mathcal{D}$ fulfilling $\mathfrak{f}(\xi_0) = 0$ and $\mathfrak{f}'(\xi_0) > 0$.

8.1.2 The Classes of Univalent and Bi-Univalent Functions

Definition 8.14 ([25], p. 26) An analytic function $\mathfrak{f}$ is named to be univalent in a domain B if it does not take the same value twice:

$$\mathfrak{f}(\xi_1) \neq \mathfrak{f}(\xi_2), \ (\xi_1; \xi_2 \in B)$$

for all pairs of distinct points $\xi_1$and ξ_2 in B.

On the other hand, $\mathfrak{f}$ is a one-to-one (injective) mapping of B onto the other domain. Alternative terms in common use are schlicht and simple. The fact that univalency is essentially preserved under uniform convexity is a fascinating situation also fundamental to the theory of univalent functions.

Univalency means that various domain points will be converted to different image domain points geometrically. Because the theory of univalent functions is very large, some simplification assumptions are essential. The most apparent is to replace the arbitrary domain B with a more convenient one, with the open unit disc $\mathcal{D}$ being the most appealing option.

As an example, let us assume the function $\mathfrak{f}(\xi) = (1+\xi)^2$ in $\mathcal{D}$. On geometric considerations, obviously this function is univalent in $\mathcal{D}$. $\mathcal{D}$ is shifted to the right by $1+\xi$, and the effect of squaring $1+\xi$ is simple to perceive. In $\mathcal{D}$, the familiar kind of argument gives that $\eta = (1+\xi)^3$ is not univalent. (see [35]).

Let we show the analytic functions in $\mathcal{D}$ by $\mathcal{H}$ and let

$$\mathcal{S} = \left\{\mathfrak{f} \in \mathcal{H} : \mathfrak{f} \text{ is normalized function by the condition } \mathfrak{f}(0) = \mathfrak{f}'(0) - 1 = 0\right\}.$$

Hence each function $\mathfrak{f}$ has a Taylor series expansion explained by

$$\mathfrak{f}(\xi) = \xi + \sum_{n=2}^{\infty} a_n \xi^n, \ (\xi \in \mathcal{D}). \tag{8.5}$$

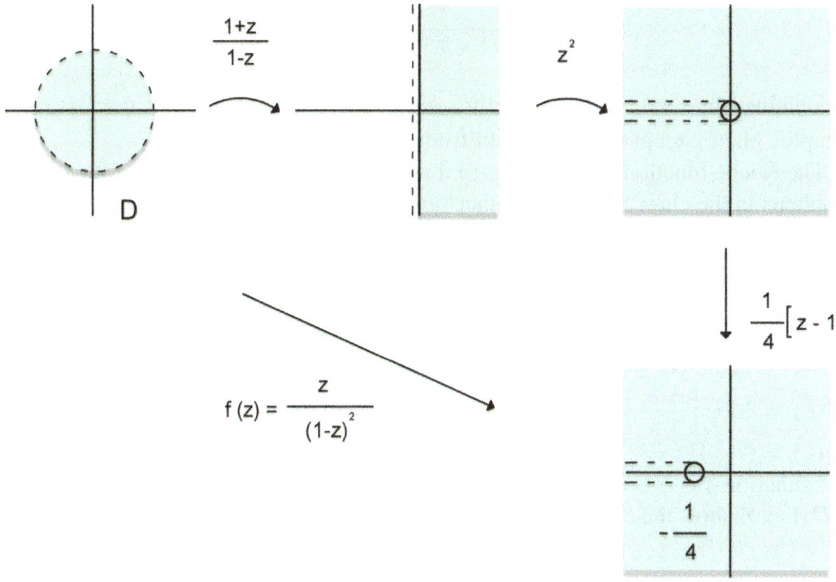

Fig. 8.3 Transformation of Koebe

Due to the angle-preserving property [25, Chaps. 1 and 2] of an analytic univalent function, it is a conformal mapping.

The Koebe function is the most famous function in $\mathcal{S}$. Since the Koebe function given by

$$\mathfrak{f}(\xi) = \frac{\xi}{(1-\xi)^2} = \xi + 2\xi^2 + 3\xi^3 + \cdots, \xi \in \mathcal{D}$$

is in the $\mathcal{S}$ and translates $\mathcal{D}$ onto the slit plane $\mathbb{C}\backslash(-\infty, \frac{1}{4}]$, it is seen that the radius $\frac{1}{4}$ in Koebe's theorem is optimal; indeed, no larger number would do. We can see this by giving

$$\mathfrak{f}(\xi) = \frac{1}{4}\left(\frac{1+\xi}{1-\xi}\right)^2 - \frac{1}{4}$$

and assuming that the function

$$\eta(\xi) = \frac{1+\xi}{1-\xi}$$

translates $\mathcal{D}$ onto the right half-plane $\mathcal{R}(\eta) > 0$ conformally.

In Fig. 8.3, it is shown that the image of the $\mathcal{D}$ under the Koebe function $\mathfrak{f}(z) = \frac{z}{(1-z)^2}, z \in \mathcal{D}$.

The rotation of the Koebe function, i.e. the function given by

$$\mathfrak{f}_\alpha(\xi) = \frac{\xi}{(1 - e^{i\alpha}\xi)^2}, (\xi \in \mathcal{D}),$$

is contained by the class $\mathcal{S}$ for each $\alpha \in \mathbb{R}$. This image of the unit disc is in the complex plane except for a radial slit from $-\infty$ to $\frac{-e^{i\alpha}}{4}$.

The Koebe function and rotations of it are the only extremal functions for various problems in the class $\mathcal{S}$. Some of other functions in $\mathcal{S}$ are as follows:

- $\mathfrak{f}(\xi) = \xi$, identity function;
- $\mathfrak{f}(\xi) = \frac{\xi}{1-\xi}$, which maps $\mathcal{D}$ conformally onto the half-plane $\mathcal{R}(\eta) > -\frac{1}{2}$;
- $\mathfrak{f}(\xi) = \frac{\xi}{(1-\xi)^2}$, which maps $\mathcal{D}$ onto the entire plane minus the two half lines $\frac{1}{2} \leq y < \infty$ and $-\infty \leq y < -\frac{1}{2}$;
- $\mathfrak{f}(\xi) = \frac{1}{2} \log \left[\frac{(1+\xi)}{1-\xi}\right]$, which maps $\mathcal{D}$ onto the horizontal strip $-\frac{\pi}{4} < \text{Im}\{\eta\} < \frac{\pi}{4}$;
- $\mathfrak{f}(\xi) = \xi - \frac{1}{2}\xi^2 = \frac{1}{2}\left[1 - (1 - \xi)^2\right]$, which maps $\mathcal{D}$ onto the interior of a cardioid. A function $\mathfrak{f} \in \mathcal{S}$ is named to be bi-univalent in $\mathcal{D}$ if both $\mathfrak{f}$ and $\mathfrak{f}^{-1}$ are univalent in $\mathcal{D}$. Let Σ show the class of bi-univalent functions estabiished in $\mathcal{D}$. The functions

$$\frac{\xi}{1-\xi}, -\log(1 - \xi), \frac{1}{2} \log \left(\frac{1+\xi}{1-\xi}\right)$$

can be given as some examples in Σ but the familiar Koebe function is not in Σ. Other well-known examples of functions in $\mathcal{S}$ such as

$$\xi - \frac{\xi^2}{2}, \frac{\xi}{1-\xi^2}$$

are also not in Σ ([25], p. 26, 27, 28).

For a brief history and significant instances of functions in Σ, see Srivastava et al.'s [68] pioneering work.

Now, we give the famous Area Theorem:

The univalence of a function

$$\mathfrak{g}(\xi) = \xi + b_0 + \sum_{n=1}^{\infty} b_n \xi^{-n}, |\xi| > 1$$

places a strong restriction on the size of the Laurent coefficients $b_n, n = 1, 2, \cdots$. This is stated by the area theorem, which is crucial for the geometric function theory of univalent functions. The source of the name is obvious from the proof. The area theorem was discovered by Gronwall in 1914 [34].

Theorem 8.15 ([25], 29) (Area Theorem) *If* $\mathfrak{g} \in \Sigma$, *then*

$$\sum_{n=1}^{\infty} n\,|b_n|^2 \leq 1,$$

with equality if and only if $\mathfrak{g} \in \widetilde{\Sigma}$.

Corollary 8.16 ([25], 30) *If* $\mathfrak{g} \in \Sigma$, *then* $|b_1| \leq 1$, *with equality if and only if* $\mathfrak{g}$ *has the form*

$$\mathfrak{g}(\xi) = \xi + b_0 + \frac{b_1}{\xi},\ |b_1| = 1.$$

This is a conformal mapping of $\{\xi : |\xi| > 1\}$, exterior to $\mathcal{D}$, onto the complement of a line segment of length 4.

It's only a simple step from this last conclusion to a Bieberach theorem [19] for estimating the second coefficient a_2 of a function of class $\mathcal{S}$. The well-known Bieberbach conjecture was founded on this theory, which was first proposed in 1916.

Theorem 8.17 ([25], (30 Bieberbach's Theorem) *If* $\mathfrak{f} \in \mathcal{S}$,

$$|a_2| \leq 2,$$

with equality if and only if $\mathfrak{f}$ *is a rotation of the Koebe function.*

At the beginning of the 1907, Koebe [44] explored that the ranges of all functions in $\mathcal{S}$ contain a common disc $|\eta| < \delta$, where δ is an absolute constant. The Koebe function indicates that $\delta \leq 1/4$, and after then Bieberbach [19] established Koebe's conjecture that δ may be chosen $1/4$.

Theorem 8.18 (Koebe One-Quarter Theorem) (see [19, 25, 44], p. 31) *For every* $\mathfrak{f} \in \mathcal{S}$, *the range* $\mathfrak{f}(\mathcal{D})$ *contains the disc* $\left\{\eta : |\eta| < \frac{1}{4}\right\}$.

Proof If a function $\mathfrak{f} \in \mathcal{S}$ omits the value $\eta \in \mathbb{C}$, then

$$\mathfrak{g}(\xi) = \frac{\eta\mathfrak{f}(\xi)}{\eta - \mathfrak{f}(\xi)} = \xi + \left(a_2 + \frac{1}{\eta}\right)\xi^2 + \cdots$$

is analytic and univalent in $\mathcal{D}$. This is the ommited-value mapping given in example 10, which is the composition of $\mathfrak{f}$ with a linear fractional mapping. To $\mathfrak{g} \in \mathcal{S}$, according to Bieberbach's theorem, we obtain

$$\left|a_2 + \frac{1}{\eta}\right| \leq 2.$$

Combining with the inequality $|a_2| \leq 2$,it is obtained that $\left|\frac{1}{\eta}\right| \leq 4$, or $|\eta| < \frac{1}{4}$.

The fact that univalence is the key to Koebe's theorem should be assumed. For instance, the analytic functions

$$\mathfrak{f}_n(\xi) = \frac{1}{n}(e^{n\xi} - 1), n = 1, 2, ...,$$

satisfy the equations $\mathfrak{f}_n(0) = 0$ and $\mathfrak{f}_n'(0) = 1$, but $\mathfrak{f}_n$ omits the value $-\frac{1}{n}$, which may be chosen arbitrarily close to origin.

According the Koebe One-Quarter Theorem [44], the image of $\mathcal{D}$ under every univalent function $\mathfrak{f} \in \mathcal{S}$ contains a disc of radius $1/4$. We know that every function $\mathfrak{f} \in \mathcal{S}$ has an inverse $\mathfrak{f}^{-1}$, satisfying

$$\mathfrak{f}^{-1}(\mathfrak{f}(\xi)) = \xi, \quad (\xi \in \mathbb{U})$$

and

$$\mathfrak{f}\left(\mathfrak{f}^{-1}(\eta)\right) = \eta, \quad \left(|\eta| < r_0(\mathfrak{f}), \; r_0(\mathfrak{f}) \geq \frac{1}{4}\right).$$

The inverse function $\mathfrak{f}^{-1}(\eta) = \mathfrak{g}(\eta)$ is established by

$$\mathfrak{g}(\eta) = \mathfrak{f}^{-1}(\eta) = \eta - a_2\eta^2 + (2a_2^2 - a_3)\eta^3 - (5a_2^3 - 5a_2a_3 + a_4)\eta^4 + \cdots. \tag{8.6}$$

8.1.3 Coefficient Estimates

As we learned up to now, each function

$$\mathfrak{f}(\xi) = \xi + \sum_{n=2}^{\infty} a_n \xi^n \tag{8.7}$$

in class $\mathcal{S}$ has the property $|a_2| \leq 2$, with equality happening only for rotations of the Koebe function. This offers the general problem to find

$$A_n = \sup_{\mathfrak{f} \in \mathcal{S}} |a_n|, n = 2, 3, \cdots.$$

Due to the fact that the Koebe function plays a vital role in every problem for the family $\mathcal{S}$, naturally it maximizes $|a_n|$ for every n.This is the eminent Bieberbach conjecture, which was first developed in 1916 [19].

Conjecture 8.19 ([19]) (Bierbach Conjecture) *The coefficients of each function* $\mathfrak{f} \in \mathcal{S}$ *satisfy*

$$|a_n| \leq n; n \geq 2.$$

With the exception of when $\mathfrak{f}$ *is the Koebe function or one of its rotations, strict inequality satisfies for all n.*

This conjecture has served as a challenge to mathematicians for many years and has sparked the creation of significant new techniques in complicated analysis. Lowner [50] and Garabedian and Schiffer [32], respectively, each proved the conjecture for the situations $n = 3$ and $n = 4$. After then, the conjecture was proven by Pederson and Schiffer [60] for $n = 5$, for $n = 6$, it was shown by Pederson [59] and Ozawa [57], independently. The Bieberbach conjecture was proved by Louis de Branges [17] for all the coefficients n in 1985. Although it took approximately 70 years for this publication to finally establish the Bieberbach conjecture, bounds for the Taylor coefficients were determined for various subclasses of univalent functions in the interim. The work of various subfamilies of analytical and univalent functions has started to take form after the Bieberbach conjecture was proven, and it is still a fascinating topic. There are numerous publications available now that focus on the study of univalent functions; Conway [22], Duren [25] Goodman [35], Miller and Mocanu [52], and Pommerenke [61]'s books come to mind. Further reading and some additional results on this subject can also be found in the wide variety of classical and modern textbooks devoted to the Univalent Function Theory and related topics, including [1, 18, 20, 21, 37, 46, 53, 64–66]. In literature, the study into Σ was initiated by Lewin [48]. It concentrated on coefficient-related problems and discovered a bound of 1.51 for the second coefficient's modulus, $|a_2|$. Afterwards, Brannan and Clunie [15] conjectured that $|a_2| \leqq \sqrt{2}$ for $\mathfrak{f} \in \Sigma$. After then, Netanyahu [54] examined that $\max |a_2| = \frac{4}{3}$ if $\mathfrak{f}(\xi) \in \Sigma$. Some subfamilies of the bi-univalent function class Σ similar to subfamilies $\mathcal{S}^{\star}(\beta)$ and $\mathcal{K}(\beta)$ of starlike and convex functions of order β $\left(0 \leqq \beta < 1\right)$ in $\mathbb{U}$ were introduced by Brannan and Taha [16]. For each of the function classes $\mathcal{S}_{\Sigma}^{\star}(\beta)$ and $\mathcal{K}_{\Sigma}(\beta)$, non-sharp estimates for the initial coefficients were found by them. In the last years, motivated substantially by the aforementioned work on this area Srivastava et al. [68], the coefficient bounds for various subclasses of bi-univalent functions (see, for example, [28, 63]) were investigated by many authors. Knowledge is not much about bounds on the general coefficient $|a_n|$ for $n \geqq 4$. Only a few articles have been published in the literature that determine the general boundaries for the coefficients $|a_n|$ for the analytic bi-univalent functions (see [36, 40]). The coefficient estimate problem for each of the coefficients $|a_n|$ $(n \in \mathbb{N} \setminus \{1, 2\};\ \mathbb{N} = \{1, 2, 3, \cdots\})$ is presumably still an open problem.

Furthermore, the Fekete–Szegö inequality is given whereby Loewner's method for the coefficients of $\mathfrak{f} \in \mathcal{S}$ is

$$\left|a_3 - \vartheta a_2^2\right| \leq 1 + 2\exp(\frac{-2\vartheta}{1-\sigma}) \text{ for } \vartheta \in [0, 1).$$

If $\vartheta \to 1^-$, then we have the basic inequality $\left|a_3 - a_2^2\right| \leq 1$. Also, the coefficient functional

$$\mathfrak{f}_{\vartheta}(\mathfrak{f}) = a_3 - \vartheta a_2^2$$

on the normalized analytic functions $\mathfrak{f}$ in $\mathcal{D}$ plays a vital role in the function theory. The problem of maximizing the absolute value of the functional $\mathfrak{f}_\vartheta(\mathfrak{f})$ is named the Fekete–Szegö problem (see [26]). The Fekete–Szego problem has continued to receive the attention of researches in Geometric Function Theory (see, for example, [7, 12, 13, 23, 24, 27, 43, 49, 55, 56, 67, 71]). Various interesting developments involving the Fekete–Szego problem can be found in [43, 49].

8.1.4 Convex and Starlike Functions

Definition 8.20 (*see* [35]) A set E is called convex if every pair of points a and b can be joined by a line segment ab such that every point in the line segment is also in the interior of E. Obviously, every convex set is necessarily connected. Thus, it follows that the intersection of two or more convex sets is convex too. If a function $\mathfrak{f}(\xi)$ maps B onto a convex domain, then $\mathfrak{f}(\xi)$ is named a convex function.

Definition 8.21 A set E in the plane is named to be starlike with respect to η_0 an interior point of E if each ray with initial point η_0 intersects the interior of E in a set that is either a line segment or a ray. If a function $\mathfrak{f}(\xi)$ maps $\mathcal{D}$ onto a domain that is starlike with respect to η_0, then we say that $\mathfrak{f}(\xi)$ is starlike with respect to η_0. In the special case $\eta_0 = 0$, we say that $\mathfrak{f}(\xi)$ is starlike function.

Examples of convex and starlike domains are shown in Figs. 8.4 and 8.5, respectively. The domain shown in Fig. 8.5 is starlike with respect to η_0 but is not starlike with respect to the origin. The special function $\eta(\xi) = \frac{1+\xi}{1-\xi}$ is a convex function due to the fact that it maps $\mathcal{D}$ onto a half-plane. The Koebe function is a starlike function. In fact the domain $k(\mathcal{D})$ is starlike with respect to each $\eta_0 > -\frac{1}{4}$.

The class $S^*(\alpha)$ of starlike functions of order α in $\mathcal{D}$ and the class $K(\alpha)$ of convex functions of order α in $\mathcal{D}$ are explained by

Fig. 8.4 Convex region

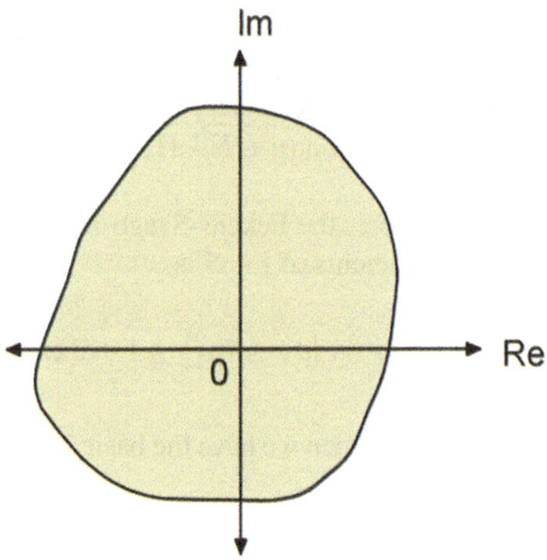

Fig. 8.5 Starlike region

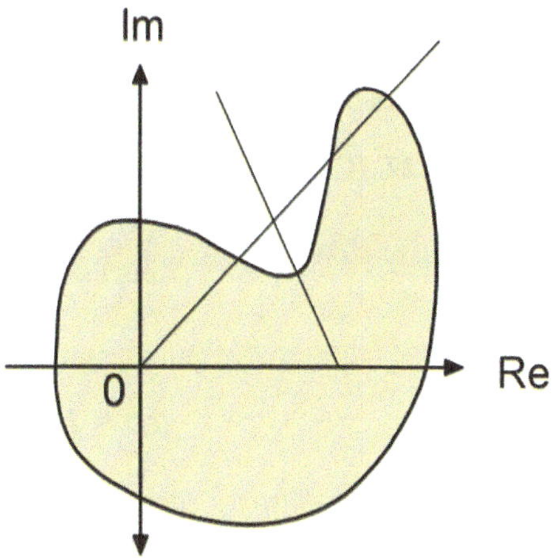

$$\mathcal{S}^*(\alpha) = \left\{ \mathfrak{f} : \mathfrak{f} \in \mathcal{S} \text{ and } \mathcal{R}e\left(\frac{\xi \mathfrak{f}'(\xi)}{\mathfrak{f}(\xi)}\right) > \alpha, \xi \in \mathcal{D} \right\}$$

for $0 \leq \alpha < 1$ and

$$\mathcal{K}(\alpha) = \left\{ \mathfrak{f} : \mathfrak{f} \in \mathcal{S} \text{ and } \mathcal{R}e\left(1 + \frac{\xi \mathfrak{f}''(\xi)}{\mathfrak{f}'(\xi)}\right) > \alpha, \xi \in \mathcal{D} \right\}$$

for $0 \leq \alpha < 1$.

We state that $\mathfrak{f}(\xi) \in \mathcal{K}(\alpha)$ if and only if $\xi \mathfrak{f}'(\xi) \in \mathcal{S}^*(\alpha)$, and that

$$\mathcal{S}^*(\alpha) \subset \mathcal{S}^*(0), \mathcal{K}(\alpha) \subset \mathcal{K}(0), \mathcal{K}(\alpha) \subset \mathcal{S}^*(\alpha)$$

for $0 \leq \alpha < 1$.

In the next section, we give the class $\mathcal{P}$, which is closely related to the families of convex functions $\mathcal{K}$ and starlike functions $\mathcal{S}^*$.

8.1.5 Functions with Positive Real Part

Very important class of functions with positive real part, consisting of all functions p analytic in $\mathcal{D}$ satisfying $p(0) = 1$ and $\mathcal{R}e\,(p(\xi)) > \alpha$, is generally shown by $\mathcal{P}$ and named the Caratheodory class. For instance, $\mathcal{P}$ contains the function

$$p(\xi) = \frac{1+\xi}{1-\xi}, (\xi \in \mathcal{D}).$$

This function states a conformal map of $\mathcal{D}$ onto the right half-plane and consequently it has an essential role in the class $\mathcal{P}$, as the Koebe function for the class $\mathcal{S}$. We state that $p(\xi)$ does not need to be univalent.

Lemma 8.22 (Caratheodory Lemma) (see [25, 61]) *If* $p \in \mathcal{P}$ *and*

$$p(\xi) = 1 + \sum_{n=1}^{\infty} c_n \xi^n,$$

then

$$|c_n| \leq 2, n = 1, 2, \cdots .$$

This inequality is sharp for each n.

Grenander and Szegö [33] obtained more refinement coefficients bounds in the class $\mathcal{P}$.

8.1.5.1 Subordination

The concept of subordination goes back to Lindelöf, but Littlewood and Rogonsky discovered the term and also introduced the main relations. For many years a substantial theory has been developed, and today, subordination has a very important role in complex analysis.

Assume that the functions $\mathfrak{f}$ and $\mathfrak{g}$ be analytic in $\mathcal{D}$, with $\mathfrak{f}(0) = \mathfrak{g}(0)$, also assume that $\mathfrak{f}$ be univalent and the range of $\mathfrak{g}$ be contained in the range of $\mathfrak{f}$. Then

$$\Psi(\xi) = \mathfrak{f}^{-1}(\mathfrak{g}(\xi))$$

is analytic in $\mathcal{D}$,

$$\Psi(0) = 0 \text{ and } |\Psi(\xi)| < 1.$$

Thus, according to the Schwarz lemma,

$$|\Psi(\xi)| < |\xi| \text{ for } 0 < |\xi| < 1$$

unless Ψ is simply a rotation of the disc.

Generally, for the functions $\mathfrak{f}, \mathfrak{g} \in \mathcal{S}$, analytic, an analytic function $\mathfrak{f}$ is named to be subordinate to the function $\mathfrak{g}$, symbolized by

$$\mathfrak{f}(\xi) \prec \mathfrak{g}(\xi),$$

if

$$\mathfrak{f}(\xi) = \mathfrak{g}(\Psi(\xi))$$

for some analytic function Ψ with $|\Psi(\xi)| \leq |\xi|$, $|\xi| < 1$. The superordinate function $\mathfrak{g}$ does not need to be univalent.

8.2 Coefficient Estimates for the Class $\mathbf{Q}_{\Sigma}(m, n; \Theta)$

8.2.1 Basic Concepts

The polynomials Lucas, Dickson, Chebyshev, Lucas–Lehmer, Lucas, Lucas-Balancing, and Fibonacci polynomials are vital in mathematics because they can be applicable to various fields of science. In very recent years, a lot of researchers have studied these polynomials from a theoretical point [29–31, 38, 39, 45, 47, 69, 70] (Also see [3–11, 14, 29, 47, 51, 58]).

Definition 8.23 [47] Let $P(\Theta)$ and $Q(\Theta)$ are polynomials with real coefficients. The (P, Q) Lucas polynomials $\mathcal{L}_{\mathcal{P},\mathcal{Q},n}(\Theta)$ are established by the recurrence relation

$$\mathcal{L}_{\mathcal{P},\mathcal{Q},n}(\Theta) = \mathcal{P}(\Theta)\mathcal{L}_{\mathcal{P},\mathcal{Q},n-1}(\Theta) + \mathcal{Q}(\Theta)\mathcal{L}_{\mathcal{P},\mathcal{Q},n-2}(\Theta) \quad (n \geq 2), \tag{8.8}$$

from which the first few Lucas polynomials can be found as

$$\begin{aligned} \mathcal{L}_{\mathcal{P},\mathcal{Q},0}(\Theta) &= 2, \\ \mathcal{L}_{\mathcal{P},\mathcal{Q},1}(\Theta) &= \mathcal{P}(\Theta), \\ \mathcal{L}_{\mathcal{P},\mathcal{Q},2}(\Theta) &= \mathcal{P}^2(\Theta) + 2\mathcal{Q}(\Theta), \\ \mathcal{L}_{\mathcal{P},\mathcal{Q},3}(\Theta) &= \mathcal{P}^3(\Theta) + 3\mathcal{P}(\Theta)\mathcal{Q}(\Theta). \end{aligned} \tag{8.9}$$

Some special situations of the $\mathcal{L}_{\mathcal{P},\mathcal{Q},n}(\Theta)$ with the chosen initial conditions are explained by following:

1. For $\mathcal{P}(\Theta) = \Theta$ and $\mathcal{Q}(\Theta) = 1$, $\mathcal{L}_{\mathcal{P},\mathcal{Q},n}(\Theta)$ are the Lucas polynomials $\mathcal{L}_n(\Theta)$;
2. For $\mathcal{P}(\Theta) = 2\Theta$ and $\mathcal{Q}(\Theta) = 1$, $\mathcal{L}_{\mathcal{P},\mathcal{Q},n}(\Theta)$ are the Pell–Lucas polynomials $\mathfrak{D}_n(\Theta)$;
3. For $\mathcal{P}(\Theta) = 1$ and $\mathcal{Q}(\Theta) = 2\Theta$, $\mathcal{L}_{\mathcal{P},\mathcal{Q},n}(\Theta)$ are the Jacobsthal–Lucas polynomials $\mathfrak{j}_n(\Theta)$;
4. For $\mathcal{P}(\Theta) = 3\Theta$ and $\mathcal{Q}(\Theta) = -2$, $\mathcal{L}_{\mathcal{P},\mathcal{Q},n}(\Theta)$ are the Fermat–Lucas polynomials $\mathfrak{f}_n(\Theta)$;
5. For $\mathcal{P}(\Theta) = 2\Theta$ and $\mathcal{Q}(\Theta) = -1$, $\mathcal{L}_{\mathcal{P},\mathcal{Q},n}(\Theta)$ are the Chebyshev polynomials first kind $T_n(\Theta)$.

Definition 8.24 [47] Let $\mathcal{G}_{\{\mathcal{L}_n(\Theta)\}}(\xi)$ be the generating function of the $(\mathcal{P}, \mathcal{Q})$ Lucas polynomial sequence $\mathcal{L}_{\mathcal{P},\mathcal{Q},n}(\Theta)$. Then

$$\mathcal{G}_{\{\mathcal{L}_n(\Theta)\}}(\xi) = \sum_{n=0}^{\infty} \mathcal{L}_{\mathcal{P},\mathcal{Q},n}(\Theta)\xi^n = \frac{2 - \mathcal{P}(\Theta)\xi}{1 - \mathcal{P}(\Theta)\xi - \mathcal{Q}(\Theta)\xi^2}. \tag{8.10}$$

We start this section by giving definition of the class $\mathcal{Q}_\Sigma(m, n; \Theta)$ and by finding the estimates on the coefficients $|a_2|$ and $|a_3|$ for functions in this class.

We know that the generating function $\mathcal{G}_{\{\mathcal{L}_n(\Theta)\}}(\xi)$ is a convergence formal series.

For a function $\mathfrak{f} \in \mathcal{H}$ given by (8.7),the differential operator $\mathcal{D}^n (n \in \mathbb{N})$ introduced by Sãlãgean [62] is

$$\mathcal{D}_\delta^0 \mathfrak{f}(\xi) = \mathfrak{f}(\xi)$$
$$\mathcal{D}^1 \mathfrak{f}(\xi) = \xi \mathfrak{f}'(\xi)$$
$$\mathcal{D}^n \mathfrak{f}(\xi) = \mathcal{D}\left(\mathcal{D}^{n-1} \mathfrak{f}(\xi)\right).$$

Thus, from these equalities, we obtain that

$$\mathcal{D}^n \mathfrak{f}(\xi) = \xi + \sum_{k=2}^{\infty} k^n \mathcal{A}_k \xi^k, \qquad n \in \mathbb{N}_0 = \{0, 1, 2, ...\} \tag{8.11}$$

with $\mathcal{D}^n \mathfrak{f}(0) = 0$.

8.2.2 The Class $\mathcal{Q}_\Sigma(m, n; \Theta)$

Definition 8.25 The function $\mathfrak{f}$ given by (8.7) is named to be in the class $\mathrm{Q}_\Sigma(m, n; \Theta)$ $(m, n \in \mathbb{N}_0, m > n)$ if the following conditions are fulfilled:

$$\frac{\mathcal{D}^m \mathfrak{f}(\xi)}{\mathcal{D}^n \mathfrak{f}(\xi)} \prec \mathcal{G}_{\{\mathcal{L}_{p,q,n}(\Theta)\}}(\xi) - 1 \tag{8.12}$$

and

$$\frac{\mathcal{D}^m \mathfrak{g}(\eta)}{\mathcal{D}^n \mathfrak{g}(\eta)} \prec \mathcal{G}_{\{\mathcal{L}_{p,q,n}(\Theta)\}}(\eta) - 1 \tag{8.13}$$

where the function $\mathcal{D}^n$ is Sãlãgean differential operator and g is given by (8.6).

Remark 8.26 There are many choices of m, n which would provide different subclasses of class $\mathrm{Q}_\Sigma(m, n; \Theta)$. For example

1. Choosing $m = 1$ and $n = 0$, we obtain the class

$$\mathcal{Q}_\Sigma(1, 0; \Theta) = \mathcal{Q}_\Sigma(1; \Theta) = S_\sigma(\Theta).$$

This class was studied by Altınkaya and Yalçın [11]. If the function $\mathfrak{f}$ is in this class, then following conditions are satisfied:

$$\frac{\xi \mathfrak{f}'(\xi)}{\mathfrak{f}(\xi)} \prec \mathcal{G}_{\{\mathcal{L}_{p,q,n}(\Theta)\}}(\xi) - 1$$

and

$$\frac{\eta \mathfrak{g}'(\eta)}{\mathfrak{g}(\eta)} \prec \mathcal{G}_{\{\mathcal{L}_{p,q,n}(\Theta)\}}(\eta) - 1.$$

2. Choosing $m = 2$ and $n = 1$, we obtain the class $Q_\Sigma(2, 1; \Theta)$. This new class consists of the functions satisfying the conditions

$$1 + \frac{\xi \mathfrak{f}''(\xi)}{\mathfrak{f}'(\xi)} \prec \mathcal{G}_{\{\mathcal{L}_{p,q,n}(\Theta)\}}(\xi) - 1$$

and

$$1 + \frac{\eta \mathfrak{g}''(\eta)}{\mathfrak{g}'(\eta)} \prec \mathcal{G}_{\{\mathcal{L}_{p,q,n}(\Theta)\}}(\eta) - 1.$$

Theorem 8.27 *Let $\mathfrak{f}$ given by (8.7) be in the class $Q_\Sigma(m, n; \Theta)$. Then,*

$$|a_2| \leq \frac{|p(\Theta)| \sqrt{|p(\Theta)|}}{\sqrt{\left|[(3^m - 3^n) - 2^m(2^m - 2^n)]\, p^2(\Theta) - 2(2^m - 2^n)^2 q(\Theta)\right|}}$$

and

$$|a_3| \leq \frac{p^2(\Theta)}{(2^m - 2^n)^2} + \frac{|p(\Theta)|}{|3^m - 3^n|}.$$

Proof Let $\mathfrak{f} \in Q_\Sigma(m, n; \Theta)$. Then from principle of subordination, for some analytic functions Ω, Λ such that

$$\Omega(0) = \Lambda(0) = 0 \text{ and } |\Omega(\xi)| < 1, |\Lambda(\eta)| < 1, \text{ for all } \xi, \eta \in U,$$

we can write

$$\frac{\mathcal{D}^m \mathfrak{f}(\xi)}{\mathcal{D}^n \mathfrak{f}(\xi)} = \mathcal{G}_{\{\mathcal{L}_{p,q,n}(\Theta)\}}(\Omega(\xi)) - 1 \tag{8.14}$$

and

$$\frac{\mathcal{D}^m \mathfrak{g}(\eta)}{\mathcal{D}^n \mathfrak{g}(\eta)} = \mathcal{G}_{\{\mathcal{L}_{p,q,n}(\Theta)\}}(\Lambda(\eta)) - 1 \tag{8.15}$$

or equivalently

$$\frac{\mathcal{D}^m \mathfrak{f}(\xi)}{\mathcal{D}^n \mathfrak{f}(\xi)} = -1 + \mathcal{L}_{p,q,0}(\Theta) + \mathcal{L}_{p,q,1}(\Theta)\Omega(\xi) + \mathcal{L}_{p,q,2}(\Theta)\Omega^2(\xi) + \cdots \tag{8.16}$$

and

$$\frac{\mathcal{D}^m \mathfrak{g}(\eta)}{\mathcal{D}^n \mathfrak{g}(\eta)} = -1 + \mathcal{L}_{p,q,0}(\Theta) + \mathcal{L}_{p,q,1}(\Theta)\Lambda(\eta) + \mathcal{L}_{p,q,2}(\Theta)\Lambda^2(\eta) + \cdots \tag{8.17}$$

From (8.16) and (8.17), we get

$$\frac{\mathcal{D}^m \mathfrak{f}(\xi)}{\mathcal{D}^n \mathfrak{f}(\xi)} = \mathcal{L}_{p,q,1}(\Theta)l_1\xi + \left[\mathcal{L}_{p,q,1}(\Theta)l_2 + \mathcal{L}_{p,q,2}(\Theta)l_1^2\right]\xi^2 + \cdots \tag{8.18}$$

and

$$\frac{\mathcal{D}^m \mathfrak{g}(\eta)}{\mathcal{D}^n \mathfrak{g}(\eta)} = \mathcal{L}_{p,q,1}(\Theta)r_1\eta + \left[\mathcal{L}_{p,q,1}(\Theta)r_2 + \mathcal{L}_{p,q,2}(\Theta)r_1^2\right]\eta^2 + \cdots \tag{8.19}$$

As we know before, if for $\xi, \eta \in U$,

$$\Omega(\xi) = \left|\sum_{i=1}^{n} l_i \xi^i\right| < 1$$

and

$$\Lambda(\eta) = \left|\sum_{i=1}^{n} r_i \eta^i\right| < 1$$

then

$$|l_i| < 1$$

and

$$\Lambda(\eta) = |r_i| < 1$$

where $i \in \mathbb{N}$.

So, if we compare the coefficients in (8.18) and (8.19), then we have

$$(2^m - 2^n)a_2 = \mathcal{L}_{p,q,1}(\Theta)l_1, \tag{8.20}$$

$$(3^m - 3^n)a_3 - 2^n(2^m - 2^n)a_2^2 = \mathcal{L}_{p,q,1}(\Theta)l_2 + \mathcal{L}_{p,q,2}(\Theta)l_1^2, \tag{8.21}$$

$$-(2^m - 2^n)a_2 = \mathcal{L}_{p,q,1}(\Theta)r_1, \tag{8.22}$$

$$(3^m - 3^n)(2a_2^2 - a_3) - 2^n(2^m - 2^n)a_2^2 = \mathcal{L}_{p,q,1}(\Theta)r_2 + \mathcal{L}_{p,q,2}(\Theta)r_1^2. \tag{8.23}$$

From (8.20) and (8.22)

$$l_1 = -r_1, \tag{8.24}$$

$$2(2^m - 2^n)^2 a_2^2 = \mathcal{L}^2_{p,q,1}(\Theta)\left(l_1^2 + r_1^2\right). \tag{8.25}$$

Adding (8.21) and (8.23), we get

$$2\left[(3^m - 3^n) - 2^n(2^m - 2^n)\right] a_2^2 = \mathcal{L}_{p,q,1}(\Theta)\,(l_2 + r_2) + \mathcal{L}_{p,q,2}(\Theta)\left(l_1^2 + r_1^2\right). \tag{8.26}$$

By using (8.25) in (8.26), we have

$$2\left[\left[(3^m - 3^n) - 2^n(2^m - 2^n)\right]\mathcal{L}^2_{p,q,1}(\Theta) - (2^m - 2^n)^2\mathcal{L}_{p,q,2}(\Theta)\right] a_2^2 = \mathcal{L}^3_{p,q,1}(\Theta)\,(l_2 + r_2) \tag{8.27}$$

which states

$$|a_2| \le \frac{|p(\Theta)|\sqrt{|p(\Theta)|}}{\sqrt{\left|\left[(3^m - 3^n) - 2^m(2^m - 2^n)\right]p^2(\Theta) - 2(2^m - 2^n)^2 q(\Theta)\right|}}.$$

Also, by subtracting (8.23) from (8.21), we get

$$2(3^m - 3^n)\left(a_3 - a_2^2\right) = \mathcal{L}_{p,q,1}(\Theta)\,(l_2 - r_2) + \mathcal{L}_{p,q,2}(\Theta)\left(l_1^2 - r_1^2\right). \tag{8.28}$$

Then, by (8.24) and (8.25), equality (8.28) becomes

$$a_3 = \frac{\mathcal{L}^2_{p,q,1}(\Theta)}{2(2^m - 2^n)^2}\left(l_1^2 + r_1^2\right) + \frac{\mathcal{L}_{p,q,1}(\Theta)}{2(3^m - 3^n)}\,(l_2 - r_2) \tag{8.29}$$

and by the help of (8.9), we conclude that

$$|a_3| \le \frac{p^2(\Theta)}{(2^m - 2^n)^2} + \frac{|p(\Theta)|}{|3^m - 3^n|}.$$

Corollaries
Letting $m = 1$ and $n = 0$ in Theorem 8.27, we get next consequence.

Corollary 8.28 *Let* $\mathfrak{f} \in \mathcal{Q}_\Sigma(1, 0; \Theta) = \mathcal{S}_\sigma(\Theta)$. *Then,*

$$|a_2| \le p(\Theta)\sqrt{\left|\frac{p(\Theta)}{2q(\Theta)}\right|}$$

and

$$|a_3| \leq p^2(\Theta) + \frac{p(\Theta)}{3^n\,(1+2\lambda)}.$$

This corollary coincides with the corollary of Theorem 2 in [11].

Corollary 8.29 *Let* $\mathfrak{f} \in Q_\Sigma(m, 0; \Theta) = Q^{\Sigma,\delta}(m; \Theta)$. *Then,*

$$|a_2| \leq p(\Theta)\sqrt{\left|\frac{p(\Theta)}{2q(\Theta)}\right|}$$

and

$$|a_3| \leq \frac{p^2(\Theta)}{(1+\lambda)^2} + \frac{|p(\Theta)|}{(1+2\lambda)}.$$

Corollary 8.30 *Let* $\mathfrak{f} \in Q_\Sigma(1, 0; \Theta) = Q^{\Sigma,\delta}(\Theta)$. *Then,*

$$|a_2| \leq \frac{|p(\Theta)|\sqrt{|p(\Theta)|}}{\sqrt{|[p^2(\Theta) + 4q(\Theta)]|}}$$

and

$$|a_3| \leq p^2(\Theta) + \frac{|p(\Theta)|}{2}.$$

8.2.3 Fekete–Szegö Inequalities

Next, we will examine Fekete–Szegö inequalities for functions in $Q_\Sigma(m, n; \Theta)$.

Theorem 8.31 *For* $\delta \in \mathbb{R}$, *let* $\mathfrak{f} \in Q_\Sigma(m, n; \Theta)$. *Then,*

$$\left|a_3 - \delta a_2^2\right| \leq \begin{cases} \frac{|p(\Theta)|}{[(3^m-3^n)-2^n(2^m-2^n)]}, & 0 \leq |t(\delta;\Theta)| \leq \frac{1}{[(3^m-3^n)-2^n(2^m-2^n)]} \\ 2\,|p(\Theta)|\,|t(\delta;\Theta)|\,, & |t(\delta;\Theta)| \geq \frac{1}{[(3^m-3^n)-2^n(2^m-2^n)]} \end{cases}$$

where

$$t(\delta;\Theta) = \frac{(1-\delta)\,\mathcal{L}^2_{p,q,1}(\Theta)}{2\left[\mathcal{L}^2_{p,q,1}(\Theta)(1+2\xi)(1+2\delta)^n - \mathcal{L}_{p,q,2}(\Theta)(1+\xi)^2(1+\delta)^{2n}\right]}.$$

Proof From the equations (8.27) and (8.28), we get

$$a_3 - \delta a_2^2 = (1-\delta)\left[\frac{\mathcal{L}^3_{p,q,1}(\Theta)\,(l_2+r_2)}{2\left[[(3^m-3^n)-2^n(2^m-2^n)]\,\mathcal{L}^2_{p,q,1}(\Theta)-(2^m-2^n)^2_{p,q,2}\mathcal{L}(\Theta)\right]}\right]$$
$$+\frac{\mathcal{L}_{p,q,1}(\Theta)\,(l_2-r_2)}{2(3^m-3^n)}$$
$$=\mathcal{L}_{p,q,1}(\Theta)\left[\left(t(\delta;\Theta)+\frac{1}{2(3^m-3^n)}\right)s_2+\left(t(\delta;\Theta)-\frac{\tau}{2(3^m-3^n)}\right)s_2\right],$$

where

$$t(\delta;\Theta)=\frac{(1-\delta)\,\mathcal{L}^2_{p,q,1}(\Theta)}{2\left[[(3^m-3^n)-2^n(2^m-2^n)]\,\mathcal{L}^2_{p,q,1}(\Theta)-(2^m-2^n)^2_{p,q,2}\mathcal{L}(\Theta)\right]}.$$

With the help of Eq. (8.9), the proof is completed.

8.2.3.1 Corollaries

An urgent and important corollary of Theorem 8.31 for different values of m and n is asserted by the following corollaries.

Corollary 8.32 *Let* $\mathfrak{f}\in Q_\Sigma(1,0;\Theta)=S_\sigma(\delta)$. *Then,*

$$\left|a_3-\delta a_2^2\right|\le\begin{cases}|p(\Theta)| &, \quad 0\le|t(\delta;\Theta)|\le 1\\ 2\,|p(\Theta)|\,|t(\delta;\Theta)| &, \quad |t(\delta;\Theta)|\ge 1\end{cases}.$$

Corollary 8.33 *Let* $\mathfrak{f}\in Q_\Sigma(2,1;\Theta)$. *Also, assume that* $\delta\in\mathbb{R}$. *Then,*

$$\left|a_3-\delta a_2^2\right|\le\begin{cases}\frac{|p(\Theta)|}{2} &, \quad 0\le|t(\delta;\Theta)|\le\frac{1}{2}\\ 2\,|p(\Theta)|\,|t(\delta;\Theta)| &, \quad |t(\delta;\Theta)|\ge\frac{1}{2}\end{cases}$$

Open Problem

For the subclass $Q_\Sigma(m,n;\Theta)\;(m,n\in\mathbb{N}_0, m>n)$, it is interesting to obtain the bounds of the Hankel determinants and to investigate Taylor–Maclaurin coefficients m-fold symmetric bi-univalent functions.

References

1. Ahlfors, L.V.: Complex Analysis, 2nd edn. McGraw-Hill, New York (1966)
2. Agarwal, R.P., Perera, K., Pinelas, S.: An Introduction to Complex Analysis. Springer Science & Business Media (2011)

3. Akgül, A., Sakar, F.M.: A certain subclass of bi-univalent analytic functions introduced by means of the q-analogue of Noor integral operator and Horadam polynomials. Turkish J. Math. **43**(5), 2275–2286 (2019)
4. Akgül, A.: A new approach to the bi-univalent analytic functions connected with q-analogue of Noor integral operator. Commun. Fac. Sci. Univ. Ank. Ser. A1 Math. Stat. **70**(2), 940–949 (2021)
5. Akgül, A.: (P, Q)-Lucas polynomial coefficient inequalities of the bi-univalent function class. Turk. J. Math. **43**, 2170–2176 (2019)
6. Akgül, A., Cotîrlă, L.I.: Coefficient estimates for a family of starlike functions endowed with quasi subordination on conic domain. Symmetry **14**(3), 582 (2022)
7. Akgül, A., Sakar, F.M.: A new characterization of $(P, Q)-$Lucas polynomial coefficients of the bi-univalent function class associated with $q-$analogue of Noor integral operator. Afrika Matematika **33**(3), 1–12 (2022)
8. Akgül, A.: Coefficient estimates of a new Bi-Univalent function class introduced by Lucas-Balancing Polynomial. Int. J. Open Probl. Compt. Math. **16**(3), 3647 (2023)
9. Akgül, A.: Initial coefficient estimates of Bi-Univalent functions linked with balancing coefficients. WSEAS Trans. Math. **22**, 792–797 (2023)
10. Akgül, A., Shaba, T.: (U, V)-Lucas polynomial coefficient relations of the bi-univalent function class. Communications Faculty of Sciences University of Ankara Series A1 Mathematics and Statistics, vol. 71(4), pp. 1121–1135 (2022)
11. Altınkaya, Ş., Yalçın, S.: On the (p, q)-Lucas polynomial coefficient bounds of the bi-univalent function class sigma. Boletín de la Sociedad Matemática Mexicana, pp. 1–9 (2019)
12. Altınkaya, Ş., Yalçın, S.: Fekete-Szego inequalities for certain classes of biunivalent functions. Internat. Scholar. Res. Notices **2014**, Article ID 327962, 1–6 (2014)
13. Altınkaya, Ş, Yalçın, S.: The Fekete-Szego problem for a general class of bi-univalent functions sattisfying subordinate conditions, Sahand. Commun. Math. Anal. **5**, 1–7 (2017)
14. Altınkaya, Ş, Yalçın, S.: Coefficient estimates for two new subclasses of Bi-univalent functions. Acta Universitatis Apulensis **43**, 53–63 (2015)
15. Brannan, D.A., Clunie, J.G. eds.: Aspects of Contemporary Complex Analysis (Proceedings of the NATO Advanced Study Institute held at the University of Durham, Durham; 1–20 July 1979). Academic Press, New York and London (1980)
16. Brannan, D.A., Taha, T.S.: On some classes of bi-univalent functions. In: Mazhar, S.M., Hamoui, A., Faour, N.S. eds.: Mathematical Analysis and Applications, Kuwait; 18–21 Feb. 1985, in: KFAS Proceedings Series, vol. 3, pp. 53–60. Pergamon Press, Elsevier Science Limited, Oxford (1988). See also Studia Univ. Babeş-Bolyai Math. **31**(2), 70–77 (1986)
17. De Branges, L.: A proof of the Bieberbach conjecture. Acta Math. **154**, 137–152 (1985)
18. Bernardi, S.D.: A survey of the development of the theory of schlicht functions. Duke Math. J. **19**, 263–287 (1952)
19. Bieberbach, L.: ¨ Uber die Koeffizienten derjenigen Potenzreihen, welche eine schlichte Abbildung des Einheitskreises vermitteln, S. B. Preuss. Akad. Wiss. **138**, 940–955 (1916)
20. Clunie, J.: On schlicht functions. Ann. Math. **69**, 511–519 (1959)
21. Clunie, J., Pommerenke, Ch.: On the coe cients of univalent functions. Michigan Math. J. **14**, 71–78 (1967)
22. Conway, J.B.: Functions of One Complex Variable, 2nd ed. Springer (1978)
23. Deniz, E., Orhan, H.: The Fekete-Szegö problem for a generalized subclass of analytic functions. Kyungpook Math. J. **50**(1), 37–47 (2010)
24. Deniz, E., Çağlar, M., Orhan, H.: The Fekete-Szegö problem for a class of analytic functions defined by Dziok-Srivastava operator. Kodai Math. J. **35**(3), 439–462 (2012)
25. Duren, P.L.: Univalent functions. Grundlehren der Mathematischen Wissenschaften, vol. 259. Springer, New York (1983)
26. Fekete, M., Szego, G.: Eine Bemerkung Uber Ungerade Schlichte Funktionen, J. London Math. Soc. [s1-8 (2)], 85–89 (1933)
27. Frasin, B.A., Darus, M.: On the Fekete-Szego problem. Int. J. Math. Math. Sci. **24**, 577–581 (2000)

28. Frasin, B.A., Aouf, M.K.: New subclasses of bi-univalent functions. Appl. Math. Lett. **24**, 1569–1573 (2011)
29. Filipponi, P., Horadam, A.F.: Second derivative sequences of Fibonacci and Lucas polynomials. Fibonacci Q **31**(3), 194–204 (1993)
30. Filipponi, P., Horadam, A.F.: Derivative sequences of Fibonacci and Lucas polynomials. Appl. Fibonacci Num. **4**, 99–108 (1991)
31. Filipponi, P., Horadam, A.F.: Second derivative sequences of Fibonacci and Lucas polynomials. Fibonacci Quart. **31**(3), 194–204 (1993)
32. Garabedian, P.R., Schiffer, M.: A proof of the Bieberbach conjecture for the fourth coe cient. J. Rational Mech. Anal. **4**, 427–465 (1955)
33. Grenander, U., Szegö, G.: Toeplitz Forms and Their Applications. California Monographs in Mathematical Sciences University, California Press, Berkeley (1958)
34. Gronwall, T.H.: Some remarks on conformal representation. Ann. Math. **16**, 72–76 (1914–1915)
35. Goodman, A.W.: Univalent Functions, vol. I. II. Polygonal Publishing House, NJ (1983)
36. Hamidi, S.G., Jahangiri, J.M.: Faber polynomial coefficient estimates for analytic bi-close-to-convex functions. C. R. Acad. Sci. Paris Ser. I **352**(1), 17–20 (2014)
37. Hayman, W.K.: On successive coe cients of univalent functions. J. London Math. Soc. **38**, 228–243 (1963)
38. Horadam, A.F., Mahon, J.M.: Pell and Pell-Lucas Polynomials. Fibonacci Q. **23**(1), 7–20 (1985)
39. Horzum, T., Koçer, E.G.: On some properties of Horadam polynomials. Int. Math. Forum. **4**, 1243–1252 (2009)
40. Jahangiri, J.M., Hamidi, S.G.: Coefficient estimates for certain classes of bi-univalent functions. Int. J. Math. Math. Sci. ArticleID 190560, 4 pp (2013)
41. Kanas, S.: An uni ed approach to the Fekete-Szego problem. Appl. Math. Comput. **218**, 8453–8461 (2012)
42. Kanas, S., Darwish, H.E.: Fekete-Szego problem for starlike and convex functions of complex order. Appl. Math. Lett. **23**, 777–782 (2010)
43. Keogh, F.R., Merkes, E.P.: A coefficient inequality for certain classes of analytic functions. Proc. Am. Math. Soc. **20**, 8–12 (1969)
44. Koebe, P.: Über die Uniformisierung beliebiger analytischer Kurven, Nachr. Akad. Wiss. Göttingen, Math.-Phys. Kl. 191–210 (1907)
45. Koshy, T.: Fibonacci and Lucas Numbers with Applications. Wiley, 17 Jan. 2019
46. Krantz, S.G.: Goemtric Function Theory. Explorations in Complex Analysis, Brikhauser, Boston (2006)
47. Lee, G., Asci, M.: Some properties of the $(p; q)$-Fibonacci and $(p; q)$-Lucas polynomials. J. Appl. Math. **2012**, 264842 (2012)
48. M. Lewin, On a coefficient problem for bi-univalent functions. Proc. Am. Math. Soc **18**, 63–68 (1967)
49. London, R.R.: Fekete-Szegö inequalities for close-to convex functions. Proc. Am. Math. Soc. **117**, 947–950 (1993)
50. Lowner, K.: Uniersuchungen uber schlichte konforme Abbildungen des Einheitskreises. I. Math. Ann. **89**, 103–121 (1923)
51. Lupas, A.: A guide of Fibonacci and Lucas polynomials. Octagon Math. Mag. **7**, 2–12 (1999)
52. Miller, S.S., Mocanu, P.T.: Differential Subordinations: Theory and Applications, Series on Monographs and Textbooks in Pure and Applied Mathematics, vol. 225. Marcel Dekker, New York, Basel (2000)
53. Nehari, Z.: Conformal Mapping. Mc. Grraw-Hill Book Comp, New York (1952)
54. Netanyahu, E.: The minimal distance of the image boundary from the orijin and the second coefficient of a univalent function in $|\xi| < 1$. Arch. Rational Mech. Anal. **32**, 100–112 (1969)
55. Olatunji, S.O., Dutta, H.: Coefficient inequalities for pseudo subclasses of analytical functions related to Petal type domains defined by error function. AIMS Math. **5**, 2526–2538 (2020)

56. Orhan, H., Deniz, E., Çağlar, M.: Fekete-Szegö problem for certain subclasses of analytic functions. Demonstratio Mathematica **45**(4), 835–846 (2012)
57. Ozawa, M.: On the Bieberbach conjecture for the sixth coe cient. Kodai Math. Sem. Rep. **21**, 97–128 (1969)
58. Özkoç, A., Porsuk, A.: A note for the (p; q)-Fibonacci and Lucas quaternion polynomials. Konuralp J. Math. **5**(2), 36–46
59. Pederson, R.W.: A proof of the Bieberbach conjecture for the sixth coefficient. Arch. Rational Mech. Anal. **31**, 331–351 (1968)
60. Pederson, R.W., Schier, M.: A proof of the Bieberbach conjecture for the nth coefficient. Arch. Rational Mech. Anal. **45**, 161–193 (1972)
61. Pommerenke, C.: Univalent Functions. Vandenhoeck & Ruprecht, Gottingen (1975)
62. Sălăgean, G.S.: Subclasses of univalent functions. In: Complex Analysis-Fifth Romanian-Finnish seminar, Part 1(Bucharest, 1981), Lecture Notes in Mathematics, vol. 1013, pp. 362–372. Springer, Berlin, Germany (1983)
63. Sakar, F.M., Aydoğan, S.M.: Initial bounds for certain subclass of generalized Sălăgean type bi-univalent functions associated with the Horadam polynomials. J. Quality Meas. Anal. **15**(1), 89–100 (2019)
64. Schaeffer, A.C., Spencer, D.C.: The coefficients of schlicht functions. Duke Math. J. **12**, 107–125 (1945)
65. Schober, G.: Univalent Functions-Selected Topics. Springer, Berlin, Heidelberg, New York (1975)
66. Srivastava, H.M., Owa, S. (eds.): Current Topics in Analytic Function Theory, World Scienti c Publishing Company, Singapore, NJ, London, and Hong Kong (1992)
67. Srivastava, H.M., Mishra, A.K., Das, M.K.: The Fekete-Szego problem for a subclass of close-to-convex functions. Complex Variables Theory Appl. **44**, 145–163 (2001)
68. Srivastava, H.M., Mishra, A.K., Gochhayat, P.: Certain subclasses of analytic and bi-univalent functions. Appl. Math. Lett. **23**(10), 1188–1192 (2010)
69. Vellucci, P., Bersani, A.M.: The class of Lucas-Lehmer polynomials. Rendiconti di Matematica **37**, 43–62 (2016)
70. Wang, T., Zhang, W.: Some identities involving Fibonacci, Lucas polynomials and their applications. Bull. Math. Soc. Sci. Math. Roum. **55**(1), 95–103 (2012)
71. Zaprawa, Z.: On Fekete-Szego problem for classes of bi-univalent functions. Bull. Belg. Math. Soc. Simon Stevin **21**, 169–178 (2014)

Chapter 9
Sufficient Conditions for Generalized Integral Operators Involving the Rabotnov Function

Sercan Kazımoğlu, Hemen Dutta, and Erhan Deniz

Abstract In this paper, our aim is to present sufficient conditions for the families of integral operators containing the normalized forms of the Rabotnov functions such that they can be univalent in the open unit disk. Moreover, we find the convexity order of the families of the investigated operators. In proof of results, we use some differential inequalities related to Rabotnov functions and well-known lemmas. Also, we give some simple results and graphical illustrations to support the results shown.

Keywords Analytic functions · Rabotnov functions · Univalent functions · Convexity · Integral operators

AMS Class 30C45 · 33C10

9.1 Introduction

Geometric function theory is one of the important branches of complex analysis. It deals with the geometric properties of analytic functions. The main foundation of geometric function theory is the theory of univalent functions, but a number of new associated areas have emerged and led to various strong results and applications. The univalent functions realize conformal representations, having applications in several fields of pure and applied mathematics such as mathematical physics, mathematical biology, fluid mechanics, electrotechnics, nuclear physics, and mathematical chemistry. Recently, several researchers have studied certain integral operators involving

S. Kazımoğlu · E. Deniz (✉)
Faculty of Science and Letters, Department of Mathematics, Kafkas University, Campus 36100, Kars, Türkiye
e-mail: edeniz36@gmail.com

S. Kazımoğlu
e-mail: srcnkzmglu@gmail.com

H. Dutta
Department of Mathematics, Gauhati University, Guwahati, India
e-mail: duttah@gauhati.ac.in

B. C. Tripathy et al. (eds.), *Operators, Inequalities and Approximation*, Industrial and Applied Mathematics, https://doi.org/10.1007/978-981-97-3238-8_9

some special functions like Bessel, Lommel, Struve, Wright and Mittag-Leffler, etc., and of their geometric properties on different classes of univalent functions. It is remarkable to note that researchers in the field are interested nowadays in obtaining new theoretical methodologies and techniques with observational results together with their several applications. Let us now recall some known definitions and results in geometric function theory.

Let $\mathcal{A}$ denote the family of functions $\hbar$ of the form

$$\hbar(\rho) = \rho + \sum_{\upsilon=2}^{\infty} a_\upsilon \rho^\upsilon$$

which are analytic in the open unit disk $\mathbb{D}$ and satisfy the usual normalization condition:

$$\hbar(0) = \hbar'(0) - 1 = 0.$$

We denote by $\mathcal{S}$ the subclass of the normalized analytic function class $\mathcal{A}$ consisting of functions which are also univalent in $\mathbb{D}$. A function $\hbar \in \mathcal{S}$ is said to be convex of order δ $(0 \leq \delta < 1)$ if it satisfies

$$\Re\left(1 + \frac{\rho\hbar''(\rho)}{\hbar'(\rho)}\right) > \delta \quad (\rho \in \mathbb{D}).$$

Let $n \in \mathbb{N} := \{1, 2, 3, \ldots\}$ and

$$\mathcal{A}^n = \{(\hbar_1, \hbar_2, \cdots, \hbar_\upsilon) : \hbar_\upsilon \in \mathcal{A}\,(\upsilon = 1, 2, \ldots, n)\}.$$

For the functions $\hbar_\upsilon \in \mathcal{A}$ $(\upsilon = 1, 2, \ldots, n)$, the parameters $\eta_\upsilon, \zeta_\upsilon \in \mathbb{C}$ $(\upsilon = 1, 2, \ldots, n)$ and $\gamma \in \mathbb{C}$, we define the following two integral operators:

$$\mathcal{J}_{\eta_1,\eta_2,\ldots,\eta_n;\zeta_1,\zeta_2,\ldots,\zeta_n;n;\gamma} : \mathcal{A}^n \longrightarrow \mathcal{A}$$

and

$$\mathcal{K}_{\eta_1,\eta_2,\ldots,\eta_n;\zeta_1,\zeta_2,\ldots,\zeta_n;n} : \mathcal{A}^n \longrightarrow \mathcal{A}$$

by

$$\mathcal{J}_{\eta_1,\eta_2,\ldots,\eta_n;\zeta_1,\zeta_2,\ldots,\zeta_n;n;\gamma}\left[\hbar_1, \hbar_2, \ldots, \hbar_n\right](\rho) := \left[\gamma \int_0^\rho t^{\gamma-1} \prod_{\upsilon=1}^{n} \left(\hbar'_\upsilon(t)\right)^{\eta_\upsilon} \left(\frac{\hbar_\upsilon(t)}{t}\right)^{\zeta_\upsilon} dt\right]^{1/\gamma} \tag{9.1}$$

and

$$\mathcal{K}_{\eta_1,\eta_2,\ldots,\eta_n;\zeta_1,\zeta_2,\ldots,\zeta_n;n}\left[\hbar_1, \hbar_2, \ldots, \hbar_n\right](\rho)$$
$$:= \left[\left(1 + \sum_{\upsilon=1}^{n} \eta_\upsilon\right) \int_0^\rho \prod_{\upsilon=1}^{n} \left(\hbar_\upsilon(t)\right)^{\eta_\upsilon} \left(e^{\hbar_\upsilon(t)}\right)^{\zeta_\upsilon} dt\right]^{1/\left(1+\sum_{\upsilon=1}^{n}\eta_\upsilon\right)}. \tag{9.2}$$

We note that, for some special parameters, the integral operators defined above have been extensively studied by many authors as follows:

(1) $\mathcal{J}_{0,0,\ldots,0;\zeta_1,\zeta_2,\ldots,\zeta_n;n;\gamma}[\hbar_1, \hbar_2, \ldots, \hbar_n] \equiv \mathcal{F}_{1/\zeta_1,1/\zeta_2,\ldots,1/\zeta_n;\gamma}$ (Seenivasagan and Breaz [26]; see also [3, 28]);
(2) $\mathcal{J}_{\eta_1,\eta_2,\ldots,\eta_n;0,0,\ldots,0;n;\gamma}[\hbar_1, \hbar_2, \ldots, \hbar_n] \equiv \mathcal{H}_{\eta_1,\eta_2,\ldots,\eta_n;\gamma}$ (Breaz and Breaz [4]);
(3) $\mathcal{J}_{\eta_1,\eta_2,\ldots,\eta_n;0,0,\ldots,0;n;1}[\hbar_1, \hbar_2, \ldots, \hbar_n] \equiv \mathcal{H}_{\eta_1,\eta_2,\ldots,\eta_n}$ (Breaz et al. [7]);
(4) $\mathcal{J}_{\eta;0;1;1}[\hbar] \equiv \mathcal{H}_{\eta}$ (Kim and Merkes [15]); see also Pfaltzgraff [23];
(5) $\mathcal{K}_{\eta,\eta,\ldots,\eta;0,0,\ldots,0;n}[\hbar_1, \hbar_2, \ldots, \hbar_n] \equiv \mathcal{G}_{n,\eta}[\hbar_1, \hbar_2, \ldots, \hbar_n]$ (Breaz and Breaz [5]; see also [6]);
(6) $\mathcal{K}_{\eta;0;1}[\hbar] \equiv \mathcal{G}_{1,\eta}[\hbar]$ (Moldoveanu and Pascu [18]).

Furthermore, the specific integral operators via an obvious parametric changes of the classical Bessel function $J_\nu(\rho)$ of order ν and of the first kind by Deniz et al. [8] were introduced and they worked on the univalence condition of the related integral operators. Also, the starlikeness, convexity, and uniform convexity of integral operators containing these equivalent forms of $J_\nu(\rho)$ were discussed by Raza et al. [25] and Deniz [9]. Recently, some sufficient conditions for univalence of various linear fractional derivative operators containing the normalized forms of the similar parametric variation of $J_\nu(\rho)$ were investigated by Al-Khrasani et al. [1]. Moreover, the theory of derivatives and integrals of an arbitrary complex or real order has been utilized not only in complex analysis but also in mathematical analysis and modeling of real-world problems in applied sciences (see for example, [30, 31]).

Inspired by the studies mentioned above, in the present paper, we work on some mappings and univalence and convexity conditions for the integral operators given by (9.1) and (9.2), associated with the Rabotnov function which is defined as follows: Consider the function $R_{\xi,\varrho}(\rho)$ defined by

$$R_{\xi,\varrho}(\rho) = \rho^{\xi} \sum_{n=0}^{\infty} \frac{\varrho^n}{\Gamma((1+\xi)(n+1))} \rho^{n(1+\xi)} \tag{9.3}$$

where Γ stands for the Euler gamma function and $\xi > -1$, $\varrho \in \mathbb{C}$ and $\rho \in \mathbb{D}$. This function was introduced by Rabotnov [24] and known today as the Rabotnov fractional exponential function or briefly Rabotnov function. In 1948, Yu. N. Rabotnov, who worked in solid mechanics including plasticity, creep theory, hereditary mechanics, failure mechanics, nonelastic stability, composites, and shell theory introduced a special function applied in viscoelasticity [24].

The function defined by (9.3) does not belong to the class $\mathcal{A}$. Therefore, we consider the following normalization of the Rabotnov function $R_{\xi,\varrho}(\rho)$: for $\rho \in \mathbb{D}$,

$$\mathbb{R}_{\xi,\varrho}(\rho) = \Gamma(1+\xi)\,\rho^{1/(1+\xi)} R_{\xi,\varrho}\left(\rho^{1/(1+\xi)}\right) = \sum_{n=0}^{\infty} \frac{\varrho^n \Gamma(1+\xi)}{\Gamma((1+\xi)(n+1))} \rho^{n+1} \tag{9.4}$$

where $\xi > -1$ and $\varrho \in \mathbb{C}$.
Recently, the geometric properties of normalized Rabotnov function have been stud-

ied by Eker and Ece [32]. Kazımoğlu and Deniz [14] obtained some results related to partial sums of the function (9.4).

Note that some special cases of $\mathbb{R}_{\xi,\varrho}(\rho)$ are

$$\begin{cases} \mathbb{R}_{0,-\frac{1}{3}}(\rho) = \rho e^{-\frac{\varrho}{3}} \\ \mathbb{R}_{1,\frac{1}{2}}(\rho) = \sqrt{2\rho}\sinh\sqrt{\frac{\varrho}{2}} \\ \mathbb{R}_{1,-\frac{1}{4}}(\rho) = 2\sqrt{\rho}\sin\frac{\sqrt{\rho}}{2} \\ \mathbb{R}_{1,1}(\rho) = \sqrt{\rho}\sinh\sqrt{\rho} \\ \mathbb{R}_{1,2}(\rho) = \frac{\sqrt{2\rho}\sinh\sqrt{2\rho}}{2} \end{cases}.$$

Let $\upsilon = 1, 2, \ldots, n$ and let $\xi_\upsilon > -1$ and $\varrho \in \mathbb{C}$. Consider the functions $\mathbb{R}_{\xi_\upsilon,\varrho}$ $(\upsilon = 1, 2, \ldots, n)$ defined by

$$\mathbb{R}_{\xi_\upsilon,\varrho}(\rho) = \Gamma(1+\xi_\upsilon)\,\rho^{1/(1+\xi_\upsilon)} R_{\xi_\upsilon,\varrho}\left(\rho^{1/(1+\xi_\upsilon)}\right).$$

Using the functions $\mathbb{R}_{\xi_\upsilon,\varrho}$ and integral operator defined by (9.1) and (9.2), we define the functions $\mathcal{J}^{\xi_1,\xi_2,\ldots,\xi_n;\varrho}_{\eta_1,\eta_2,\ldots,\eta_n;\zeta_1,\zeta_2,\ldots,\zeta_n;n;\gamma}$ and $\mathcal{K}^{\xi_1,\xi_2,\ldots,\xi_n;\varrho}_{\eta_1,\eta_2,\ldots,\eta_n;\zeta_1,\zeta_2,\ldots,\zeta_n;n} : \mathbb{D} \longrightarrow \mathbb{C}$ as follows:

$$\begin{aligned} \mathcal{J}^{\xi_1,\xi_2,\ldots,\xi_n;\varrho}_{\eta_1,\eta_2,\ldots,\eta_n;\zeta_1,\zeta_2,\ldots,\zeta_n;n;\gamma}(\rho) &:= \mathcal{J}_{\eta_1,\eta_2,\ldots,\eta_n;\zeta_1,\zeta_2,\ldots,\zeta_n;n;\gamma}\left[\mathbb{R}_{\xi_1,\varrho}, \mathbb{R}_{\xi_2,\varrho}, \ldots, \mathbb{R}_{\xi_n,\varrho}\right](\rho) \\ &= \left[\gamma\int_0^\rho t^{\gamma-1}\prod_{\upsilon=1}^n \left(\mathbb{R}'_{\xi_\upsilon,\varrho}(t)\right)^{\eta_\upsilon}\left(\frac{\mathbb{R}_{\xi_\upsilon,\varrho}(t)}{t}\right)^{\zeta_\upsilon} dt\right]^{1/\gamma} \end{aligned} \tag{9.5}$$

and

$$\begin{aligned} \mathcal{K}^{\xi_1,\xi_2,\ldots,\xi_n;\varrho}_{\eta_1,\eta_2,\ldots,\eta_n;\zeta_1,\zeta_2,\ldots,\zeta_n;n}(\rho) &:= \mathcal{K}_{\eta_1,\eta_2,\ldots,\eta_n;\zeta_1,\zeta_2,\ldots,\zeta_n;n}\left[\mathbb{R}_{\xi_1,\varrho}, \mathbb{R}_{\xi_2,\varrho}, \ldots, \mathbb{R}_{\xi_n,\varrho}\right](\rho) \\ &= \left[\left(1+\sum_{\upsilon=1}^n \eta_\upsilon\right)\int_0^\rho \prod_{\upsilon=1}^n \left(\mathbb{R}_{\xi_\upsilon,\varrho}(t)\right)^{\eta_\upsilon}\left(e^{\mathbb{R}_{\xi_\upsilon,\varrho}(t)}\right)^{\zeta_\upsilon} dt\right]^{1/(1+\sum_{\upsilon=1}^n \eta_\upsilon)} \end{aligned} \tag{9.6}$$

In this paper, we drive some sufficient conditions for the following operators:

$$\mathcal{J}^{\xi_1,\xi_2,\ldots,\xi_n;\varrho}_{\eta_1,\eta_2,\ldots,\eta_n;\zeta_1,\zeta_2,\ldots,\zeta_n;n;\gamma}$$

and

$$\mathcal{K}^{\xi_1,\xi_2,\ldots,\xi_n;\varrho}_{\eta_1,\eta_2,\ldots,\eta_n;\zeta_1,\zeta_2,\ldots,\zeta_n;n},$$

defined by (9.5) and (9.6), respectively, to be univalent in $\mathbb{D}$. We also determine the order of convexity of the functions defined by using the above-mentioned integral operators (9.5) and (9.6).

Recently, many mathematicians have set the univalence criteria for several those integral operators which preserve the class $\mathcal{S}$. By using a variety of different analytic techniques, operators, and special functions, several authors have studied the univalence criterion, a few of them are as mentioned below.

In 2010, Baricz and Frasin [3] studied some integral operators involving Bessel functions. These integral operators were defined by using the normalized Bessel functions of the first kind. Frasin [12] and Arif and Raza [2] studied the convexity and strong convexity of the integral operators defined in [3]. There is an extensive literature in geometric function theory that deals with the geometric properties of the integral operators defined by different kinds of special functions like Bessel function [2, 3, 8, 9, 12], Struve function [10, 16, 17], Lommel function [11, 20], Wright function [19], and Mittag-Leffler function [29]. Motivated by the work of these above authors, we contribute to this univalence theory by studying the univalence and convexity of integral operators involving the Rabotnov function.

9.2 A Set of Lemmas

The following lemmas will be required in our present investigations.

Lemma 9.2.1 (see Pescar [22]) *Let α and β be a complex number such that*

$$\Re(\alpha) > 0 \text{ and } |\beta| \leqq 1 \quad (\beta \neq -1).$$

If the function $h \in \mathcal{A}$ satisfies the following inequality:

$$\left| \beta |\rho|^{2\alpha} + \left(1 - |\rho|^{2\alpha}\right) \frac{\rho h''(\rho)}{\alpha h'(\rho)} \right| \leqq 1$$

for all $\rho \in \mathbb{D}$, then the function $F_\alpha \in \mathcal{A}$ defined by

$$F_\alpha(\rho) = \left(\alpha \int_0^\rho t^{\alpha-1} h'(t)dt\right)^{1/\alpha} \tag{9.7}$$

is in the class $\mathcal{S}$.

Lemma 9.2.2 (see Pascu [21]) *Let $\varpi \in \mathbb{C}$ such that $\Re(\varpi) > 0$. If $h \in \mathcal{A}$ satisfies the following inequality:*

$$\left(\frac{1 - |\rho|^{2\Re(\varpi)}}{\Re(\varpi)}\right) \left|\frac{\rho h''(\rho)}{h'(\rho)}\right| \leqq 1$$

for all $\rho \in \mathbb{D}$. Then, for all $\alpha \in \mathbb{C}$ such that

$$\Re(\alpha) \geqq \Re(\varpi),$$

the function F_α defined by (9.7) is in the class $\mathcal{S}$.

Lemma 9.2.3 *Let $\xi > -1$ and $\varrho \in \mathbb{C}$. Then, $\forall\ \rho \in \mathbb{D}$, the function $\mathbb{R}_{\xi,\varrho}$ identified by (9.4) provides the following inequalities:*

$$\left| \mathbb{R}'_{\xi,\varrho}(\rho) - \frac{\mathbb{R}_{\xi,\varrho}(\rho)}{\rho} \right| \leqq \left(\frac{|\varrho|}{1+\xi} \right) e^{\frac{|\varrho|}{1+\xi}}, \tag{9.8}$$

$$\left| \frac{\rho\mathbb{R}'_{\xi,\varrho}(\rho)}{\mathbb{R}_{\xi,\varrho}(\rho)} - 1 \right| \leqq \frac{\left(\frac{|\varrho|}{1+\xi} \right) e^{\frac{|\varrho|}{1+\xi}}}{2 - e^{\frac{|\varrho|}{1+\xi}}}, \quad (|\varrho| < (1+\xi)\ln 2), \tag{9.9}$$

$$\left| \rho\mathbb{R}'_{\xi,\varrho}(\rho) \right| \leqq \left(\frac{1+\xi+|\varrho|}{1+\xi} \right) e^{\frac{|\varrho|}{1+\xi}}, \tag{9.10}$$

$$\left| \frac{\rho\mathbb{R}'_{\xi,\varrho}(\rho)}{\mathbb{R}_{\xi,\varrho}(\rho)} \right| \leqq \frac{\left(\frac{1+\xi+|\varrho|}{1+\xi} \right) e^{\frac{|\varrho|}{1+\xi}}}{2 - e^{\frac{|\varrho|}{1+\xi}}}, \quad (|\varrho| < (1+\xi)\ln 2) \tag{9.11}$$

and

$$\left| \frac{\rho\mathbb{R}''_{\xi,\varrho}(\rho)}{\mathbb{R}'_{\xi,\varrho}(\rho)} \right| \leqq \frac{\left(\frac{|\varrho|(2+2\xi+|\varrho|)}{(1+\xi)^2} \right) e^{\frac{|\varrho|}{1+\xi}}}{2 - \left(\frac{1+\xi+|\varrho|}{1+\xi} \right) e^{\frac{|\varrho|}{1+\xi}}}, \quad \left(|\varrho| < (1+\xi)\ln\left(\frac{2(1+\xi)}{1+\xi+|\varrho|} \right) \right). \tag{9.12}$$

Proof By using the triangle inequality and the following inequality [32]:

$$\frac{\Gamma(1+\xi)}{\Gamma((1+\xi)(v+1))} \leq \frac{1}{(1+\xi)^v (v)!}, \quad v \in \mathbb{N}, \tag{9.13}$$

we have

$$\begin{aligned} \left| \mathbb{R}'_{\xi,\varrho}(\rho) - \frac{\mathbb{R}_{\xi,\varrho}(\rho)}{\rho} \right| &= \left| \sum_{v=1}^{\infty} \frac{v\varrho^v \Gamma(1+\xi)}{\Gamma((1+\xi)(v+1))} \rho^v \right| \\ &\leq \sum_{v=1}^{\infty} \frac{v|\varrho|^v \Gamma(1+\xi)}{\Gamma((1+\xi)(v+1))} \\ &\leq \sum_{v=1}^{\infty} \frac{v|\varrho|^v}{v!(1+\xi)^v} = \left(\frac{|\varrho|}{1+\xi} \right) e^{\frac{|\varrho|}{1+\xi}} \end{aligned} \tag{9.14}$$

and

$$\begin{aligned}\left|\frac{\mathbb{R}_{\xi,\varrho}(\rho)}{\rho}\right| &= \left|1+\sum_{\upsilon=1}^{\infty}\frac{\varrho^{\upsilon}\Gamma\left(1+\xi\right)}{\Gamma\left(\left(1+\xi\right)\left(\upsilon+1\right)\right)}\rho^{\upsilon}\right| \\ &\geq 1-\sum_{\upsilon=1}^{\infty}\frac{\left|\varrho\right|^{\upsilon}\Gamma\left(1+\xi\right)}{\Gamma\left(\left(1+\xi\right)\left(\upsilon+1\right)\right)} \\ &\geq 1-\sum_{\upsilon=1}^{\infty}\frac{\left|\varrho\right|^{\upsilon}}{\upsilon!\left(1+\xi\right)^{\upsilon}} = 2-e^{\frac{|\varrho|}{1+\xi}}, \quad \left(|\varrho|<(1+\xi)\ln 2\right).\end{aligned} \tag{9.15}$$

From (9.14) and (9.15), we get

$$\left|\frac{\rho\mathbb{R}'_{\xi,\varrho}(\rho)}{\mathbb{R}_{\xi,\varrho}(\rho)}-1\right| = \left|\frac{\mathbb{R}'_{\xi,\varrho}(\rho)-\frac{\mathbb{R}_{\xi,\varrho}(\rho)}{z}}{\frac{\mathbb{R}_{\xi,\varrho}(\rho)}{\rho}}\right| \leq \frac{\left(\frac{|\varrho|}{1+\xi}\right)e^{\frac{|\varrho|}{1+\xi}}}{2-e^{\frac{|\varrho|}{1+\xi}}}, \quad \left(|\varrho|<(1+\xi)\ln 2\right).$$

Thus, (9.8) and (9.9) are proved.

Now, to prove (9.10) inequality, by using again the well-known triangle inequality and (9.13), we have

$$\begin{aligned}\left|\rho\mathbb{R}'_{\xi,\varrho}(\rho)\right| &= \left|\rho+\sum_{\upsilon=1}^{\infty}\frac{(\upsilon+1)\,\varrho^{\upsilon}\Gamma\left(1+\xi\right)}{\Gamma\left(\left(1+\xi\right)\left(\upsilon+1\right)\right)}\rho^{\upsilon+1}\right| \\ &\leq 1+\sum_{\upsilon=1}^{\infty}\frac{(\upsilon+1)\left|\varrho\right|^{\upsilon}\Gamma\left(1+\xi\right)}{\Gamma\left(\left(1+\xi\right)\left(\upsilon+1\right)\right)} \\ &\leq 1+\sum_{\upsilon=1}^{\infty}\frac{(\upsilon+1)\left|\varrho\right|^{\upsilon}}{\upsilon!\left(1+\xi\right)^{\upsilon}} = \left(\frac{1+\xi+|\varrho|}{1+\xi}\right)e^{\frac{|\varrho|}{1+\xi}}.\end{aligned} \tag{9.16}$$

From (9.15) and (9.16), we obtain

$$\left|\frac{\rho\mathbb{R}'_{\xi,\varrho}(\rho)}{\mathbb{R}_{\xi,\varrho}(\rho)}\right| \leq \frac{\left(\frac{1+\xi+|\varrho|}{1+\xi}\right)e^{\frac{|\varrho|}{1+\xi}}}{2-e^{\frac{|\varrho|}{1+\xi}}}, \quad \left(|\varrho|<(1+\xi)\ln 2\right).$$

Finally, using the same inequalities to prove (9.12), we arrive at the following:

$$\begin{aligned}\left|\rho\mathbb{R}''_{\xi,\varrho}(\rho)\right| &= \left|\sum_{\upsilon=1}^{\infty}\frac{(\upsilon+1)\,\upsilon\varrho^{\upsilon}\Gamma\left(1+\xi\right)}{\Gamma\left(\left(1+\xi\right)\left(\upsilon+1\right)\right)}\rho^{\upsilon}\right| \\ &\leq \sum_{\upsilon=1}^{\infty}\frac{(\upsilon+1)\,\upsilon\left|\varrho\right|^{\upsilon}\Gamma\left(1+\xi\right)}{\Gamma\left(\left(1+\xi\right)\left(\upsilon+1\right)\right)} \\ &\leq \sum_{\upsilon=1}^{\infty}\frac{(\upsilon+1)\,\upsilon\left|\varrho\right|^{\upsilon}}{\upsilon!\left(1+\xi\right)^{\upsilon}} = \left(\frac{|\varrho|\left(2+2\xi+|\varrho|\right)}{(1+\xi)^{2}}\right)e^{\frac{|\varrho|}{1+\xi}}.\end{aligned} \tag{9.17}$$

Moreover, if we use reverse triangle inequality, then we have

$$\begin{aligned}\left|\mathbb{R}'_{\xi,\varrho}(\rho)\right| &= \left|1+\sum_{\upsilon=1}^{\infty}\frac{(\upsilon+1)\,\varrho^{n}\Gamma(1+\xi)}{\Gamma((1+\xi)(\upsilon+1))}\rho^{\upsilon}\right| \\ &\geq 1-\sum_{\upsilon=1}^{\infty}\frac{(\upsilon+1)\,|\varrho|^{\upsilon}\,\Gamma(1+\xi)}{\Gamma((1+\xi)(\upsilon+1))} \\ &\geq 1-\sum_{\upsilon=1}^{\infty}\frac{(\upsilon+1)\,|\varrho|^{\upsilon}}{\upsilon!\,(1+\xi)^{\upsilon}}=2-\left(\frac{1+\xi+|\varrho|}{1+\xi}\right)e^{\frac{|\varrho|}{1+\xi}},\ \left(|\varrho|<(1+\xi)\ln\left(\frac{2(1+\xi)}{1+\xi+|\varrho|}\right)\right).\end{aligned} \tag{9.18}$$

Next, by combining the inequalities (9.17) with (9.18), we instantly determine that

$$\left|\frac{\rho\mathbb{R}''_{\xi,\varrho}(\rho)}{\mathbb{R}'_{\xi,\varrho}(\rho)}\right| \leqq \frac{\left(\frac{|\varrho|(2+2\xi+|\varrho|)}{(1+\xi)^2}\right)e^{\frac{|\varrho|}{1+\mu}}}{2-\left(\frac{1+\xi+|\varrho|}{1+\xi}\right)e^{\frac{|\varrho|}{1+\xi}}},\qquad \left(|\varrho|<(1+\xi)\ln\left(\frac{2(1+\xi)}{1+\xi+|\varrho|}\right)\right).$$

This completes the proof. □

The inequalities (9.8), (9.9) and (9.15) were proved by Eker and Ece [32] for $\varrho\in\mathbb{R},\ \varrho>0$.

9.3 Univalence and Convexity Conditions for the Integral Operator in (9.5)

Firstly, we take into account the integral operator identified by (9.5).

Theorem 9.3.1 *Let* $\upsilon=1,2,\ldots,n,\quad \xi_\upsilon>-1,\quad \varrho\in\mathbb{C}$ *and* $|\varrho|<(1+\xi_\upsilon)\ln\left(\frac{2(1+\xi_\upsilon)}{1+\xi_\upsilon+|\varrho|}\right)$. *Also, let* γ, β, η_υ, *and* ζ_υ *be in* $\mathbb{C}$ *such that*

$$\Re(\gamma)>0,\quad |\beta|\leq 1\quad (\beta\neq -1).$$

Assume that follows inequality satisfied

$$|\beta|+\frac{1}{|\gamma|}\left(\frac{\left(\frac{|\varrho|(2+2\xi+|\varrho|)}{(1+\xi)^2}\right)e^{\frac{|\varrho|}{1+\xi}}}{2-\left(\frac{1+\xi+|\varrho|}{1+\xi}\right)e^{\frac{|\varrho|}{1+\xi}}}\sum_{\upsilon=1}^{n}|\eta_\upsilon|+\frac{\left(\frac{|\varrho|}{1+\xi}\right)e^{\frac{|\varrho|}{1+\xi}}}{2-e^{\frac{|\varrho|}{1+\xi}}}\sum_{\upsilon=1}^{n}|\zeta_\upsilon|\right)\leqq 1, \tag{9.19}$$

where $\xi=\min\{\xi_1,\xi_2,\ldots,\xi_n\}$. *Then the function:*

$$\mathcal{J}^{\xi_1,\xi_2,\ldots,\xi_n;\varrho}_{\eta_1,\eta_2,\ldots,\eta_n;\zeta_1,\zeta_2,\ldots,\zeta_n;n;\gamma},$$

defined by (9.5) is in the class $\mathcal{S}$.

Proof Let us define the function φ as follows:

$$\varphi(\rho) = \mathcal{J}^{\xi_1,\xi_2,\ldots,\xi_n;\varrho}_{\eta_1,\eta_2,\ldots,\eta_n;\zeta_1,\zeta_2,\ldots,\zeta_n;n;1}(\rho) = \int_0^\rho \prod_{\upsilon=1}^{n} \left(\mathbb{R}'_{\xi_\upsilon,\varrho}(t)\right)^{\eta_\upsilon} \left(\frac{\mathbb{R}_{\xi_\upsilon,\varrho}(t)}{t}\right)^{\zeta_\upsilon} dt. \tag{9.20}$$

First of all, we observe that $\mathbb{R}_{\xi_\upsilon,\varrho}(0) = \mathbb{R}'_{\xi_\upsilon,\varrho}(0) - 1 = 0$, since $\mathbb{R}_{\xi_\upsilon,\varrho} \in \mathcal{A}$ for all $\upsilon = 1, 2, \ldots, n$. Thus, it is obvious that $\varphi(\rho) = \varphi'(\rho) - 1 = 0$. However, we have

$$\varphi'(\rho) = \prod_{\upsilon=1}^{n} \left(\mathbb{R}'_{\xi_\upsilon,\varrho}(\rho)\right)^{\eta_\upsilon} \left(\frac{\mathbb{R}_{\xi_\upsilon,\varrho}(\rho)}{\rho}\right)^{\zeta_\upsilon}. \tag{9.21}$$

By separating both sides of (9.21), we get

$$\frac{\rho\varphi''(\rho)}{\varphi'(\rho)} = \sum_{\upsilon=1}^{n} \eta_\upsilon \frac{\rho\mathbb{R}''_{\xi_\upsilon,\varrho}(\rho)}{\mathbb{R}'_{\xi_\upsilon,\varrho}(\rho)} + \sum_{\upsilon=1}^{n} \zeta_\upsilon \left(\frac{\rho\mathbb{R}'_{\xi_\upsilon,\varrho}(\rho)}{\mathbb{R}_{\xi_\upsilon,\varrho}(\rho)} - 1\right) \tag{9.22}$$

and, from (9.9) and (9.12), we have

$$\begin{aligned}
\left|\frac{\rho\varphi''(\rho)}{\varphi'(\rho)}\right| &\leqq \sum_{\upsilon=1}^{n} \left(|\eta_\upsilon| \left|\frac{\rho\mathbb{R}''_{\xi_\upsilon,\varrho}(\rho)}{\mathbb{R}'_{\xi_\upsilon,\varrho}(\rho)}\right| + |\zeta_\upsilon| \left|\frac{\rho\mathbb{R}'_{\xi_\upsilon,\varrho}(\rho)}{\mathbb{R}_{\xi_\upsilon,\varrho}(\rho)} - 1\right|\right) \\
&\leqq \sum_{\upsilon=1}^{n} \left(|\eta_\upsilon| \frac{\left(\frac{|\varrho|(2+2\xi_\upsilon+|\varrho|)}{(1+\xi_\upsilon)^2}\right) e^{\frac{|\varrho|}{1+\xi_\upsilon}}}{2 - \left(\frac{1+\xi_\upsilon+|\varrho|}{1+\xi_\upsilon}\right) e^{\frac{|\varrho|}{1+\xi_\upsilon}}} + |\zeta_\upsilon| \frac{\left(\frac{|\varrho|}{1+\xi_\upsilon}\right) e^{\frac{|\varrho|}{1+\xi_\upsilon}}}{2 - e^{\frac{|\varrho|}{1+\xi_\upsilon}}}\right) \\
&\leqq \sum_{\upsilon=1}^{n} \left(|\eta_\upsilon| \frac{\left(\frac{|\varrho|(2+2\xi+|\varrho|)}{(1+\xi)^2}\right) e^{\frac{|\varrho|}{1+\xi}}}{2 - \left(\frac{1+\xi+|\varrho|}{1+\xi}\right) e^{\frac{|\varrho|}{1+\xi}}} + |\zeta_\upsilon| \frac{\left(\frac{|\varrho|}{1+\xi}\right) e^{\frac{|\varrho|}{1+\xi}}}{2 - e^{\frac{|\varrho|}{1+\xi}}}\right)
\end{aligned} \tag{9.23}$$

where $\rho \in \mathbb{D}$; $|\varrho| < (1+\xi_\upsilon) \ln\left(\frac{2(1+\xi_\upsilon)}{1+\xi_\upsilon+|\varrho|}\right)$, $(\upsilon = 1, 2, \ldots, n)$. We have also used the fact that the functions $\Theta_1, \Theta_2 : (-1, \infty) \longrightarrow \mathbb{R}$, defined by

$$\Theta_1(x) = \frac{\left(\frac{|\varrho|(2+2x+|\varrho|)}{(1+x)^2}\right) e^{\frac{|\varrho|}{1+x}}}{2 - \left(\frac{1+x+|\varrho|}{1+x}\right) e^{\frac{|\varrho|}{1+x}}} \quad \text{and} \quad \Theta_2(x) = \frac{\left(\frac{|\varrho|}{1+x}\right) e^{\frac{|\varrho|}{1+x}}}{2 - e^{\frac{|\varrho|}{1+x}}},$$

are decreasing and, consequently, we have

$$\frac{\left(\frac{|\varrho|(2+2\xi+|\varrho|)}{(1+\xi)^2}\right) e^{\frac{|\varrho|}{1+\xi}}}{2 - \left(\frac{1+\xi+|\varrho|}{1+\xi}\right) e^{\frac{|\varrho|}{1+\xi}}} > \frac{\left(\frac{|\varrho|(2+2\xi_\upsilon+|\varrho|)}{(1+\xi_\upsilon)^2}\right) e^{\frac{|\varrho|}{1+\xi_\upsilon}}}{2 - \left(\frac{1+\xi_\upsilon+|\varrho|}{1+\xi_\upsilon}\right) e^{\frac{|\varrho|}{1+\xi_\upsilon}}}$$

and

$$\frac{\left(\frac{|\varrho|}{1+\xi}\right)e^{\frac{|\varrho|}{1+\xi}}}{2-e^{\frac{|\varrho|}{1+\xi}}} > \frac{\left(\frac{|\varrho|}{1+\xi_v}\right)e^{\frac{|\varrho|}{1+\xi_v}}}{2-e^{\frac{|\varrho|}{1+\xi_v}}}.$$

Therefore, we obtain

$$\begin{aligned} &\left|\beta\,|\rho|^{2\gamma} + \left(1-|\rho|^{2\gamma}\right)\frac{\rho\varphi''(\rho)}{\gamma\varphi'(\rho)}\right| \\ &\leqq |\beta| + \frac{1}{|\gamma|}\left(\frac{\left(\frac{|\varrho|(2+2\xi+|\varrho|)}{(1+\xi)^2}\right)e^{\frac{|\varrho|}{1+\xi}}}{2-\left(\frac{1+\xi+|\varrho|}{1+\xi}\right)e^{\frac{|\varrho|}{1+\xi}}}\sum_{v=1}^{n}|\eta_v| + \frac{\left(\frac{|\varrho|}{1+\xi}\right)e^{\frac{|\varrho|}{1+\xi}}}{2-e^{\frac{|\varrho|}{1+\xi}}}\sum_{v=1}^{n}|\zeta_v|\right) \qquad (9.24) \\ &\leqq 1. \end{aligned}$$

So, applying that Lemma 9.2.1 we obtain the desired result. □

Theorem 9.3.2 *Let* $v = 1, 2, \ldots, n$, $\xi_v > -1$, $\varrho \in \mathbb{C}$ *and* $|\varrho| < (1+\xi_v)\ln\left(\frac{2(1+\xi_v)}{1+\xi_v+|\varrho|}\right)$. *Also, let* γ, η_v *and* ζ_v *be in* $\mathbb{C}$ *such that* $\Re(\gamma) > 0$. *Assume that follows inequality satisfied*

$$\Re(\gamma) \geqq \sum_{v=1}^{n}|\eta_v|\frac{\left(\frac{|\varrho|(2+2\xi+|\varrho|)}{(1+\xi)^2}\right)e^{\frac{|\varrho|}{1+\xi}}}{2-\left(\frac{1+\xi+|\varrho|}{1+\xi}\right)e^{\frac{|\varrho|}{1+\xi}}} + \sum_{v=1}^{n}|\zeta_v|\frac{\left(\frac{|\varrho|}{1+\xi}\right)e^{\frac{|\varrho|}{1+\xi}}}{2-e^{\frac{|\varrho|}{1+\xi}}}, \qquad (9.25)$$

where $\xi = \min\{\xi_1, \xi_2, \ldots, \xi_n\}$. *Then the function:*

$$\mathcal{J}^{\xi_1,\xi_2,\ldots,\xi_n;\varrho}_{\eta_1,\eta_2,\ldots,\eta_n;\zeta_1,\zeta_2,\ldots,\zeta_n;n;\gamma},$$

defined by (9.5) is in the class $\mathcal{S}$.

Proof Let us consider the function φ as in (9.20). By using inequalities (9.21) and (9.22), we get

$$\begin{aligned} &\frac{1-|\rho|^{2\Re(\gamma)}}{\Re(\gamma)}\left|\frac{\rho\varphi''(\rho)}{\varphi'(\rho)}\right| \\ &\leqq \frac{1-|\rho|^{2\Re(\gamma)}}{\Re(\gamma)}\sum_{v=1}^{n}\left(|\eta_v|\frac{\left(\frac{|\varrho|(2+2\xi_v+|\varrho|)}{(1+\xi_v)^2}\right)e^{\frac{|\varrho|}{1+\xi_v}}}{2-\left(\frac{1+\xi_v+|\varrho|}{1+\xi_v}\right)e^{\frac{|\varrho|}{1+\xi_v}}} + |\zeta_v|\frac{\left(\frac{|\varrho|}{1+\xi_v}\right)e^{\frac{|\varrho|}{1+\xi_v}}}{2-e^{\frac{|\varrho|}{1+\xi_v}}}\right) \\ &\leqq \frac{1}{\Re(\gamma)}\sum_{v=1}^{n}\left(|\eta_v|\frac{\left(\frac{|\varrho|(2+2\xi+|\varrho|)}{(1+\xi)^2}\right)e^{\frac{|\varrho|}{1+\xi}}}{2-\left(\frac{1+\xi+|\varrho|}{1+\xi}\right)e^{\frac{|\varrho|}{1+\xi}}} + |\zeta_v|\frac{\left(\frac{|\varrho|}{1+\xi}\right)e^{\frac{|\varrho|}{1+\xi}}}{2-e^{\frac{|\varrho|}{1+\xi}}}\right) \qquad (9.26) \\ &\leq 1. \end{aligned}$$

Then, the desired result follows from Lemma 9.2.2. □

Theorem 9.3.3 *Let* $v = 1, 2, \ldots, n,$ $\xi_v > -1,$ $\varrho \in \mathbb{C}$ *and* $|\varrho| < (1+\xi_v)\ln\left(\frac{2(1+\xi_v)}{1+\xi_v+|\varrho|}\right)$. *Also, let* η_v *and* ζ_v *be in* $\mathbb{C}$. *Assume that follows inequality satisfied*

$$0 < \sum_{v=1}^{n}\left(|\eta_v|\frac{\left(\frac{|\varrho|(2+2\xi+|\varrho|)}{(1+\xi)^2}\right)e^{\frac{|\varrho|}{1+\xi}}}{2-\left(\frac{1+\xi+|\varrho|}{1+\xi}\right)e^{\frac{|\varrho|}{1+\xi}}} + |\zeta_v|\frac{\left(\frac{|\varrho|}{1+\xi}\right)e^{\frac{|\varrho|}{1+\xi}}}{2-e^{\frac{|\varrho|}{1+\xi}}}\right) \leq 1, \tag{9.27}$$

where $\xi = \min\{\xi_1, \xi_2, \ldots, \xi_n\}$. *Then the function:*

$$\mathcal{J}^{\xi_1,\xi_2,\ldots,\xi_n;\varrho}_{\eta_1,\eta_2,\ldots,\eta_n;\zeta_1,\zeta_2,\ldots,\zeta_n;n;1},$$

defined by (9.5) with $\gamma = 1$, *is convex of order* δ *given by*

$$\delta = 1 - \sum_{v=1}^{n}\left(|\eta_v|\frac{\left(\frac{|\varrho|(2+2\xi+|\varrho|)}{(1+\xi)^2}\right)e^{\frac{|\varrho|}{1+\xi}}}{2-\left(\frac{1+\xi+|\varrho|}{1+\xi}\right)e^{\frac{|\varrho|}{1+\xi}}} + |\zeta_v|\frac{\left(\frac{|\varrho|}{1+\xi}\right)e^{\frac{|\varrho|}{1+\xi}}}{2-e^{\frac{|\varrho|}{1+\xi}}}\right). \tag{9.28}$$

Proof Let us consider the function φ by (9.20). By using (9.23) and (9.28), we have

$$\begin{aligned}\left|\frac{\rho\varphi''(\rho)}{\varphi'(\rho)}\right| &\leq \sum_{v=1}^{n}|\eta_v|\frac{\left(\frac{|\varrho|(2+2\xi+|\varrho|)}{(1+\xi)^2}\right)e^{\frac{|\varrho|}{1+\xi}}}{2-\left(\frac{1+\xi+|\varrho|}{1+\xi}\right)e^{\frac{|\varrho|}{1+\xi}}} + \sum_{v=1}^{n}|\zeta_v|\frac{\left(\frac{|\varrho|}{1+\xi}\right)e^{\frac{|\varrho|}{1+\xi}}}{2-e^{\frac{|\varrho|}{1+\xi}}}\\ &= 1-\delta.\end{aligned} \tag{9.29}$$

As a result, the function φ is convex of order

$$\delta = 1 - \sum_{v=1}^{n}\left(|\eta_v|\frac{\left(\frac{|\varrho|(2+2\xi+|\varrho|)}{(1+\xi)^2}\right)e^{\frac{|\varrho|}{1+\xi}}}{2-\left(\frac{1+\xi+|\varrho|}{1+\xi}\right)e^{\frac{|\varrho|}{1+\xi}}} + |\zeta_v|\frac{\left(\frac{|\varrho|}{1+\xi}\right)e^{\frac{|\varrho|}{1+\xi}}}{2-e^{\frac{|\varrho|}{1+\xi}}}\right). \tag{9.30}$$

□

In Theorem 9.3.1 with $n = 1$, $\xi_1 = 0$ and $\varrho = -1/3$, we can write the following corollary (Fig. 9.1).

Corollary 9.3.1 *Let* γ, β, η *and* β *be in* $\mathbb{C}$ *such that* $\Re(\gamma) > 0$, $|\beta| \leq 1$ $(\beta \neq -1)$. *If the inequality*

$$|\beta| + \frac{1}{|\gamma|}\left(\frac{7e^{1/3}}{18-12e^{1/3}}|\eta| + \frac{e^{1/3}}{6-3e^{1/3}}|\zeta|\right) \leq 1$$

then the function:

$$\left[\frac{\gamma}{3^\eta}\int_0^\rho t^{\gamma-1}(3-t)^\eta\left(e^{-t/3}\right)^{\eta+\zeta}dt\right]^{1/\gamma}$$

is in the class $\mathcal{S}$.

Example 9.3.1 From Corollary 9.3.1;

$$f_1(\rho) = \frac{1}{3}\int_0^\rho (3-t)\, e^{-2t/3} dt \in \mathcal{S}.$$

Setting $n = 1, \xi_1 = 1$ and $\varrho = 1/2$ in Theorem 9.3.1, we can get result below.

Corollary 9.3.2 *Let* γ, β, η *and* ζ *be in* $\mathbb{C}$ *such that* $\Re(\gamma) > 0, |\beta| \leq 1\ (\beta \neq -1)$. *If the inequality:*

$$|\beta| + \frac{1}{|\gamma|}\left(\frac{9e^{1/4}}{32 - 20e^{1/4}}|\eta| + \frac{e^{1/4}}{8 - 4e^{1/4}}|\zeta|\right) \leqq 1$$

holds, then the function:

$$\left[2^{\left(\frac{\zeta}{2}-\eta\right)}\gamma\int_0^\rho t^{\frac{-\zeta}{2}+\gamma-1}\left(\cosh\sqrt{\frac{t}{2}} + \frac{\sqrt{2}\sinh\sqrt{\frac{t}{2}}}{\sqrt{t}}\right)^\eta \left(\sinh\sqrt{\frac{t}{2}}\right)^\zeta dt\right]^{1/\gamma}$$

is in the class $\mathcal{S}$.

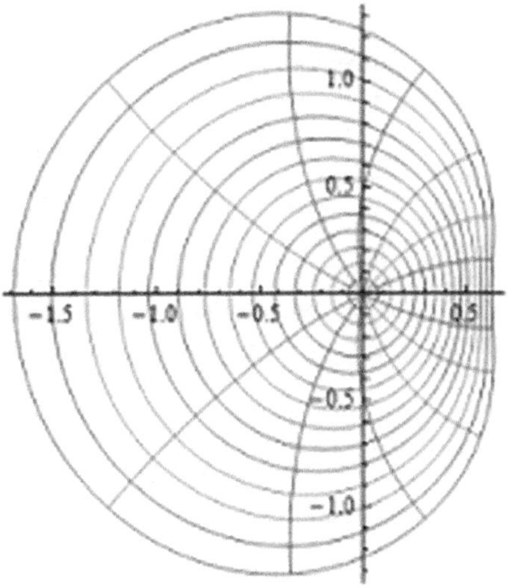

Fig. 9.1 The image of f_1 on $\mathbb{D}$

From Theorem 9.3.3 with $n = 1, \xi_1 = 0$ and $\varrho = -1/3$, we can get result below.

Corollary 9.3.3 *Let η and ζ be complex numbers such that*

$$0 < \frac{7e^{1/3}}{18 - 12e^{1/3}} |\eta| + \frac{e^{1/3}}{6 - 3e^{1/3}} |\zeta| \leqq 1.$$

Then the function:

$$3^{-\eta} \int_0^{\rho} t^{\gamma-1} (3-t)^{\eta} \left(e^{-t/3}\right)^{\eta+\zeta} dt$$

is convex of order δ given by

$$\delta = 1 - \frac{7e^{1/3}}{18 - 12e^{1/3}} |\eta| - \frac{e^{1/3}}{6 - 3e^{1/3}} |\zeta| .$$

Let $n = 1, \xi_1 = 1$ and $\varrho = 1/2$ in the Theorem 9.3.3, then we get following result.

Corollary 9.3.4 *Let η and ζ be complex numbers such that*

$$0 < \frac{9e^{1/4}}{32 - 20e^{1/4}} |\eta| + \frac{e^{1/4}}{8 - 4e^{1/4}} |\zeta| \leqq 1.$$

Then the function:

$$2^{\left(\frac{\zeta}{2} - \eta\right)} \int_0^{\rho} t^{\frac{-\zeta}{2}} \left(\cosh\sqrt{\frac{t}{2}} + \frac{\sqrt{2}\sinh\sqrt{\frac{t}{2}}}{\sqrt{t}}\right)^{\eta} \left(\sinh\sqrt{\frac{t}{2}}\right)^{\zeta} dt$$

is convex of order δ given by

$$\delta = 1 - \frac{9e^{1/4}}{32 - 20e^{1/4}} |\eta| - \frac{e^{1/4}}{8 - 4e^{1/4}} |\zeta| .$$

9.4 Univalence and Convexity Conditions for the Integral Operator in (9.6)

We examine the univalence and convexity requirements for the integral operator defined by (9.6) in this section.

Theorem 9.4.1 *Let $v = 1, 2, \ldots, n,\ \xi_v > -1,\ \zeta_v,\ \varrho \in \mathbb{C}$ and $|\varrho| < (1 + \xi_v) \ln 2$. Also, let β and η_v be in $\mathbb{C}$ such that*

$$|\beta| \leq 1 \ (\beta \neq -1) \quad \text{and} \quad \Re\left(1 + \sum_{v=1}^{n} \eta_v\right) > 0.$$

Assume that follows inequality satisfied

$$|\beta| + \frac{1}{\left|1+\sum_{\upsilon=1}^{n}\eta_\upsilon\right|}\sum_{\upsilon=1}^{n}\left[|\eta_\upsilon|\frac{\left(\frac{|\varrho|}{1+\xi}\right)e^{\frac{|\varrho|}{1+\xi}}}{2-e^{\frac{|\varrho|}{1+\xi}}} + |\zeta_\upsilon|\left(\frac{1+\xi+|\varrho|}{1+\xi}\right)e^{\frac{|\varrho|}{1+\xi}}\right] \leqq 1, \tag{9.31}$$

where $\xi = \min\{\xi_1, \xi_2, \ldots, \xi_n\}$. *Then the function:*

$$\mathcal{K}^{\xi_1,\xi_2,\ldots,\xi_n;\varrho}_{\eta_1,\eta_2,\ldots,\eta_n;\zeta_1,\zeta_2,\ldots,\zeta_n;n},$$

defined by (9.6) is in the class $\mathcal{S}$.

Proof Let us define the functions ψ by

$$\psi(\rho) = \int_0^\rho \prod_{\upsilon=1}^{n}\left(\frac{\mathbb{R}_{\xi_\upsilon,\varrho}(t)}{t}\right)^{\eta_\upsilon}\left(e^{\mathbb{R}_{\xi_\upsilon,\varrho}(t)}\right)^{\zeta_\upsilon} dt. \tag{9.32}$$

Then $\psi(0) = \psi'(0) - 1 = 0$. Differentiating both sides of (9.32) logarithmically, we get

$$\frac{\rho\psi''(\rho)}{\psi'(\rho)} = \sum_{\upsilon=1}^{n}\eta_\upsilon\left(\frac{\rho\mathbb{R}'_{\xi_\upsilon,\varrho}(\rho)}{\mathbb{R}_{\xi_\upsilon,\varrho}(\rho)} - 1\right) + \sum_{\upsilon=1}^{n}\zeta_\upsilon\left(\rho\mathbb{R}'_{\xi_\upsilon,\varrho}(\rho)\right) \tag{9.33}$$

and, from (9.9) and (9.10) in Lemma 9.2.3, we have

$$\begin{aligned}
\left|\frac{\rho\psi''(\rho)}{\psi'(\rho)}\right| &\leqq \sum_{\upsilon=1}^{n}\left(|\eta_\upsilon|\left|\frac{\rho\mathbb{R}'_{\xi_\upsilon,\varrho}(\rho)}{\mathbb{R}_{\xi_\upsilon,\varrho}(\rho)} - 1\right| + |\zeta_\upsilon|\left|\rho\mathbb{R}'_{\xi_\upsilon,\varrho}(\rho)\right|\right)\\
&\leqq \sum_{\upsilon=1}^{n}\left[|\eta_\upsilon|\frac{\left(\frac{|\varrho|}{1+\xi_\upsilon}\right)e^{\frac{|\varrho|}{1+\xi_\upsilon}}}{2-e^{\frac{|\varrho|}{1+\xi_\upsilon}}} + |\zeta_\upsilon|\left(\frac{1+\xi_\upsilon+|\varrho|}{1+\xi_\upsilon}\right)e^{\frac{|\varrho|}{1+\xi_\upsilon}}\right]\\
&\leqq \sum_{\upsilon=1}^{n}\left[|\eta_\upsilon|\frac{\left(\frac{|\varrho|}{1+\xi}\right)e^{\frac{|\varrho|}{1+\xi}}}{2-e^{\frac{|\varrho|}{1+\xi}}} + |\zeta_\upsilon|\left(\frac{1+\xi+|\varrho|}{1+\xi}\right)e^{\frac{|\varrho|}{1+\xi}}\right],
\end{aligned} \tag{9.34}$$

where $(\rho \in \mathbb{D};\ |\varrho| < (1+\xi_\upsilon)\ln 2,\ (\upsilon = 1, 2, \ldots, n))$, we have also used the fact that the functions

$$\Theta_3, \Theta_4 : (-1, \infty) \longrightarrow \mathbb{R},$$

defined by

$$\Theta_3(x) = \frac{\left(\frac{|\varrho|}{1+x}\right)e^{\frac{|\varrho|}{1+x}}}{2-e^{\frac{|\varrho|}{1+x}}} \quad \text{and} \quad \Theta_4(x) = \left(\frac{1+x+|\varrho|}{1+x}\right)e^{\frac{|\varrho|}{1+x}},$$

are decreasing and, consequently,

$$\frac{\left(\frac{|\varrho|}{1+\xi_v}\right) e^{\frac{|\varrho|}{1+\xi_v}}}{2-e^{\frac{|\varrho|}{1+\xi_v}}} < \frac{\left(\frac{|\varrho|}{1+\xi}\right) e^{\frac{|\varrho|}{1+\xi}}}{2-e^{\frac{|\varrho|}{1+\xi}}}$$

and

$$\left(\frac{1+\xi_v+|\varrho|}{1+\xi_v}\right) e^{\frac{|\varrho|}{1+\xi_v}} < \left(\frac{1+\xi+|\varrho|}{1+\xi}\right) e^{\frac{|\varrho|}{1+\xi}}.$$

Therefore, we obtain

$$\begin{aligned}
&\left|\beta\,|\rho|^{2\left(1+\sum_{v=1}^{n}\eta_v\right)} + \left(1-|\rho|^{2\left(1+\sum_{v=1}^{n}\eta_v\right)}\right)\frac{\rho\psi''(\rho)}{\left(1+\sum_{v=1}^{n}\eta_v\right)\psi'(\rho)}\right| \\
&\leq |\beta| + \left|\frac{\rho\psi''(\rho)}{\left(1+\sum_{v=1}^{n}\eta_v\right)\psi'(\rho)}\right| \qquad (9.35)\\
&\leq |\beta| + \frac{1}{\left|1+\sum_{v=1}^{n}\eta_v\right|}\sum_{v=1}^{n}\left[|\eta_v|\frac{\left(\frac{|\varrho|}{1+\xi}\right) e^{\frac{|\varrho|}{1+\xi}}}{2-e^{\frac{|\varrho|}{1+\xi}}} + |\zeta_v|\left(\frac{1+\xi+|\varrho|}{1+\xi}\right) e^{\frac{|\varrho|}{1+\xi}}\right] \\
&\leq 1.
\end{aligned}$$

By Lemma 9.2.1 with

$$\alpha = 1 + \sum_{v=1}^{n}\eta_v,$$

the desired result is obtained. □

Theorem 9.4.2 *Let* $v = 1, 2, \ldots, n,\ \xi_v > -1,\ \zeta_v,\ \varrho \in \mathbb{C}$ *and* $|\varrho| < (1+\xi)\ln 2$. *Also, let* η_v *be in* $\mathbb{C}$ *such that* $\Re(\eta_v) \geqq 0$. *Assume that follows inequality satisfied*

$$\sum_{v=1}^{n}\left[|\eta_v|\frac{\left(\frac{|\varrho|}{1+\xi}\right) e^{\frac{|\varrho|}{1+\xi}}}{2-e^{\frac{|\varrho|}{1+\xi}}} + |\zeta_v|\left(\frac{1+\xi+|\varrho|}{1+\xi}\right) e^{\frac{|\varrho|}{1+\xi}}\right] \leqq 1, \qquad (9.36)$$

where $\xi = \min\{\xi_1, \xi_2, \ldots, \xi_n\}$. *Then the function:*

$$\mathcal{K}^{\xi_1,\xi_2,\ldots,\xi_n;\varrho}_{\eta_1,\eta_2,\ldots,\eta_n;\zeta_1,\zeta_2,\ldots,\zeta_n;n},$$

defined by (9.6) is in the class $\mathcal{S}$.

Proof Let us define the function ψ as in (9.32). Therefore, from (9.34), we have

$$\begin{aligned}
&\left(1-|\rho|^2\right)\left|\frac{\rho\psi''(\rho)}{\psi'(\rho)}\right| \\
&\leqq \sum_{\upsilon=1}^{n}\left[|\eta_\upsilon|\frac{\left(\frac{|\varrho|}{1+\xi}\right)e^{\frac{|\varrho|}{1+\xi}}}{2-e^{\frac{|\varrho|}{1+\xi}}}+|\zeta_\upsilon|\left(\frac{1+\mu+|\varrho|}{1+\xi}\right)e^{\frac{|\varrho|}{1+\xi}}\right]. \qquad (9.37)\\
&\leq 1.
\end{aligned}$$

By Lemma 9.2.2, with $\varpi = 1$ and $\alpha = 1 + \sum_{\upsilon=1}^{n} \eta_\upsilon$, the desired result is obtained. □

Theorem 9.4.3 *Let* $\upsilon = 1, 2, \ldots, n,\ \xi_\upsilon > -1,\ \zeta_\upsilon,\ \varrho \in \mathbb{C}$ *and* $|\varrho| < (1+\xi)\ln 2$. *Also, let* η_υ *be in* $\mathbb{C}$ *such that*

$$\Re\left(1+\sum_{\upsilon=1}^{n}\eta_\upsilon\right) > 0.$$

Assume that follows inequality satisfied

$$0 < \sum_{\upsilon=1}^{n}\left[|\eta_\upsilon|\frac{\left(\frac{|\varrho|}{1+\xi}\right)e^{\frac{|\varrho|}{1+\xi}}}{2-e^{\frac{|\varrho|}{1+\xi}}}+|\zeta_\upsilon|\left(\frac{1+\xi+|\varrho|}{1+\xi}\right)e^{\frac{|\varrho|}{1+\xi}}\right] \leqq 1, \qquad (9.38)$$

where $\xi = \min\{\xi_1, \xi_2, \ldots, \xi_n\}$. *Then the function:* ψ *defined by (9.32), is convex of order* δ *given by*

$$\delta = 1 - \sum_{\upsilon=1}^{n}\left[|\eta_\upsilon|\frac{\left(\frac{|\varrho|}{1+\xi}\right)e^{\frac{|\varrho|}{1+\xi}}}{2-e^{\frac{|\varrho|}{1+\xi}}}+|\zeta_\upsilon|\left(\frac{1+\xi+|\varrho|}{1+\xi}\right)e^{\frac{|\varrho|}{1+\xi}}\right]. \qquad (9.39)$$

Proof Let us define the function ψ as given by (9.32). Using similar methods in the proof of Theorem 9.4.1, we get

$$\begin{aligned}
\left|\frac{\rho\psi''(\rho)}{\psi'(\rho)}\right| &\leqq \sum_{\upsilon=1}^{n}\left[|\eta_\upsilon|\left|\frac{\rho\mathbb{R}'_{\xi_\upsilon,\varrho}(\rho)}{\mathbb{R}_{\xi_\upsilon,\varrho}(\rho)}-1\right|+|\zeta_\upsilon|\left|\rho\mathbb{R}'_{\xi_\upsilon,\varrho}(\rho)\right|\right]\\
&\leqq \sum_{\upsilon=1}^{n}\left[|\eta_\upsilon|\frac{\left(\frac{|\varrho|}{1+\xi}\right)e^{\frac{|\varrho|}{1+\xi}}}{2-e^{\frac{|\varrho|}{1+\xi}}}+|\zeta_\upsilon|\left(\frac{1+\xi+|\varrho|}{1+\xi}\right)e^{\frac{|\varrho|}{1+\xi}}\right] \qquad (9.40)\\
&= 1-\delta.
\end{aligned}$$

Therefore, the function ψ is convex of order

$$\delta = 1 - \sum_{\upsilon=1}^{n} \left[|\eta_\upsilon| \frac{\left(\frac{|\varrho|}{1+\xi}\right) e^{\frac{|\varrho|}{1+\xi}}}{2 - e^{\frac{|\varrho|}{1+\xi}}} + |\zeta_\upsilon| \left(\frac{1+\xi+|\varrho|}{1+\xi}\right) e^{\frac{|\varrho|}{1+\xi}} \right].$$

□

From Theorem 9.4.1 with $n = 1, \xi_1 = 0$ and $\varrho = -1/3$, we can get result below.

Corollary 9.4.1 *Let η and β be in $\mathbb{C}$ such that $\Re(1+\eta) > 0$ and $|\beta| \leq 1\ (\beta \neq -1)$. If these numbers satisfy the inequality below:*

$$|\beta| + \frac{1}{|1+\eta|} \left(\frac{e^{1/3}}{6 - 3e^{1/3}} |\eta| + \frac{4e^{1/3}}{3} |\zeta| \right) \leqq 1,$$

then the function:

$$\left[(1+\eta) \int_0^\rho \left(te^{-t/3}\right)^\eta e^{\zeta\left(te^{-t/3}\right)} dt \right]^{1/(1+\eta)} \tag{9.41}$$

is in the class $\mathcal{S}$.

Example 9.4.1 From Corollary 9.4.1 with $\eta = 1$ and $\zeta = 1$, we have the following result:

$$\hbar_2(\rho) = \left[2 \int_0^\rho \left(te^{-t/3}\right) e^{\left(te^{-t/3}\right)} dt \right]^{1/2} \in \mathcal{S}.$$

In reality, by a simple calculation, we get

$$\frac{\rho \hbar_2'(\rho)}{\hbar_2(\rho)} + 1 + \frac{p \hbar_2''(\rho)}{\hbar_2'(\rho)} = 1 + \frac{(3-\rho)\left(e^{\rho/3} + \rho\right)}{3e^{\rho/3}} = g(\rho).$$

It also holds true that $\Re(g(\rho)) > 0$ for all $\rho \in \mathbb{D}$ (see Fig. 9.2). Therefore, $\hbar_2$ is a $1/2-$convex function [13, Vol. I, p. 142]. Thus it follows from [13, Vol. I, p. 142] that $\hbar_2$ belongs to the class $\mathcal{S}$.

From Theorem 9.4.1 with $n = 1, \xi_1 = 1$ and $\varrho = 1/2$, we can get result below.

Corollary 9.4.2 *Let β, η and ζ be in $\mathbb{C}$ such that $\Re(1+\eta) > 0$ and $|\beta| \leq 1\ (\beta \neq -1)$. If these numbers satisfy the inequality below:*

$$|\beta| + \frac{1}{|1+\eta|} \left(\frac{e^{1/4}}{8 - 4e^{1/4}} |\eta| + \frac{5e^{1/4}}{4} |\zeta| \right) \leqq 1,$$

then the function:

$$\left[(1+\eta) \int_0^\rho \left(\sqrt{2t} \sinh \sqrt{\frac{t}{2}}\right)^\eta e^{\zeta\left(\sqrt{2t} \sinh \sqrt{\frac{t}{2}}\right)} dt \right]^{1/(1+\eta)} \tag{9.42}$$

is in the class $\mathcal{S}$.

From Theorem 9.4.3 with $n = 1, \xi_1 = 0$ and $\varrho = -1/3$, we can get result below.

Corollary 9.4.3 *Let η and ζ be complex numbers such that*

$$\Re(1+\eta) > 0 \quad \text{and} \quad 0 < \frac{e^{1/3}}{6-3e^{1/3}}|\eta| + \frac{4e^{1/3}}{3}|\zeta| \leqq 1.$$

Then the function $\int_0^\rho \left(e^{-t/3}\right)^\eta e^{\zeta\left(te^{-t/3}\right)}dt$ is convex of order δ given by

$$\delta = 1 - \frac{e^{1/3}}{6-3e^{1/3}}|\eta| - \frac{4e^{1/3}}{3}|\zeta|.$$

From Theorem 9.4.3 with $n = 1, \xi_1 = 1$ and $\varrho = 1/2$, we can get result below.

Corollary 9.4.4 *Let α and ζ be complex numbers such that*

$$\Re(1+\eta) > 0 \quad \text{and} \quad 0 < \frac{e^{1/4}}{8-4e^{1/4}}|\eta| + \frac{5e^{1/4}}{4}|\zeta| \leqq 1.$$

Then the function $\int_0^\rho \left(\sqrt{\frac{2}{t}}\sinh\sqrt{\frac{t}{2}}\right)^\eta e^{\zeta\left(\sqrt{2t}\sinh\sqrt{\frac{t}{2}}\right)}dt$ is convex of order δ given by

$$\delta = 1 - \frac{e^{1/4}}{8-4e^{1/4}}|\eta| - \frac{5e^{1/4}}{4}|\zeta|.$$

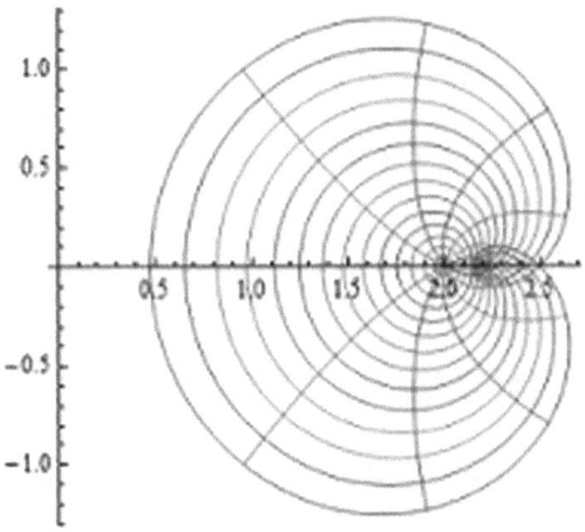

Fig. 9.2 The image of g on $\mathbb{D}$

9.5 Conclusions

In the present investigation, we have first introduced a family of integral operators and the Rabotnov function which, in particular, plays a very important role in the study of pure and applied mathematical sciences. We have then successfully obtained various interesting mapping and geometric properties, such as univalence and convexity conditions, for the integral operators, introduced in the paper and associated with the Rabotnov function, by using the known techniques which were used in the literature. Finally, we have a number of corollaries and examples together with the associated graphical illustrations that are potentially useful for motivating further researches on this subject and on other related topics.

The newly introduced operators could be used for many purposes, just as operators have generated interesting outcomes in geometric function theory during time being studied in many aspects. Hopefully, the original results contained here would stimulate researchers' imagination and inspire them just as all the operators introduced before in studies related to functions of a complex variable have done. Other properties related to them could be investigated, and also they could prove useful in introducing special classes of functions based on those properties.

References

1. Al-Kharsani, H.A., Al-Zahrani, A.M., Al-Hajri, S.S., Pogány, T.K.: Univalence criteria for linear fractional differential operators associated with a generalized Bessel function. Math. Commun. **21**, 171–188 (2016)
2. Arif, M., Raza, M.: Some properties of an integral operator defined by Bessel functions. Acta Universitatis Apulensis **26**, 69–74 (2011)
3. Baricz, Á., Frasin, B.A.: Univalence of integral operators involving Bessel functions. Appl. Math. Lett. **23**, 371–376 (2010)
4. Breaz, D., Breaz, N.: Univalence conditions for certain integral operators. Stud. Univ. Babeş-Bolyai Math. **47**, 9–15 (2002)
5. Breaz, D., Breaz, N.: Univalence of an integral operator. Mathematica (Cluj) **47**, 35–38 (2005)
6. Breaz, D., Breaz, N., Srivastava, H.M.: An extension of the univalent condition for a family of integral operators. Appl. Math. Lett. **22**, 41–44 (2009)
7. Breaz, D., Owa, S., Breaz, N.: A new integral univalent operator. Acta Univ. Apulensis Math. Inform. **16**, 11–16 (2008)
8. Deniz, E., Orhan, H., Srivastava, H.M.: Some sufficient conditions for univalence of certain families of integral operators involving generalized Bessel functions. Taiwanese J. Math. **15**, 883–917 (2011)
9. Deniz, E.: Convexity of integral operators involving generalized Bessel functions. Integral Transform. Spec. Funct. **24**, 201–216 (2013)
10. Din, M.U., Srivastava, H.M., Raza, M.: Univalence of certain integral operators involving generalized Struve functions. Hacettepe J. Math. Stat. **47**(4), 821–833 (2018)
11. Din, M.U., Yalcin, S.: Certain geometric properties of modified Lommel Functions. Honam Math. J. **42**(4), 719–731 (2020)
12. Frasin, B.A.: Sufficient conditions for integral operator defined by Bessel functions. J. Math. Ineq. **4**(2), 301–306 (2010)

13. Goodman, A.W.: Univalent Functions, Mariner Publishing Company Incorporated, vols. I and II. Tampa, FL (1983)
14. Kazımoğlu, S., Deniz, E.: Partial sums of the Rabotnov function, Acta Universitatis Sapientiae. Mathematica **14**(2), 250–261 (2022)
15. Kim, Y.J., Merkes, E.P.: On an integral of powers of a spirallike function. Kyungpook Math. J. **12**, 249–252 (1972)
16. Mahmood, S., Mahroz, S., Rafiq, A., Malik, S.N., Raza, M.: Convexity of certain integral operators defined by Struve functions. J. Funct. Spaces, Article ID 6327132 (2018)
17. Mahmood, S., Srivastava, H.M., Malik, S.N., Raza, M., Shahzadi, N., Zainab, S.: A certain family of integral operators associated with the Struve functions. Symmetry **11**(4), 463 (2019)
18. Moldoveanu, S., Pascu, N.N.: Integral operators which preserve the univalence. Mathematica (Cluj) **32**, 159–166 (1990)
19. Mustafa, N.: Univalence of certain integral operators involving normalized Wright functions. Commun. Faculty Sci. Univ. Ankara Ser. A1 Math. Stat. **66**(1), 19–28 (2017)
20. Park, J.H., Srivastava, H.M., Cho, N.E.: Univalence and convexity conditions for certain integral operators associated with the Lommel function of the first kind. AIMS Math. **6**(10), 11380–11402 (2021)
21. Pascu, N.N.: An improvement of Becker's univalence criterion. In: Proceedings of the Commemorative Session: Simion Stoilow (Brasov), pp. 43–48 (1987)
22. Pescar, V.: A new generalization of Ahlfor's and Becker's criterion of univalence. Bull. Malaysian Math. Soc. **19**, 53–54 (1996)
23. Pfaltzgraff, J.A.: Univalence of the integral of $(f'(z))^{\lambda}$. Bull. London Math. Soc. **7**, 254–256 (1975)
24. Rabotnov, Y.N.: Equilibrium of elastic media with an aftereffect. Prikl. Matem. Mekh. **12**, 53–62 (1948)
25. Raza, R., Noreen, S., Malik, S.N.: Geometric properties of integral operators defined by Bessel functions. J. Inequal. Spec. Funct. **7**, 34–48 (2016)
26. Seenivasagan, N., Breaz, D.: Certain sufficient conditions for univalence. Gen. Math. **15**, 7–15 (2007)
27. Srivastava, H.M.: Operators of basic (or q-) calculus and fractional q-calculus and their applications in geometric function theory of complex analysis. Iran. J. Sci. Technol. Trans. A: Sci. **44**, 327–344 (2020)
28. Srivastava, H.M., Deniz, E., Orhan, H.: Some general univalence criteria for a family of integral operators. Appl. Math. Comput. **215**, 3696–3701 (2010)
29. Srivastava, H.M., Frasin, B.A., Pescar, V.: Univalence of integral operators involving Mittag-Leffler functions. Appl. Math. Inform. Sci. **11**, 635–641 (2017)
30. Srivastava, H.M., Jan, R., Jan, A., Deebai, W., Shutaywi, M.: Fractional-calculus analysis of the transmission dynamics of the dengue infection, Chaos, 31. Article ID **53130**, 1–18 (2021)
31. Srivastava, H.M., Owa, S.: Univalent Functions; Fractional Calculus; and Their Applications, Halsted Press (Ellis Horwood Limited, Chichester), John Wiley and Sons, New York. Brisbane, and Toronto, Chichester (1989)
32. Sümer Eker, S., Ece, S.: Geometric properties of normalized Rabotnov function. Hacet. J. Math. Stat. **51**(5), 1248–1259 (2022)

SPRINGER NATURE

GPSR Compliance

The European Union's (EU) General Product Safety Regulation (GPSR) is a set of rules that requires consumer products to be safe and our obligations to ensure this.

If you have any concerns about our products, you can contact us on ProductSafety@springernature.com

In case Publisher is established outside the EU, the EU authorized representative is:

Springer Nature Customer Service Center GmbH
Europaplatz 3
69115 Heidelberg, Germany

The manufacturer's authorised representative in the EU is Springer Nature Customer Service Centre GmbH, Europaplatz 3, 69115 Heidelberg, Germany. If you have any concerns regarding our products, please contact ProductSafety@springernature.com

Printed and bound by CPI Group (UK) Ltd, Croydon, CR0 4YY
30/09/2026
02225300-0002